Saas-Fee Advanced Course 32
2002

AF181211

Springer
Berlin
Heidelberg
New York
Hong Kong
London
Milan
Paris
Tokyo

Physics and Astronomy ONLINE LIBRARY

springeronline.com

A. W. Blain F. Combes B. T. Draine

The Cold Universe

Saas-Fee Advanced Course 32
2002

Swiss Society for Astrophysics and Astronomy

Edited by D. Pfenniger and Y. Revaz

With 129 Figures

Springer

Andrew W. Blain
California Institute of Technology
Astronomy 105-24
Pasadena, CA 91125, USA

Bruce T. Draine
Princeton University
Department of Astrophysical Sciences
Peyton Hall 108
Princeton, NJ 08544-1011, USA

Françoise Combes
Observatoire de Paris, DEMIRM
Place de l'Observatoire 61
75014 Paris, France

Volume Editors:

Daniel Pfenniger
Yves Revaz
Observatoire de Genève
ch. des Maillettes 51
1290 Sauverny, Switzerland

This series is edited on behalf of the Swiss Society for Astrophysics and Astronomy:
Société Suisse d'Astrophysique et d'Astronomie
Observatoire de Genève, ch. des Maillettes 51, 1290 Sauverny, Switzerland

Cover picture: Copyright 2002, Didier Greusard, Geneva Observatory. All rights reserved.

Library of Congress Cataloging-in-Publication Data
Blain, A. (Andrew)
The cold universe / A. Blain, F. Combes, B. Draine ; edited by D. Pfenniger and Y.Revaz. p. m. - - (Saas-Fee advanced course 32, lecture notes 2002, Swiss Society for Astrophysics and Astronomy)
Includes bibliographical references and index.
ISBN 3-540-40838-X (alk. paper)
1. Cosmic dust–Congresses. 2. Astronomical spectroscopy–Congresses. 3. Galaxies–Formation–Congresses. I. Combes, F. II. Draine, Bruce T., 1947– III. Pfenniger, D. IV. Revaz, Y. (Yves) V. Title. VI. Saas-Fee advanced course ... lecture notes ; 2002.
QB791.B53 2003 523.1'125–dc22 2003059117

ISBN 978-3-642-07407-3 e-ISBN 978-3-540-31636-7

This work is subject to copyright. All rights are reserved, whether the whole or part of the material is concerned, specifically the rights of translation, reprinting, reuse of illustrations, recitation, broadcasting, reproduction on microfilm or in any other way, and storage in data banks. Duplication of this publication or parts thereof is permitted only under the provisions of the German Copyright Law of September 9, 1965, in its current version, and permission for use must always be obtained from Springer-Verlag. Violations are liable for prosecution under the German Copyright Law.

Springer-Verlag is a part of Springer Science+Business Media

springeronline.com

© Springer-Verlag Berlin Heidelberg 2010
Printed in Germany

The use of general descriptive names, registered names, trademarks, etc. in this publication does not imply, even in the absence of a specific statement, that such names are exempt from the relevant protective laws and regulations and therefore free for general use.

Cover design: *design & production*, Heidelberg

Preface

Modern astronomy has stretched its domains of exploration tremendously. Not only objects at very large distances and very old states of the Universe can be examined, but also all kinds of radiations and phenomena are now accessible. Astronomers constantly move from considerations about very diluted to very dense systems.

Hot and energetic systems, being the easiest to observe, have attracted a lot of attention. However the cold and low energetic states have been somewhat neglected, either because being harder to observe they appear unexciting, or because being less well known they tend to be ignored. However the Universe background radiation has now been determined as the most perfect known case of a black-body spectrum, a substantial fraction of matter spends some time close to the temperature of this universal thermal bath, before being transformed into stars or planets. Some objects, such as rapidly expanding gas shells in planetary nebulae, may even succeed in reaching a temperature well below the background radiation temperature through the mere action of adiabatic expansion. In view of the highly dynamical and turbulent state of the interstellar medium, hot *and cold* temperature fluctuations must be expected, while the clear observational bias is to observe the hot rather than the cold fluctuations. Fortunately with the accessibility of far-infrared and sub-millimetric instruments such as SCUBA, WMAP, Planck or ALMA, we can expect in the coming years continuous advances in our understanding of these harder to observe cold stages of matter.

For these reasons the members of the Swiss Society for Astrophysics and Astronomy voted in favor of the organisation of a winter Saas-Fee course on the Cold Universe. The aim of this course was to cover as well as possible the current knowledge in observational and theoretical astrophysics dealing with the coldest objects in the Universe. The selected lecturers, Andrew Blain, Françoise Combes and Bruce Draine have succeeded in bringing to the 70 participants a very rich and interesting amount of information, that the reader can enjoy now in the following pages. We are very grateful to the lecturers for their invaluable live lectures as well for their presently available written versions.

We thank also the course Secretary, Ms Irène Scheffre, and our colleagues Paul Bartholdi, Veruska Muccione, Yves Revaz, and Nunos Santos, for all their help in the practical organisation of the course.

As as reminder the course took place in the picturesque village of Grimentz in the south Alps of Switzerland. As a practical introduction to the Cold Universe, the participants could taste ice-cream made on the spot with a mixture of rasberries, strawberries, liquid cream and liquid nitrogen! The picture shows from left to right Yves Revaz, Paul Bartholdi, Veruska Muccione and Françoise Combes enjoying the cooling process.

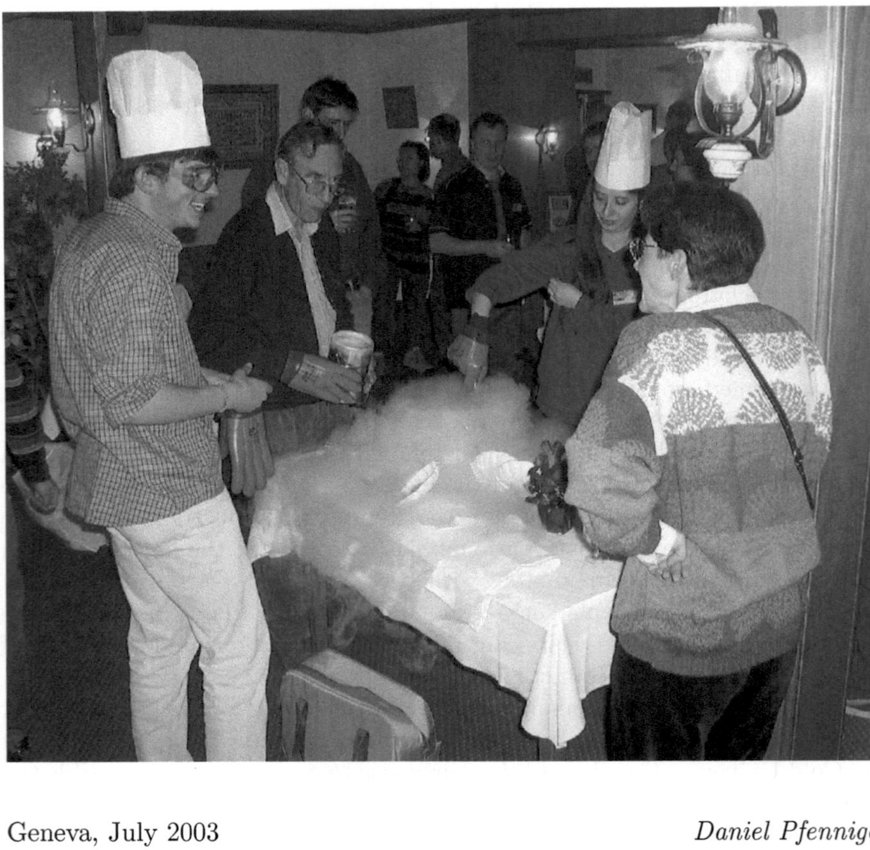

Geneva, July 2003

Daniel Pfenniger
Arnold Benz
Thierry Courvoisier
(The Course Organisers)

Contents

Galaxy Formation and Evolution in the Cold Universe

Andrew W. Blain

California Institute of Technology, Astronomy 105-24, Pasadena, CA 91125, USA

1 Introduction

This part of the proceedings discusses our knowledge of the way that galaxies formed, and the information that we can gain about the process by observing the radiation from the cold components of the Universe. The physical processes involving the molecular and atomic gas and solid-phase components of the interstellar medium (ISM) in galaxies are discussed in great detail in the other contributions. Here we will concentrate on the continuum emission processes that are relevant to studies of distant galaxies, mentioning briefly the importance of molecular studies. Little new ground will be broken in describing the processes at work as galaxies form. This is an extremely large and active research area, and one that has been described comprehensively elsewhere, including at this school in past years, sometimes touching upon the cold Universe theme [183]. Some current issues in the study of galaxy formation at long wavelengths will be raised, and the detailed physics and chemistry being discussed alongside [84,103] will be put into the context of our rapidly developing *observationally driven* understanding of the way in which galaxies formed. In addition, the discussion will be focused on how observations of the thermal emission from the constituents of the Universe at temperatures from a few 10's to a few 100's of degrees are moving into the mainstream of observational cosmology, and the illumination of some of the directions that these studies will take in the years ahead.

There are two main ways to probe the distant cold Universe. First, observing the radiation from galaxies at mid-infrared (mid-IR) and longer wavelengths ($\lambda > 10\,\mu$m) that is emitted by atomic and molecular gas and grains of dust at temperatures of order $10 - 1000$ K. Line emission from molecular and atomic gas plays an important role in cooling dense gas clouds in the ISM, although little observational information is available about these phenomena early in the history of the Universe. In terms of bolometric luminosity, thermal emission from dust is certainly the dominant process at far-IR wavelengths out to moderate and high redshifts, $z \simeq 1-2$ and $z > 2-3$ respectively. Unless extreme redshifts are being investigated, the emission from stars makes only a very small contribution to the signal detected at wavelengths longer than $10\,\mu$m. Non-thermal synchrotron emission from high-energy electrons is also unlikely to be significant, apart from in very rare powerful radio-loud active galactic nuclei (AGN). Dust grains, the solid-phase of the ISM, and atomic and molecular ISM gas all radiate at wavelengths between a few microns and several millimetres. Grains are heated radiatively by optical and UV radiation from young stars and AGN. They could also be excited by mechanical energy deposited by shocks generated by interactions between galaxies, clouds within galaxies and supernova, but this is likely to provide only a very small fractional correction to the luminosity inferred from radiative heating.

Second, observing the effects of absorption by neutral and low-excitation gas along the line of sight to distant luminous objects. While this is predominantly a technique for optical observers, absorption line measurements of molecules backlit by intense radio sources have been made, revealing very detailed information about the properties of the ISM in galaxies at moderate redshift [268]. As the simultaneous spectral coverage of submillimetre-wave instruments increases and their sensitivity steadily improves, absorption line studies offer to become an important part of the armoury for studying the cool parts of the Universe.

Additional probes can be provided by exploiting features that are visible against the Cosmic Microwave Background (CMB) radiation at millimetre wavelengths by scattering along the line of sight. Fluctuations in the CMB, both spatial and spectral, typically at the $10^{-5} \to 10^{-4}$ level, are critical probes of cosmology. Spatial anisotropies in the CMB emission encode direct evidence about the galaxy formation process. Primordial features in the spectrum of the CMB trace activity in the early Universe, while foreground objects, which potentially include the very earliest galaxies and clusters of galaxies, can impose a variety of detectable characteristic 'secondary' anisotropies to the CMB signal. Both primary and secondary features of the CMB are already being detected. The most prominent is the Sunyaev–Zeldovich (SZ) effect [255,70], produced by electrons at X-ray-emitting temperatures, of order $10^8\,\mathrm{K}/10\,\mathrm{keV}$, which has been clearly detected along the line of sight to clusters of galaxies, and may develop into an important probe of less massive structures at high redshifts. The intensity of the CMB increases with redshift z as $(1+z)^4$, exactly counteracting the reduction in surface brightness with increasing redshift that affects all cosmological observations.

Direct observations of the submillimetre and far-IR emission from high-redshift galaxies are amongst the most recent techniques to be applied to the study of galaxy formation, beginning in 1997 [57]. This is largely because the technology to detect distant galaxies has matured later at submillimetre to mid-IR wavelengths than at radio and optical wavelengths [215]. The results discussed are generally all less than 5 years old, and so most of the relevant information was gathered since the extremely deep optical Hubble Deep Field (HDF) image was taken in 1995 [269]. Hopefully, the descriptions presented here will remain valid for many years to come; however, this is a young, very active field, and some of the results may be overturned when more data, secure multiwavelength galaxy identifications, and more time to contemplate the results are available. Observational capabilities continue to develop rapidly, in almost all the windows into the distant Universe. Some of the most relevant developments outside of the far-IR/submillimetre sphere are at radio, near-IR, optical and X-ray wavelengths. At radio wavelengths, the Very Large Array (VLA) will soon be upgraded to provide significantly improved sensitivity and resolution, and drastically improved spectroscopic capabilities. The UK-based MERLIN radio interferometer is also being upgraded dramatically, and the Indian Giant Metrewave Radio Telescope (GMRT) is being commissioned with a very large collecting area at the longest radio wavelengths. The number of 8-m-class optical/near-IR telescopes in service is increasing steadily, and full suites of instruments are being delivered for them. Wide-field multi-object spectroscopic capability at the Keck, Gemini, Subaru and VLT telescopes are on the verge of generating a new level of detail in deep galaxy redshift surveys, while the shallower, wide-area 2-degree-field (2dF) redshift survey is complete with about 230,000 redshifts [83], and the Sloan Digital Sky Survey (SDSS) [58] is well underway, generating galaxy redshifts at the rate of over $10^5\,\mathrm{yr}^{-1}$. The *Chandra* X-ray observatory has provided the first arcsec-resolution images of the very opposite of the cold Universe, and together with the greater collecting area, but coarser resolution provided by the *XMM-Newton* telescope, both offer excellent spectroscopic imaging of the high-energy Universe.

All of these facilities will have an impact on our understanding of the distant Universe and the genesis of galaxies. The importance of observing over as wide a range of wavelengths as possible, and the ways in which existing observations and theories can be extended and improved by adding information from the cooler parts of galaxies, will be emphasized. In the final section we will report on developments in instrumentation that should lead to a clear high-resolution view of the formation of galaxies in the years ahead.

The other parts of this book describe the details of the radiation processes that take place in the ISM. When studying galaxy evolution at long wavelengths, it is generally difficult to make very detailed observations, both because of the restrictions of spatial resolution, and due to there being only a few flux density measurements available from an entire distant galaxy. It is therefore important to remember that extragalactic astronomy at thermal IR to millimetre wavelengths lacks the precision of studies of molecular clouds in the Milky Way and nearby galaxies.

A practical long-wavelength limit to the spectral window in which galaxy evolution can be probed by observing the ISM is imposed by the CMB, which dominates the energy density in radiation in the Universe at the present epoch by about two orders of magnitude. As the total energy output of galaxies appears to be declining steadily with cosmic time at present, and it is likely that even as the Universe expands further and redshifts away the CMB energy density, CMB photons will not only remain the most numerous photons in the Universe but will also always dominate the energy density in the diffuse intergalactic cosmic radiation field, away from individual galaxies.

At the present epoch, the CMB radiation spectrum peaks at a wavelength of 2 – 3 mm. The CMB is the most accurate blackbody known. Averaged over the whole sky, its temperature $T_0 = 2.726 \pm 0.001$ (at a 95% confidence limit)[190], and increases with redshift z as $T_0(1+z)$. This temperature sets a lower limit to the temperature of any constituents of the Universe that interact strongly with radiation, such as electrons, dust grains and polar molecules. Because only Universal contents with this property are likely to radiate efficiently, then sources of thermal radiation at wavelengths longer than about 3 mm need not be considered.

In the context of galaxies and their constituent stars and gas clouds, the CMB is usually a relatively insignificant source of heat. The temperature associated with the interstellar radiation field (ISRF) in galaxies and the temperatures equivalent to the velocities associated with outflows, winds and galactic dynamics are usually much higher. For example, the Milky Way has a rotation velocity of order $220\,\mathrm{km\,s^{-1}}$, and the dispersion in velocity across a molecular cloud is typically $10-20\,\mathrm{km\,s^{-1}}$. The typical temperature of dust grains throughout the ISM is only 15 – 20 K, still much greater than the CMB temperature. However, given that the volume of space occupied by galaxies is much less than 10^{-6}, and the CMB is all pervading, the energy density in the CMB is large, corresponding to $4.0 \times 10^{-31}\,\mathrm{kg\,m^{-3}}$, as compared with the $1.88 \times 10^{-26}\,\Omega_\mathrm{m} h^2\,\mathrm{kg\,m^{-3}}$ in the form of regular matter. Since the density parameter $\Omega_\mathrm{m} \simeq 0.3$ and Hubble's constant $h \simeq 0.65$ (see later), the CMB contains 0.017% of the energy density in matter in the Universe at the present epoch.

This discussion will concentrate on thermal emission at wavelengths between the mid-IR and millimetre wavebands. The mid-IR spans wavelengths from about $5 - 40\,\mu\mathrm{m}$, starting at wavelengths where stellar emission no longer dominates the energy output of galaxies; the far-IR waveband covers from about $40 - 200\,\mu\mathrm{m}$; and the submillimetre and millimetre wavebands span from about $200 - 1000\,\mu\mathrm{m}$ and longer than $1000\,\mu\mathrm{m}$ respectively. This wavelength range is bounded at the long-wavelength side by the CMB. Observational capabilities throughout this spectral region are developing rapidly.

1.1 The Difficulties of Observing Faint and Distant Cold Objects

In comparison with the oldest branches of observational cosmology that probe the distant Universe by detecting galaxies at optical and radio wavelengths, studies at infrared wavelengths were slow to develop. This reflects the need to overcome several physical challenges to imaging faint objects between infrared and submillimetre wavelengths.

The first challenge is one of resolution. We are accustomed to seeing spectacular, high-resolution, multicolour views of the most distant galaxies obtained by the *Hubble Space Telescope (HST)*, including the Northern and

Southern HDFs (HDF-N & HDF-S[270]).[1] At longer wavelengths, the resolution of images from a telescope of fixed aperture D falls steadily, and images cease to be arresting as soon as the ability to resolve the sub-arcsecond scale structures within galaxies at moderate redshifts is lost. In the absence of seeing this occurs at wavelengths longer than about $4[D/\mathrm{m}]^{-1}\,\mu$m. At near-IR wavelengths diffraction-limited images retain excellent spatial information content; however, at wavelengths longer than 10 μm, the aperture required to resolve distant galaxies grows towards 10 m. At still longer wavelengths, arcsec resolution cannot be achieved without the use of interferometry with baselines in the 100–1000 m range. The largest apertures flown in space to date are just 0.6 m aboard both the *Infrared Astronomy Satellite (IRAS)* in 1984 and the *Infrared Space Observatory (ISO)* in 1995. The *Space Infra-Red Telescope Facility (SIRTF)*, scheduled to fly early in April 2003, has a 0.85 m aperture. The largest planned space aperture at long wavelengths is currently the 3.5 m *Herschel Space Observatory* (formerly *FIRST*) ESA/NASA far-IR/submillimetre mission scheduled for launch in 2008, and the proposed Japanese 4-m class cold-aperture *SPICA*[2] mission. $10-20$ m-class single-aperture mid- and far-IR space telescopes, such as *SAFIR*, and interferometers with longer baselines [191] have been proposed.

If the resolution is coarsened further, to levels insufficient to separate adjacent galaxies on the sky, then the ultimate depth of observations is limited not by the ability to integrate longer and suppress photon noise, but by the unknown distribution of faint unresolved sources on the sky that cannot be detected individually. This problem of 'confusion noise' dogged early radio and X-ray observations. It is also a significant potential problem for observations using *SIRTF*, especially for the deepest surveys to probe the high-redshift Universe. As detectors become more sensitive, confusion registers as an ever more pressing problem: if the relevant depth cannot be reached, due to noise in the instrument, then confusion is not an issue! At submillimetre wavelengths from the ground, it currently takes many 10's of hours of integration to reach the confusion limit. With a cooled telescope in space it could potentially take only seconds. At the longest wavelength probed by *SIRTF* – 160 μm – confusion noise is likely to dominate other sources of noise in only about 10 s; although the time required will be much greater at shorter wavelengths.

A second challenge is a physical one in the detectors themselves. There are two major relevant regimes. At the longest wavelengths, it is natural to make 'heterodyne' electronic detections of the electric field of radiation as it interferes with a very stable reference 'local oscillator' (LO) signal in a non-linear detecting element of comparable size to the wavelength of detected

[1] The Advanced Camera for Surveys (ACS) has just been installed on *HST* to provide even more sensitive and larger images than those from the existing Wide Field and Planetary Camera (WFPC-II).

[2] `www.isas.ac.jp`

radiation, or in a smaller detector coupled by a waveguide. This produces a beat 'intermediate frequency' (IF), whose processing dramatically reduces the frequency at which the receiver electronics are required to operate. The full spectral information is recovered by auto-correlating the IF output in time. This works extremely well at frequencies less than of order several hundred GHz. Efficient detection becomes more and more difficult as the frequency increases: it becomes challenging to manufacture the millimetre-scale components necessarily, to simulate the electromagnetic fields in the detector, and to generate precise LO signals at the relevant frequencies. Cryogenic heterodyne receivers fitted to radio, millimetre and submillimetre telescopes currently perform at close to the theoretical quantum efficiency limit set by the photon energy at these frequencies. Heterodyne devices at frequencies greater than 1 THz have been tested, but it is still very challenging to design and manufacture low-noise receivers for submillimetre wavelengths.

At the shortest wavelengths it is more natural to detect the energy deposited by a photon in the detector. This typically requires the promotion of electrons above a band gap in some semiconductor device, or the detection of the temperature change caused by deposited energy using a bolometer. At optical wavelengths this is a mature technology: CCD devices operate with quantum efficiencies over 90%. At longer wavelengths the technology is more challenging. It is essential to cool the detectors so that the bandgap remains much larger than the typical thermal energy of the electrons in the device. At wavelengths longer than about $10\,\mu$m this requires cryogenic temperatures, and very low cryogenic temperatures for wavelengths longer than $200\,\mu$m. Self-stabilizing superconducting sensors operating at their transition temperature due to thermo-electric feedback, and other innovative detection devices, offer dramatic future gains for bolometer sensitivities at submillimetre wavelengths. The advantage of stability is that the bolometer current/voltage that measures the intensity of illuminating radiation does not need to be read out constantly, but just twice per thermal response time, thus reducing the number of wiring connections required. This both limits the conduction of heat into the detectors and simplify their manufacture. Some novel devices are read out using microwave radiation, allowing their designers to exploit advanced yet inexpensive mobile telecoms technology to address the detectors, thus eliminating wiring heat load entirely.

The reasons for the difficulties of detecting $10-1000\,\mu$m sources, and why the first deep surveys at these wavelengths were made only in the late 1990's is now hopefully clear.

A third challenge is the effects of the atmosphere and optics. Putting a large cold aperture in space has not so far been practical, either financially or technically, and so observations have been carried out from small space telescopes or from the ground. Thermal radiation from the atmosphere and the telescope optics mask mid-IR signals. Molecules in the atmosphere, especially O_2, OH and H_2O, both emit and absorb powerfully, ruling out observations in specific bands. With the exception of wavelengths shortward of the far-ultraviolet(UV) atmospheric cutoff, all the opaque parts of the atmosphere are found between the submillimetre and near-IR wavebands.

These factors – technical difficulty of reaching sensitive detection limits, absorption and masking emission by the atmosphere and limited resolution – have all conspired to keep the submillimetre and infrared windows closed to sensitive observations long after the optical and radio windows had been substantially investigated. There is much to learn from this previously hidden window.

2 Radiation from the Cool ISM in Galaxies and Its Detection

2.1 Galaxy Structure

The ISM of galaxies includes a variety of phases in a very complex interlinked geometry. The stellar populations of large low-redshift galaxies can usually be divided easily into an old, evolved, passive, pressure-supported spheroidal component at the centre, with a scale height of up to a few kiloparcsecs (kpc)[3]; and a more extended, rotationally-supported disk component with a scale length of typically several kpc, that contains stars with a much broader age distribution, and a much greater proportion of gas. In elliptical galaxies the spheroid is dominant; whereas spiral galaxies are dominated by the disk component. Spiral and elliptical galaxies dominate optical atlases of the nearby Universe. However, these relatively serene and static objects are not the galaxies most likely to be found when investigating the evolution and formation of galaxies [25]. They are the almost fossilized end-points of the galaxy formation process, and so while any model or theory must explain their existence, it is the more dramatically active and luminous objects that are most important at infrared wavelengths.

Irregular galaxies do not fit into the elliptical/spiral description. They are relatively rare in the local Universe and were relegated by Hubble to an extra box, disconnected from the main tuning fork of his galaxy morphology diagram. Their disorganized structure is taken to be evidence of either a recent strong tidal disturbance, or the debris remaining from a merger or interaction with a neighbour. By stirring up the ISM, these processes could easily lead to an enhanced rate of star formation: irregular galaxies are typically blue, and contain some very young stars. Interacting galaxies, for example the Antennae [196,271] have very disturbed ISMs. The disks of the two spiral galaxies involved in this interaction are deeply entwined. A large density of gas has been funneled into the region between the galactic nuclei, and a minority of the mass of the disks has been flung out in long (100 kpc), thin tails of tidal debris, extending far beyond the extent of the disks prior to the interaction.

[3] $1\,\mathrm{kpc} = 3.09 \times 10^{19}\,\mathrm{m}$

When observed at infrared and submillimetre wavelengths, irregular and interacting galaxies often show the most powerful emission [198]. They are the objects that should come to mind immediately when powerful infrared galaxies are discussed.

In addition, a small fraction of low-redshift galaxies contain AGN, with very bright, point-like nucleus that is naturally interpreted as the accretion disk surrounding a massive blackhole. AGN detected at optical wavelengths in the form of extremely bright, blue/UV quasars/QSOs, and at radio wavelengths as radio galaxies with extended double lobes of synchrotron emission, were the first high-redshift objects to be discovered and identified. The detected increase in their space density with increasing redshift to $z \simeq 2$, combined with the eventual discovery of a convincing decline in their space density at still greater redshifts [236,104] provided the first evidence for strong evolution of galaxy populations over the Universe's history.

In some cases, AGN could result from intense, central clusters of OB stars in the nuclei of galaxies, while strong star-formation activity is often likely to be associated with the accumulation of gas that goes on to feed a blackhole. This feeding process is likely to be most efficient during galaxy interactions, as gas is driven inwards by tidal forces, and the orbits of gas and stars will be disrupted, refilling orbits that cross the accretion disk of the blackhole. Periodic blackhole accretion/AGN activity matched to enhanced star-formation activity during galaxy interactions seems likely to be an important part of the story of galaxy evolution, and one that is likely to excite emission from the ISM in regions heavily enshrouded by dust, and thus easiest to study at long wavelengths.

2.2 AGN Accretion and Star Formation

The question of the relative roles of heating by AGN and young, high-mass stars is an important one, as the most luminous inferred galaxies could be powered in almost any proportion by accretion onto an AGN or the formation of massive stars. The efficiency of conversion of mass to energy in star formation is limited to the $\sim 0.7\%$ of rest-mass energy that can be released by fusion between hydrogen and iron, but for accretion to the radius of the last stable orbit of a supermassive blackhole the efficiency could reach about 30%, and so much less mass is required to generate luminosity by fueling an AGN than by forming stars.

The recent discovery of an almost linear correlation between the mass or velocity dispersion of the evolved stellar bulge component in galaxies and the inferred mass of the central blackhole [188,120,135] can inform this issue. A few parts per thousand of the mass of the most homogeneous, oldest and tightly-bound stellar populations in most galaxies is in the form of a blackhole. This is certainly true of the Milky Way, in which a $2.9\times10^6\,\mathrm{M}_\odot$ compact object lurks at the Galactic Centre within the $\simeq 10^9\,\mathrm{M}_\odot$ bulge [137,111].

The processes setting these stellar and blackhole masses would seem to be connected, with a clear role for feedback between accretion and star-formation in the centres of galaxies [239]. Accretion could perhaps regulate star formation in the bulge. With an abundant fuel supply, a blackhole can accrete at the Eddington rate, gaining mass exponentially with radiation pressure on electrons balancing gravitational forces on protons. Direct or scattered radiation from the AGN extends its influence further and further into the galaxy, far into low-ionization regions of the bulge, producing an increasingly powerful, radiation-pressure driven outflow, clearing out gas from the central regions of the galaxy. Beyond a threshold strength depending on the size and mass of the host galaxy, this would prevent both further star-formation and limit accretion as fueling gas is expelled. An alternative picture in which an efficient burst of star formation in the bulge consumes all the available gas, terminating accretion on the same timescale as the star-formation is also possible [12]. Seed blackholes could form by the direct gravitational collapse of primordial overdense regions [145], from core collapse in an early generation of supermassive stars, or from the collapse of dense star clusters [110]. Seed blackholes could gradually accrete baryonic and dark matter from the surrounding halo inefficiently [186], invisibly generating the observed correlation, if the properties of the dark matter halo are tuned correctly. Independent star-formation activity would take place both earlier and later, to form a surrounding bulge. To arrive at the observed correlation by the present epoch, would require all relaxed stellar bulges to have similar velocity and density structure, and for the process of blackhole accretion having proceeded for a similar time in all cases, which does not seem very likely. The limited scatter in the blackhole–bulge 'Magorrian relation' implies that the bulge and blackhole formation processes are linked.

This general discussion of the structure of galaxies and their possible formation histories is intended to illustrate the importance of their infrared emission for studying galaxy formation. However, only the closest galaxies can be resolved at these wavelengths, and only the disk components and nuclei of galaxies are detectable: ongoing star formation or AGN activity is essential for heating dust. Furthermore, there is no way to resolve dispersed star-formation activity from a much smaller AGN (not that this is necessarily easy in the case of exquisite angular resolution [262]).

2.3 The Structure of Infrared Emission

Without angular resolution, infrared observations provide a single measurement of the integrated energy emitted by all parts of a galaxy. This is in marked contrast to the resolved observations possible for star-forming clouds in the Milky Way. Hopefully, this situation will improve, with 0.1 – arcsec resolution becoming available with the commissioning of the SubMillimeter Array (SMA) [148] on Mauna Kea, with a 500 – m maximum baseline and the Atacama Large Millimeter Array (ALMA)[272] with a maximum baseline longer than 10 km. Limits to the angular size of very luminous infrared galaxies at millimetre wavelengths are available, from the existing OVRO Millimeter Array (MMA)[157] and IRAM Plateau de Bure (PdBI) interferometers [225,100].

However, spatially resolved submillimetre images are not currently available, and none are foreseeable at far-IR wavelengths, prior to a space-borne telescope with an aperture or baseline larger than about 10 m. As a result, it is important to be aware that all sources of infrared emission from distant galaxies are superimposed, and can be discriminated only in the spectral domain.

To investigate the spectral energy distribution (SEDs) of infrared-luminous galaxies at optical or infrared wavelengths, it is not presently either possible to constrain adequately or necessary to use radiative transfer techniques to determine the dust distribution in distant galaxies, due to the lack of detailed information about its spatial structure. Simplifying assumptions can be made when considering radiative transfer at longer wavelengths, however. The main one is that, once emitted, a far-IR photon is likely to escape from the galaxy unhindered, without scattering or absorption. The opacity of dusty regions of galaxies can be extremely high at the optical and UV wavelengths where the sources heating the dust are most luminous, and the absorption efficiency of dust is greatest: there are up to 20 magnitudes of V-band extinction to the Galactic Centre, and at least 50 magnitudes to the centres of the most heavily-enshrouded AGN. However, even at 2.2 μm the opacity has reduced enough for the most luminous stars in the Galactic Center can be discerned [111,137]. The optical depth at wavelengths of order 20 μm is expected to be relatively small, while at wavelengths longer than of order 100 μm the galaxy the opacity of the overlying ISM is less, by of order 100, and so general sightlines through galaxies are optically thin or only marginally optically thick. Only the very densest clouds are likely to remain optically thick at longer wavelengths. Hence, illuminating star-forming regions and AGN can all be considered as point sources, heating dust locally, and contributing to the general ISRF that heats dust on larger galaxy-wide scales. At longer wavelengths, radiation escapes from the galaxy without further interactions, and so this part of the radiative transfer leading to the far-IR SED is easy to deal with.

Existing radiative transfer studies have generally attempted to discriminate between point-like AGN and star-formation regions distributed throughout the galaxy [141], or to predict expected SEDs for star-forming galaxies from semi-analytic galaxy formation models [96]. Note, however, that these models are restricted to very constrained geometries: cylindrical or spherical distributions, and sources embedded within disks. It is unlikely that the most luminous and interesting galaxies will conform with these simple geometries, and so care is still required when interpreting the SEDs of these objects.

Predicting the SED of a dusty galaxy at long wavelengths is much more straightforward than at optical wavelengths, where scattering and attenuation occurs throughout the ISM as a photon makes its way out of a galaxy. In contrast, the reprocessed SED of each dust cloud as detected on Earth at long wavelengths should be very similar to the spectrum emitted by the cloud within the galaxy. The difficult step is treating the processing of optical and UV photons within the probably turbulent and inhomogeneous regions around AGN and young stars, and the neighbouring dusty molecular clouds. For AGN, scattering and absorption from clouds at considerable distances from the nucleus could be important – especially in a merging system, if the intense escaping emission from the AGN in one galaxy shines on the disk of its neighbour. If predicting the far-IR properties of a dusty galaxy is relatively easy, apart from the issue of geometry, predicting its optical properties, and the fraction of energy that will be absorbed by dust is effectively impossible. This may change when data with comparable resolution from ALMA at submillimetre wavelengths and the *James Webb (Next Generation) Space Telescope (JWST)* in the near-IR are available for large numbers of galaxies.

The far-IR emission produced by a point-like AGN/nuclear star-forming region, and hot stars distributed more widely through the ISM can be described by a distribution of dust temperatures and relative luminosities at each temperature. Almost all the hot dust within the galaxy will be contained in these regions, but their intense radiation field and strong outflows are likely to disperse the dust on timescale comparable to the lifetime of a high-mass star. Hence the youngest star-formation regions are likely to start out invisible at optical and UV wavelengths, and to be detectable only in the infrared. Understanding the optical/UV appearance of the galaxy will be further complicated by this time-dependent appearance of different dust-enshrouded star-forming regions. The optical luminosity of a dusty galaxy is thus likely to increase at later times as the dust is dispersed [160]. The ISM that fills the space between high-luminosity regions will be illuminated by the general ISRF of the galaxy, contributed both by scattered and escaping optical/UV radiation from the densest luminous regions and from longer-lived low-mass stars throughout the galaxy, and the CMB.

2.4 Dust Emission at the Coarse Level Relevant to Cosmology

The complex details of heating of the solid-phase ISM are explained in much more detail elsewhere [103]. Here we discuss the issues that are most relevant to studying and understanding unresolved infrared galaxies. This is not to deny that a deeper understanding of the physics and chemistry of the ISM is essential; rather it expresses the difficulty of constraining these details from existing observations. It is important to retain a healthy scepticism about reports that a certain galaxy contains dust grains with a precise dust temperature or temperature distribution, or that any specified dust mass in a galaxy is quoted with an error less than 200%.

On a grain-by-grain basis, the thermal emission from dust has a Planck function SED B_ν, defined by dust temperature T, modified by an emissivity function ϵ_ν, while at the longest wavelengths, scattering theory indicates that $\epsilon \propto \nu^\beta$ with $\beta = 2$, while at the shortest wavelengths, $\beta \simeq 1$. The observed value from fitting the SEDs of galaxies at $1 - 0.1$ mm is of order $1.3 - 1.5$ [107]. This single composite value reflects complex geometry, variable optical properties of the grains, and an inevitable distribution of dust temperatures. At shorter wavelengths, grains at temperatures less than 100 K emit on the Wien side of the spectrum, and an exponential SED is expected. This will be modified by small grains heated out of equilibrium by absorbing single high-energy photons before cooling radiatively back to an equilibrium temperature of $10 - 100$ K. Also, the small fraction of grains very close to hot stars and AGN will be heated to higher temperatures than grains in the general ISRF, and emit at shorter wavelengths. Galaxies are observed to have mid-IR SEDs closer to a power-law than an exponential function, due to the summed emission from these hotter components (Fig. 1). Dust sublimes when heated to temperatures of order 2000 K. At the corresponding wavelength of about 2 μm, low-mass starlight starts to dominate the SED, and so the inevitable Wien tail of the near-IR dust spectrum from the hottest grains present at higher frequencies is masked by stellar emission.

The far-IR SED of a galaxy is the sum of many different components. However, as most galaxies are described only by up to ten data points all the way from mid-IR to millimetre wavelengths, it is reasonable at present to reduce the number of parameters and components involved to about three. If these parameters are well chosen, then the results will be physically relevant. We adopt the simplest parametrization permitted by the data here. At long wavelengths, the SED is described by $f_\nu \propto \nu^\beta B_\nu(T)$; in the mid-IR waveband, a power-law spectrum with $f_\nu \propto \nu^\alpha$ is assumed. These functions are grafted together at the frequency where the logarithmic gradient of the long-wavelength SED f_ν, $\mathrm{d}\ln(\nu^\beta B_\nu)/\mathrm{d}\nu$, steepens beyond α. As shown in Fig. 1, this can provide a reasonable description of diverse observed galaxy SEDs. Without including the power-law spectrum in the mid-IR, predicted SEDs are too steep to agree with observations.

In order to reflect the general features of the mid-IR SED of a galaxy made up of dust clouds at various temperatures, or to describe the summed emission from many galaxies at wavelengths from about 200–20 μm it is reasonable to describe the spectra of the individual components as δ functions, which makes the process of integrating over the different components very easy. A more detailed description is only called for at wavelengths that extend beyond the range of peak wavelengths of the SED [44].

The value of temperature T has a clear meaning. It determines the position of the peak of the SED, and reflects a luminosity-weighted temperature for the dust in the galaxy. β includes some information about the physics of dust grains, from their long-wavelength spectra, and about the relative

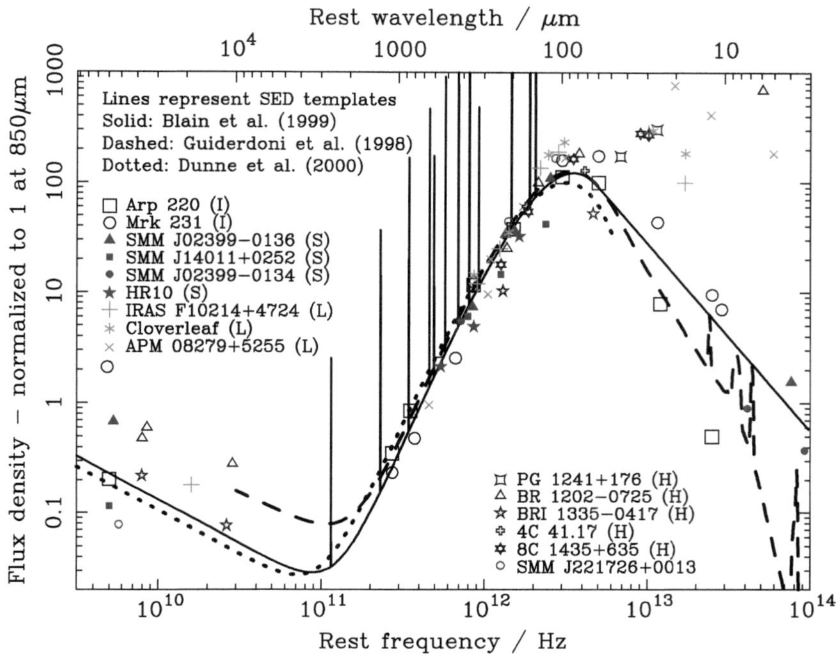

Fig. 1. Various observed multi-wavelength SEDs of the dust spectra of luminous galaxies, both nearby *IRAS* galaxies (I), distant submillimetre-selected galaxies (S), high-redshift lensed galaxies (L), and high-redshift AGN (H). Three model SEDs are shown for comparison [56,143,107], derived from fits to submillimetre and far-IR data, from fits to composite *IRAS* galaxy SEDs [91], and from submillimetre observations of low-redshift *IRAS* galaxies respectively. With the exception of some of the mainly lensed AGN, the templates provide a reasonable description. One of the models (*solid line*) includes CO line emission expected at submillimetre wavelengths, and another (*dashed line*) includes polyaromatic hydrocarbon (PAH) emission/absorption features at mid-IR wavelengths. Note, however, that the less luminous Milky Way has a temperature of order 20 K, approximately half of the 40 K assumed in the templates of luminous galaxies, while the SED of the extremely luminous lensed QSO APM 08279+5255 [152] has a thermal spectrum with $T \simeq$ 110 K. A wide range of dust temperatures is clearly present in galaxies.

amounts of dust present at temperatures much less than T. α reflects relatively directly the fractional contribution from much warmer grains. Some very luminous AGN at high redshift clearly have a very flat mid-IR spectrum [175]. This implies that they have a very significant contribution from hot dust close to an AGN, or possibly an intense, but very inhomogeneous ISRF.

Similar quality fits to data can be obtained by representing the SED of a galaxy using 2 components with different dust temperatures [108]; however, it is unclear how best to determine these temperatures and whether they have any direct physical relevance to a two-phase ISM [53].

2.5 Dust Mass and Temperature

The mass of dust in a galaxy is relevant for understanding its chemical evolution prior to the observation. The creation, destruction and recycling of dust are important for understanding both the process of star formation, and the feedback of processed metal-rich material into the ISM, in all galaxies. This is a potential advantage of studying galaxy evolution from dust emission. First, the currently active star-forming regions are picked out clearly by their powerful dust emission, without any delay as stars to burn out of their dust-enshrouded nurseries to become visible at optical wavelengths. Secondly, the dust mass should give a direct measurement of the amount of heavy elements that have formed earlier, subject to few uncertainties about stellar astrophysics or the fraction of metals sequestered in stars, which can be probed using stellar spectroscopy.

The geometry and size of dust grains is very significant for mass determinations, but less so for the general appearance of the galaxy SED. Dust grains, regardless of size or composition, have a high optical/UV opacity, and thus absorb efficiently, re-emitting the radiation at similar temperatures, with the caveat of small grains emitting briefly out of equilibrium in the mid-IR.

It is not clear whether dust is composed of spherical, filled grains, or more exotic strongly ellipsoidal objects, whiskers, miniature rubble piles, or colloidal 'snowflake' structures. All but the first of these geometries could significantly reduce the dust mass required to produce a certain optical effect. If, for example, high-redshift dust is composed of delicate networks of carbon nanotubes in 'snowflake' grains, then the mass of carbon required to generate sufficient absorption to obscure early generation of stars could be very modest, as compared with the expectation for filled spherical 'hailstone' grains at low redshifts in the Milky Way. This would have a very significant effect on calculations of the prior cumulative amount of star formation activity required to generate the observed opacity in high-redshift QSOs [32,153,201]. Uncertainties in the geometry of dust grains can also affect the likelihood of their destruction, for example by electrostatic stress built up by photoelectric emission by X-rays [129].

It is likely that moderate spectral-resolution observations of grains in the $10\,\mu$m wavelength range in the Milky Way and nearby galaxies could detect signatures of vibrational and loosely-bound electronic transitions in grains, perhaps revealing this type of information. However, it is unlikely that the evolution of dust properties could be studied at high-redshifts, as the spectral signatures would be too faint to detect.

In any case, it is important to be aware of the potential for determining some estimate of the dust mass from multi-wavelength observations of a galaxy SED between about 1 mm and $10\,\mu$m. As a galaxy is likely to be almost transparent to dust radiation, the total luminosity should be linked to dust mass, subject to the size of the grains a. The emissivity of a grain is linked to its area, which depends on a^2. Mass depends on a^3, and so the specific emissivity $\epsilon \propto a^{-1}$: a cloud of dust is a better emitter than a cloud of pebbles with the same mass. It is also a better absorber. The interaction of radiation and dust is much more complex [103], but this is the level at which it is typically treated by extragalactic observers.

The detected flux S_ν from a cloud of dust with a certain dust grain size distribution should depend on the total mass M and the SED f_ν of the cloud as $S \propto f_\nu(T)M$, while the luminosity $L \propto M \int f_\nu \, d\nu \propto MT^{4+\beta}$. In the Rayleigh–Jeans regime, typical of submillimetre-wave observations, $f_\nu \propto \nu^{2+\beta}T$ and so $S \propto \nu^{2+\beta}MT$. This means that for a certain measured long-wavelength flux density, the mass is determined as accurately as the temperature. However, the luminosity $L \propto ST^{3+\beta}$, and is very poorly determined without a measurement of the flux near the peak of the SED to fix the temperature accurately [109]. An example of the values of the parameters T and β required to fit a real dataset are shown in Fig. 2. Near the peak, the form of the SED depends weakly on temperature, and so mass and luminosity

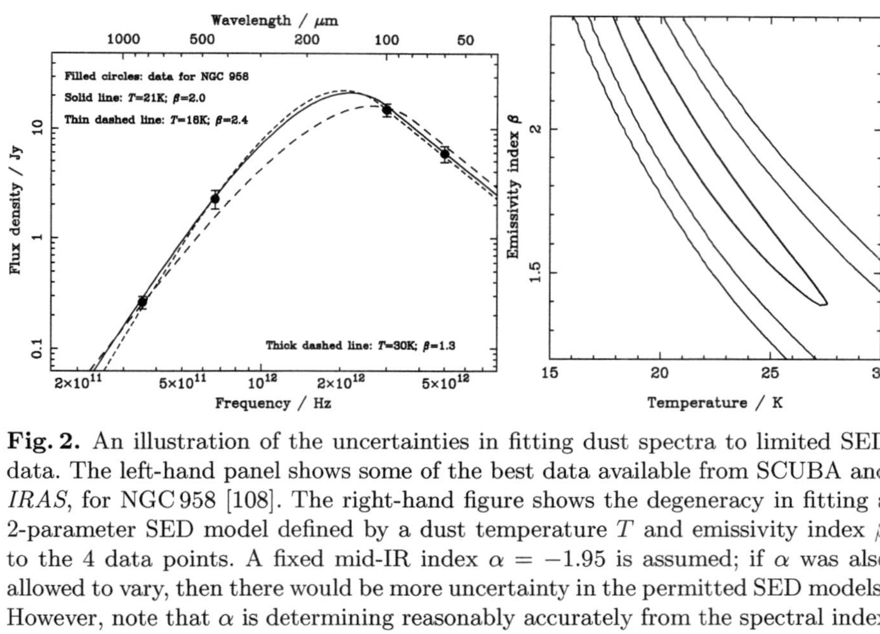

Fig. 2. An illustration of the uncertainties in fitting dust spectra to limited SED data. The left-hand panel shows some of the best data available from SCUBA and *IRAS*, for NGC 958 [108]. The right-hand figure shows the degeneracy in fitting a 2-parameter SED model defined by a dust temperature T and emissivity index β to the 4 data points. A fixed mid-IR index $\alpha = -1.95$ is assumed; if α was also allowed to vary, then there would be more uncertainty in the permitted SED models. However, note that α is determining reasonably accurately from the spectral index between the 60− and $100 - \mu$m *IRAS* points. There is a great deal of freedom to exchange hotter T values and lower β values. The SEDs that correspond to both ends of the probability contours in the right-hand panel are shown on the left by the solid and thin dashed lines respectively. The $450 - \mu$m data point was obtained most recently. Without it, the thick dashed curve provided the best fit to the data [107]. Adding additional data points near the peak of the SED would reduce the width of the probability contours, but not remove the degeneracy between T and β (see [53]).

can be constrained more accurately. The effects of redshifting the spectrum are discussed below.

In Fig. 3 the effects of assuming different functional forms for the SED are compared. An SED that includes an opacity [32], two different dust cloud components [108], and a power-law distribution of dust masses and temperatures [91] are compared with the SED model introduced above. As can be seen from the caption, the value of temperature required to obtain a good fit to the data differs by a small factor. However, the simplest model can provide

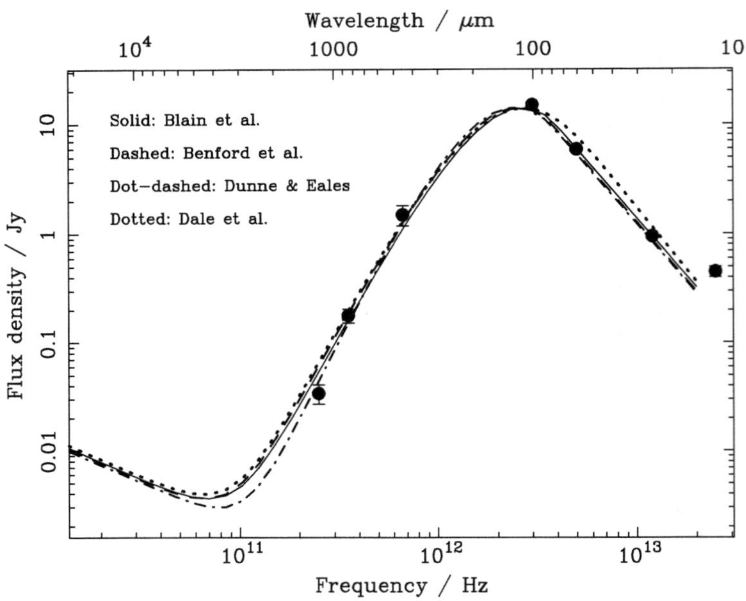

Fig. 3. A comparison of the best-fitting models for the SED of NGC 958 (see Fig. 2), extended to include additional data from the NED database. Four template SEDs are included. The one described above [25], with $T = 29\,\mathrm{K}$, $\beta = 1.5$ and $\alpha = -2.1$. The value of β is assumed to be fixed at 1.5, and the other two parameters were fitted. One includes an optically thick dust cloud [32] with an emissivity function $1 - \exp(\nu/\nu_{\mathrm{opt}})^{\beta}$. ν_{opt} is assumed to be $2.4 \times 10^{12}\,\mathrm{Hz}$, $\beta = 1.5$ and the best-fit temperature $T = 33\,\mathrm{K}$. The choice of the value of ν_{opt} within reasonable bounds has very little effect on the result. Another model includes two clouds of dust at different temperatures [108]: the cooler cloud is assumed to be at 20 K, emissivity indices $\beta = 2$ are imposed on each cloud, and the best fit is obtained when the hotter cloud has $T = 30\,\mathrm{K}$ and contains 12.6% of the dust mass of the cooler one. A mid-IR SED slope $\alpha = -2.1$ is assumed in all three models. The final model includes a power-law distribution of dust temperatures [91], with $m(T) \propto T^{\gamma}$ with $\gamma = -7.75$ between temperatures of $T_{\mathrm{min}} = 22\,\mathrm{K}$ and 2000 K, an emissivity index $\beta = 1.5$ with T_{min} and γ determined by the data. Assuming the standard form of the far-IR–radio correlation [85] all models accurately reproduce the observed 1.4-GHz flux density of NGC 958.

a good representation of the data, and it is not necessary, given the current data with low spectral resolution to be concerned with the effects of opacity and different dust components on the results.

In order to investigate the SEDs of dusty galaxies in detail, spatially resolved observations, capable of revealing all the star-forming regions with various ages, and the contribution from AGN, complex geometries and opacity distributions are required. Since these are not yet available, a reasonable description of the SED of a dusty galaxy can be provided by a dust temperature. That sets the position of the peak of the SED, and a distribution of dust clouds at hotter temperatures that contribute to an approximately power-law spectrum defined by α at shorter wavelengths.

2.6 Long-Wavelength Observations of Low-Redshift Galaxies

The only galaxies for which we have accurate and detailed SEDs that extend over the complete spectral range from millimetre-wave to near-IR wavelengths are at very low redshifts. This includes the Milky Way, which was mapped in detail by *IRAS*, and the FIRAS and DIRBE instruments aboard the *COBE* satellite [219]. The importance of dust emission as a contributor to the luminosity of typical low-redshift galaxies was first established by *IRAS* at 12, 25, 60 and 100 μm, yet the most luminous galaxies detected by *IRAS* lie at redshifts of only about 0.2 or 0.3. The SEDs of galaxies like the Milky Way typically peak at around 200 μm, while hotter galaxies more typical of those found in the *IRAS* survey peak at wavelengths between 60 and 100 μm (Fig. 1). Hence, the temperature of the dust in *IRAS* galaxies can be determined only from the 60– and 100 – μm channels. In fact, because these bands both lie either close to the peak of the SED or on its short wavelength side, it is difficult to extract detailed information about dust temperature, especially for cooler galaxies (Fig. 2). The *IRAS* data imposes a maximum permitted temperature, to avoid generating excess flux in the 25 – μm band. The temperature of cooler galaxies with SEDs that peak at wavelengths longer than the 100 – μm *IRAS* band are difficult to constrain without data at longer wavelengths.

A subset of galaxies detected by *IRAS* were subsequently selected for observation in pointed observations using the *ISO* satellite in the mid 1990's. These observations were made using a variety of imaging and spectroscopic modes, at wavelengths over a wider range than *IRAS*, covering from 5 to 200 μm. Deeper imaging surveys using *ISO* [164,218,7,114] found new populations of dusty galaxies at greater distances than *IRAS*, providing significant information about the evolution of relatively low-redshift dusty galaxies.

At longer submillimetre wavelengths, information is still relatively scarce. Using the first generation of single-element submillimetre-wave bolometer photometers, *IRAS* galaxies were detected in the 1990's [10], providing information about their SEDs at longer wavelengths. The commissioning of the SCUBA 450/850 – μm camera with a 5-arcmin^2 field of view at the JCMT provided opportunities for studies of the long-wavelength emission from a larger sample of *IRAS* galaxies [107,179,108]. The typical temperature of this *IRAS*-selected sample [107] is $T = 38 \pm 3$ K, with an emissivity index $\beta \simeq 1.3 \pm 0.2$. Observations of more luminous low-redshift galaxies [179] tend not to indicate a dramatic luminosity dependence of the SED, other than to note that spiral galaxies with infrared luminosities of order several $10^{10}\,L_\odot$ have SEDs with $T \simeq 20$ K, whereas more luminous $10^{12} - L_\odot$ galaxies have $T \simeq 40$ K. 40 K also seems to be the typical temperature of many high-redshift submillimetre-detected AGN [32,201,153].

However, there are exceptions. The nearby irregular galaxy M82 has a well-determined dust temperature $T \simeq 40$ K, despite a relatively low luminosity of about $3 \times 10^{10}\,L_\odot$ [53]. It is clearly involved in a short-lived burst of star-formation activity, however, and so conditions within are perhaps likely to be more similar to those in more luminous galaxies. Also, some moderate and high-redshift AGN have much higher dust temperatures [93,175]. This is likely to represent a significant fraction of the luminosity of these galaxies being generated either close to the accretion disk of an AGN, or from a region of a galaxy in which the ISRF is at least an order of magnitude more intense than in cooler 40-K objects. It is not yet possible to be sure of the true distribution of dust SEDs in high-redshift galaxies, and some $175 - \mu$m *ISO*-selected galaxies have been found with very modest dust temperatures close to 30 K [76]. Hopefully, the large-format, very sensitive images expected from the MIPS instrument on the forthcoming *SIRTF* satellite will provide the necessary information at wavelengths of 24, 70 and 160 μm, spanning the peak wavelength of any reasonable galaxy SED. The SINGS *SIRTF* Legacy Project[4] will generate detailed information for about low-redshift galaxies, while a variety of deeper surveys will investigate the evolution of galaxies and enable cross-correlation studies with a large sample of optically-selected galaxies.

2.7 Absorption Studies

Optical: The first technique to study ordinary galaxies in the high-redshift Universe was observations of absorption lines against the intense optical radiation from QSOs. While the emission from these galaxies, which can be optically thick and very faint, is difficult to detect, they can be clearly detected against bright background sources by way of saturated absorption features, in some cases with exquisite signal-to-noise ratios [117]. These studies began as soon as the first high-redshift QSOs were identified. Gunn & Peterson [144] noted in 1965 that the existence of detectable emission bluewards of redshifted Lyman-α in the spectra of QSOs indicated that the intergalactic medium (IGM) along the line of sight was almost entirely ionized, presenting

[4] Information about *SIRTF* Legacy Projects can be found at `http://www.sirtf.caltech.edu` .

no resonant scattering cross section. Only this year has the first evidence for such absorption been presented, at $z > 6$ [30,99]. The IGM along the line of sight to distant galaxies is ionized; while discrete dense, neutral systems provide Lyman-α absorption lines in QSO spectra, with the most spectacular results coming from the brightest illuminating sources [117]. In principle, the abundance and density of the absorbers can be used to trace the evolution of galactic systems that contain cool neutral gas, subject to possible selection effects, such as bias against including QSOs with significant foreground extinction in surveys [118], which now seems relatively unimportant based on samples selected with no foreground extinction in the radio [116].

The evolution of the cosmic density of neutral gas that results from integrating the whole population of Lyman-α absorbers is consistent with most estimates of the rate of gas consumption. However, it is difficult to disentangle the role of time-dependent effects: the consumption of gas in stars, its dispersion in stellar winds, the formation of new galaxies and the assembly of reservoirs of cold absorbing gas in pre-existing dark-matter haloes. At all observed redshifts, the abundance of the most massive, gas-rich systems, the damped Lyman-α systems, with hydrogen column densities greater than $10^{20}\,\mathrm{cm}^{-2}$ seems to be rather similar [213]. The details of these optically-based studies are far beyond the scope of this paper, but it is important to remember that optical studies of cold gas in the high-redshift Universe can be made.

Gamma-Ray Bursts: One of the most exciting prospects is extending studies of absorption line systems to include objects silhouetted against the intense short-lived optical flash/afterglow emission of high-redshift γ-ray bursts (GRBs). GRBs are thought to trace the formation rate of high-mass stars, on the grounds of their apparent association with iron X-ray absorption features, typical of the debris from supernovae, supernova-like features in some of their light curves [220], and their common association with irregular blue host galaxies [98]. As a result of their exceptional brightness, GRBs should be detectable out to the highest redshifts, and as γ-rays are not subject to absorption by gas or dust, GRBs even at $z \gg 6$ should be readily detectable at MeV–GeV energies, regardless of foreground structures, including galaxies with a dense neutral ISM. The associated optical/UV afterglow emission from such a GRB will then illuminate the full redshift range over which galaxy formation has taken place. Particularly promising is the prospect of using the emission from the extremely bright, very-short-lived, prompt GRB optical flash that sometimes occurs ($V \simeq 9$ for 15 s from $z = 1.2$ GRB 990123 [227]), to probe absorbers with column densities too low for detection otherwise.

Unlike QSOs, which ionize the surrounding IGM and modify the ionization state of the ISM in galaxies up to about 1 Mpc away, causing a deficit of absorption lines at redshifts close to the QSO, GRBs would not suffer a 'proximity effect' [231]. The proximity effect is useful for estimating the intensity of the extragalactic radiation field at the redshift of the target QSO, but prevents detailed studies of nearby galaxies, that could perhaps be members of a group or cluster surrounding the QSO. Only the physical conditions in the very local pc-scale environment within the host galaxy of the GRB would be affected by a stellar progenitor.

The drawback of GRB absorption studies is the single opportunity to observe them, with no prospects of a later confirmation. Instrumental setup and acquisition must be rapid and reliable in order to avoid missing the action, an important consideration when minutes are required to read out data from the largest CCD cameras. A few GRBs per thousand are expected to be gravitationally lensed into multiple images by foreground galaxies, groups and clusters. Each image would be magnified by several times. With time delays from minutes to months, these lensed GRBs could provide a great opportunity for observers to be waiting for the arrival of each subsequent image, so that the light curve of the same GRB can be studied again, including its earliest rise. Lensing deflections of photons are sufficiently small that even the most tightly beamed GRB should still be visible along the different lines of sight.

Studies of the submillimetre and far-infrared emission from the host galaxies of GRBs also offer interesting prospects for investigating the evolution of galaxies: see later.

Millimetre/Submillimetre Absorption Studies: A comparable absorption probe has been used at millimetre wavelengths to detect absorption by sometimes exotic molecules against the strong radio continuum emission of background radio-loud AGN. Only a handful of candidates have been detected, along the line of sight to gravitationally lensed AGN where the lensing galaxy is within about 1 arcsec of the background radio source [268]. The small number of detections is due to a combination of the need for deep integrations and bright background sources. Because the molecular gas in a typical galaxy is much more centrally concentrated than neutral hydrogen – the scale lengths differ by about an order of magnitude – the area on the sky intercepted by molecular absorption systems is smaller than for Lyman-α systems. Additionally, only a single line in an absorbing object can be observed at once, as the spectral bandwidth of existing instruments covers only a narrow spectral region $< 1\,$GHz wide, corresponding to a a very narrow redshift interval at observing frequencies about 100 GHz. This is in contrast to the ability to resolve lines from about 40 to 110 nm, $\Delta\nu/\nu \sim 1$, in an optical QSO Lyman-α absorption spectrum.

When a wider band is available using ALMA, and wide-band spectrographs designed for detecting molecular-line redshifts are in service on single-antenna millimetre-wave telescopes, it may be possible to observe a 'forest' of intervening CO-line absorbers against the radio continuum emission of more distant radio-loud AGN, while searching for CO emission/absorption from the AGNs themselves. The limited spatial extent of the absorber, and the relatively narrow spectral range is still likely to render millimetre-wave CO absorption line studies less significant than optical absorption studies for understanding galaxy formation; however, they are sure to be valuable for studying low-excitation, representative, high-redshift galaxies, and ordinary regions of the ISM in fortuitously aligned galaxies.

3 Evolution of Low-Redshift Galaxies

3.1 Description of Galaxy Surface Densities

At this point it is necessary to discuss some of the descriptive terms required to analyse the evolution of galaxies. First, a statistical luminosity function $\Phi_\nu(L, z)$ at an observing frequency ν, which describes the space density of galaxies with luminosities between L and $L + \mathrm{d}L$ in the redshift interval between z and $z + \mathrm{d}z$. It is possible to define L as the total luminosity of a galaxy, in W or $\mathrm{L}_\odot$, or as the luminosity in a frequency range centred on ν, L_ν in $\mathrm{W\,Hz^{-1}}$. As *IRAS* $60-\mu$m observations were made at a frequency close to the peak of the SED of a typical dusty galaxy, the luminosity of this class of galaxy is often described by νL_ν, a quantity with dimensions of power, that differs from the true luminosity $L = \int L_\nu \,\mathrm{d}\nu$ by a factor of order unity.

Luminosity functions can typically be described by a Schechter function form [229],

$$\Phi(L) = \Phi_0 \left(\frac{L}{L^*}\right)^\alpha \exp\left(-\frac{L}{L^*}\right) . \tag{1}$$

Whether the luminosity function is defined per unit linear or logarithmic interval is largely a matter for personal taste. As luminosity functions span many orders of magnitude it is usually sensible to use a logarithmic form in numerical integrations to yield observational quantities. If logarithmic definitions are used, then it is conventional to divide by a factor of L in the definition above. The density of galaxies as a whole is such that $\Phi_0 L^* \sim 0.1\,\mathrm{Mpc^{-3}}$, $L^* \sim 10^{11}\,\mathrm{L}_\odot$ and $\alpha \simeq -1.1$.

The form of the Schechter function is intrinsically interesting. There is little reason to expect a power-law tail at low luminosities – galaxies could have a preferred luminosity. The low-luminosity power-law behaviour, with $\alpha > -2$ at small L to ensure that $\int L\Phi(L)\,\mathrm{d}L$ converges, could be indicative of strong interactions and feedback in the star-formation process. Power-law distributions describe self-similar, self-regulating behaviour observed in complex interacting natural systems, such as the distributions of wave heights or the dimensions of forest clearings. The exponential cutoff could then clearly indicate a natural scale that the process has yet to reach, and on which luminous/massive objects have yet to form. The luminosity function of dusty $60-\mu$m *IRAS* galaxies [247,228] (Fig. 4) is unusual, having a quadratic not exponential high-luminosity cutoff. This is unlike any other population of

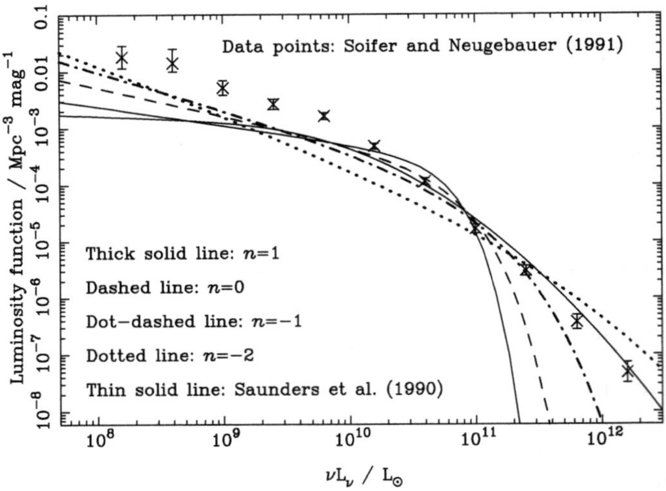

Fig. 4. The luminosity function of low-redshift galaxies detected by *IRAS* at 60 μm. The data points and thin solid line correspond to different determinations [247,228]. The difference is only significant at the faint end. The observed $\Phi \propto L^{-2}$ high-luminosity tail of this luminosity function is unusual, and almost certainly reflects the appearance of a population of strongly evolving luminous galaxies that are the first sign of the abundance of such objects in the high-redshift Universe. This luminosity function, consisting of both very luminous merging galaxies at the high-luminosity end, and ordinary spiral galaxies at the low-luminosity end is an important benchmark that successful models of the properties of high-redshift dusty galaxies must reproduce. The overplotted curves show Press–Schechter luminosity functions, assuming a constant mass-to-light ratio and different values of the fluctuation spectral index n or γ (Sect. 4.7).

galaxies, and seems to indicate that the most luminous of all local galaxies are relatively brighter using *IRAS* than at other wavelengths, presumably due to their greater tendency to be obscured by dust at optical wavelengths.

The luminosity function can be the sum of a variety of different galaxy types, or a grand multivariate function describing the whole population; however, the subpopulations also seem to be described individually by Schechter functions. To determine a luminosity function requires knowledge of the redshifts of the galaxies in a survey, so that their detected flux densities S_ν, in units $\mathrm{W\,m^{-2}\,Hz^{-1}}$ can be matched to their luminosities,[5] and the volume in which they could be detected can be calculated to construct the function.

The observational quantities that follow from the luminosity function are defined in terms of distances. We will return to define cosmological distances in Sect. 4.4.

[5] $1\,\mathrm{W\,m^{-2}\,Hz^{-1}}$ is equivalent to 10^{26} Jy (Jansky) in radio-based units.

Counts of Galaxies: The cumulative counts of galaxies on the sky brighter than a flux density S_ν, in units of sr^{-1}, are given by

$$N(>S_\nu) = \int_0^{r_{\max}} \int_{L_{\min}(r')}^{\infty} \Phi D(r')^2 \, \mathrm{d}L \, \mathrm{d}r' \; . \tag{2}$$

$\mathrm{d}r'$ is the radial distance element in the Universe in comoving coordinates, and D is the comoving distance parameter, with a functional form that depends on the geometry of the Universe. $r_{\max}$ is the distance to the most distant galaxy. The equation can easily be transformed into an expression for redshift z, as $r(z)$ is easy to determine (see Sect. 4.4). Strong observational evidence now constrains the geometry of the Universe [94], removing much of the capacity for different authors to use different cosmological parameters when interpreting observed results. The integral includes all the galaxies that are brighter than a flux density

$$S_\nu = \frac{L}{4\pi D^2(1+z)} \frac{f_{\nu(1+z)}}{\int f_{\nu'} \, \mathrm{d}\nu'} \; , \tag{3}$$

where L is the bolometric luminosity and f_ν represents the SED. The factor of $1+z$ in the denominator takes account of the redshifted observing bandwidth, and the reductions in both photon energy and arrival rate due to cosmology. The sampled restframe frequency of emission $\nu(1+z)$ is blueshifted from the observed frequency. $L_{\min}$ in (2) can be found by evaluating L in (3), and may differ between sub-populations, depending on the SED. It is possible to impose multiple constraints on the properties of galaxies that contribute to a count. For example, if the count of galaxies that exceed chosen flux densities at two different wavelengths are required, then the values of $L_{\min}$ for each can be calculated, and the most restrictive then applied to calculate the integral.

At low redshifts and for bright sources, the integrals can be evaluated immediately: $N(>S_\nu) \propto S_\nu^{-3/2}$ for any luminosity function. As the counts are pushed to greater depths, the change in their power-law slope indicates the presence of evolution in the galaxy population. If they rise above the $-3/2$ slope, then evolution is positive, if they fall below, then it is negative (see Fig. 5, in which the derivative of a count is plotted with this slope subtracted). When large numbers of sources are available, this differential count $\mathrm{d}N(>S_\nu)/\mathrm{d}S_\nu$ can be determined (Fig. 5), avoiding the cumulative, non-independent nature of the integral count $N(>S_\nu)$.

Background Radiation Intensity: Another important observable quantity is the total amount of background light I_ν contributed by a population of galaxies. This can be obtained at any wavelength by integrating over the count distribution, or directly from a luminosity function,

$$I_\nu = \int S_\nu \, \mathrm{d}N(>S_\nu) = \int_0^{r_{\max}} \int_0^{\infty} \frac{L\Phi}{4\pi(1+z)} \frac{f_{\nu(1+z)}}{\int f_{\nu'} \, \mathrm{d}\nu'} \, \mathrm{d}L \, \mathrm{d}r' \; , \tag{4}$$

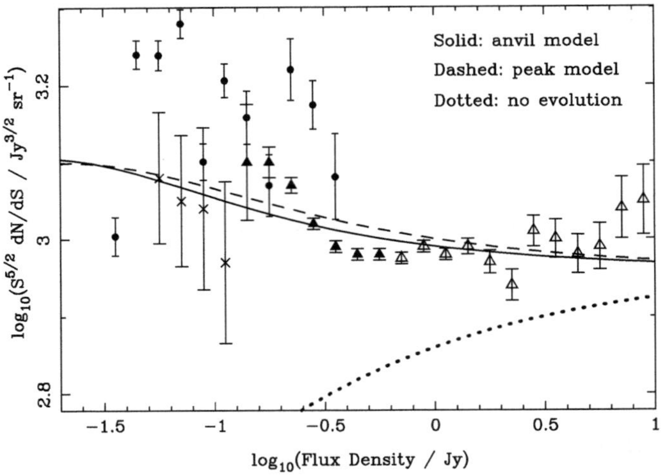

Fig. 5. Counts of galaxies determined at 60 μm from various *IRAS* surveys, summarized most recently by Bertin et al. [40]. Deep surveys at the different wavelengths of 15, 90 and 175 μm using *ISO* have extended the evolution baseline in redshift from $z = 0.1$ to $z \sim 1$, but prior to *SIRTF* at 60 μm, the *IRAS* results remain the deepest survey of the Universe. The plotted form of the counts ensures that Euclidean counts $N(> S_\nu) \propto S_\nu^a$ with $a = -3/2$, expected in cases where there is no evolution of galaxies, no cosmological factors in distances and a geometrically flat Universe, lie horizontally in the figure. That the faint counts rise significantly is clear evidence for evolution in the population. The two curves for evolving models (solid and dashed lines) correspond to the best-fitting values shown in Fig. 6. Cosmological factors and the effect of redshifting the SED naturally leads to a downturn in non-evolving counts in this plotting format, as shown by the dotted line.

in units of W m^{-2} Hz^{-1} sr^{-1}. A lower limit to the background radiation intensity, which may or may not be useful, or comprise a significant fraction of the total, can always be obtained by summing over a measured count. For the deepest optical, near-IR and submillimetre-wave surveys, these limits are interesting, and comparable to absolute background measurements determined from the brightness of the sky after subtracting the emission from known Zodiacal and Galactic components [259,38,55,88]. If the observations are made from within the atmosphere, then the atmospheric emission must also be subtracted.

From *IRAS* the limit to the background density that can be obtained by summing over the 60 − μm counts brighter than 30 mJy (Fig. 6) is about 6.3×10^{-10} W m^{-2} sr^{-1}, only 5% of the total emission from all external galaxies at this wavelength. Hence, it provides a very weak limit on evolution, other than confirming that most of the background radiation at this wavelength is generated by galaxies too distant for *IRAS* to have detected. Since *IRAS* flew, constraints that are tighter by a factor of about 3−10 have been imposed by *ISO* surveys at longer and shorter wavelengths (see Fig. 16).

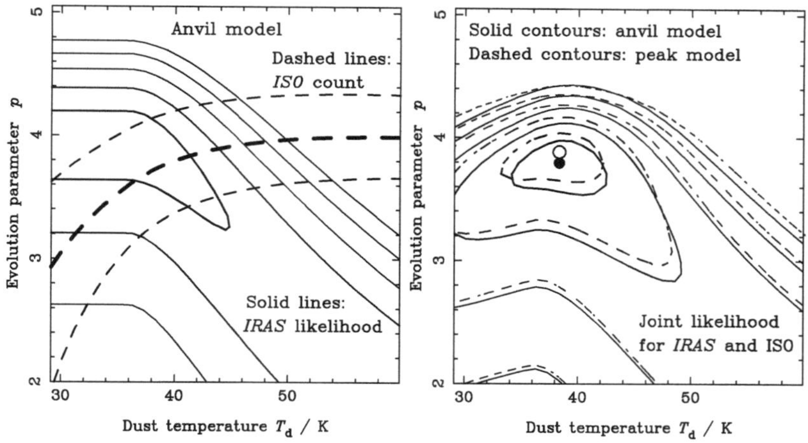

Fig. 6. Constraints on the form of the evolution of the luminosity function of galaxies at $z < 1$ from *IRAS* [40] and *ISO* $175 - \mu$m data [164,165]. The left-hand panel shows the results of fits to *IRAS* data, represented by solid contours, showing fit qualities of $\chi^2 = 1, 2, 3, 4$ and 5, and *ISO* data, represented by dashed lines corresponding to reproducing the data [164], and for missing by factors of 0.5 and 2. This is carried out in an 'anvil model' [56] in which luminosity evolution has a $(1 + z)^p$ form out to a redshift greater than that sampled by these populations of galaxies. The right hand figure shows the joint constraints imposed, and compares the effect of a 'peak model', an alternative parametrization in which the $(1 + z)^p$ luminosity evolution flattens off smoothly at $z \simeq 1.5$. The effect of choosing these slightly different forms of the evolution functions is not significant. The evolution of low-redshift dusty galaxies out to $z \simeq 1$ is strong, and matches the form derived from optical data [177].

Redshift Distribution: The redshift distribution of galaxies brighter than a specified flux density or subject to some other chosen condition can be determined by evaluating (2) without integrating over z. The redshift distribution of a carefully selected sample of galaxies is the most useful quantity for determining the form of evolution. In general, a redshift distribution provides more information than the count from which it is drawn, while a deep count in turn provides more information than a determination of the background intensity, which is undifferentiated by galaxy flux density. In terms of observational effort, finding a redshift distribution is demanding: accurate positions must be determined and follow-up spectroscopy carried out. However, the need for the careful analysis of systematic effects and the accurate determination of the contribution from unrelated sources of emission can also make background observations challenging [217,38].

Fossil Records: Alongside the radiation produced by luminous galaxies, remnants of the activity are also generated. For example, metals are generated by high-mass star formation activity, and distributed into the ISM of the host galaxy with a $\sim$ 40% efficiency at the end of stellar evolution. The number of low-mass stars still burning that formed at high redshifts can be calculated, assuming a stellar initial mass function (IMF) [171], and models of stellar evolution. When compared with the observed abundance of stars in the Milky Way, or the integrated colours and densities of local galaxies, then this provides a useful but weak constraint on the high-redshift IMF in advance of huge 100-m-class near-IR/optical telescopes that can resolve individual stars directly in a representative sample of galaxies extending as far as the Virgo cluster. Accretion power onto AGN instead generates massive remnant blackholes in galactic centres.

The density ρ of any remnant from the galaxy formation process, either long-lived low-mass stars, heavy elements or stellar remnants – white dwarves, neutron stars and low-mass blackholes – can be found by integrating the energy production rate or 'luminosity density' $\rho_{\mathrm{L}} = \int L\Phi\,\mathrm{d}L$ within a fixed comoving volume over cosmic time,

$$\rho = K \int_0^{t_0} \rho_{\mathrm{L}}(t)\,\mathrm{d}t = \frac{1}{c}\int_0^{z_0} \tau(z) \frac{1}{1+z}\frac{\mathrm{d}r}{\mathrm{d}z}\,\mathrm{d}z\ , \tag{5}$$

where the value of K determines the link between luminosity of the sources ρ_{L} and the generated material.

For low-mass stars, K depends on the stellar IMF, with

$$K \propto \int_{M_{\mathrm{min}}}^{M_{\mathrm{ex}}} IM\,\mathrm{d}M \bigg/ \int_{M_{\mathrm{min}}}^{M_{\mathrm{max}}} IM\,\mathrm{d}M\ , \tag{6}$$

where M_{min} and M_{max} are the lower and upper mass limits of the whole IMF, and M_{ex} is the mass of the most massive star still burning at the present epoch, which depends on its formation epoch and theories of stellar evolution. For nucleosynthesis, $K \simeq (0.007c^2)^{-1}$, reflecting the efficiency of fusion energy conversion, which is about 0.5–0.7% of the rest-mass energy of hydrogen when converted into helium, or iron respectively. For AGN, $K \simeq \epsilon^{-1}$, where ϵ is the potentially much greater accretion efficiency ($\simeq$ 30%). The typical IMF and the accretion efficiency could be quite different at high and low redshifts, and so the results might be difficult to interpret.

This type of integral constraint can, however, impose useful limits on the amount of activity at high redshifts. For example, the ratio between blackhole and stellar bulge mass in elliptical galaxies and spiral bulges implies that the relative amount of energy emitted by accretion as compared with star-formation activity was of order 1:5 [51], regardless of cosmology, if the processes are assumed to be coincident in time, and so $\rho_{\mathrm{L}}(z)$ has the same form in both cases. Of course, this need not be true. The link between blackhole and star formation is very important. If this Magorrian relation is the result of a series of coincidences, then it is an important goal to list them all and to determine their respective roles in the 'conspiracy' to fix this ratio.

It seems unlikely that no deeper link exists, reflecting an important part of the galaxy formation process [239]. However, another remarkable result, the existence of the fundamental plane of elliptical galaxy properties [97], has remained an interesting but rather ill-explained feature of our understanding of the formation of elliptical galaxies for many years. The realization that elliptical galaxies have an unexpectedly narrow range of properties has not allowed the process of their formation to be determined absolutely.

3.2 Detecting Evolution in the *IRAS* Survey

From the *IRAS* survey it was clear that comparable amounts of energy were being released by both nearby galaxies and the Milky Way, at optical and far-IR wavelengths. Hence, it was clear that far-IR luminous galaxies could make a significant contribution to the energy budget of the Universe, if this continued out to high redshifts. However, *IRAS* did not reach out to cosmological redshifts[6]. It is interesting to review the evidence for strong galaxy evolution available from *IRAS* results in the late 1980's, but we must first put the studies into context.

Although deep galaxy surveys had been made at optical wavelengths since the advent of CCD detectors around 1980, the type of evolution of the luminosity function was not evaluated effectively until large galaxy redshift surveys were carried out in the early/mid 1990's [177]. The results from optical redshift surveys at $z < 1$ were obtained at almost the same time as the deepest optical HDF-N image of the distant Universe was taken. Analysis of the colours of the faint galaxies provided approximate 'photometric' redshift information, backed up by campaigns of spectroscopy at the 10-m Keck telescope, confirmed that the luminosity released by galaxies in the high-redshift Universe is greater by an order of magnitude than at low redshift [187,86,273].

These results tend to be supported by subsequent observations, but a less dramatic form of evolution has been claimed from redshift surveys that select galaxies at several optical wavelengths, not just the I-band [87]. Observations of AGN had already demonstrated strong evolution of the number of double-lobed radio galaxies [104], and optically-selected QSOs [119], conclusions based on count data and backed up by reasonably complete redshift distributions for the sources. At the time of their publication, details of the evolution of *IRAS*-selected galaxies, were consistent with the evolution inferred from AGNs to much greater redshifts [200].

How can the observed counts and backgrounds be linked to the form of evolution of galaxies? There are two conventional forms of evolution, which empirically describe the evolution of luminosity functions ((1)). One is 'pure density evolution', in which Φ_0 is a function of redshift but L^* is constant. This ensures that the evolution of both the space and luminosity density of

---

[6] Even in 1985, when only $z = 0.5$ being a high redshift, and the only more distant objects with redshifts were radio- and optically-selected AGN.

galaxies, $\int \Phi \, \mathrm{d}L$ and $\int L\Phi \, \mathrm{d}L$ respectively, is described by $n(z) = \Phi_0(z)/\Phi_0(0)$. This simple scaling feeds through to boost both counts and backgrounds. An alternative is 'pure luminosity evolution', in which Φ_0 is fixed and L^* is a function of redshift, with the evolution function $g = L^*(z)/L^*(0)$. This introduces the same dependence of the luminosity density on redshift, when the entire luminosity function is integrated, but leads to an enhanced luminosity density of galaxies brighter than a certain detection threshold. As a result, the two forms of evolution affect the counts in different ways. The additional factor that shifts the detection threshold means that a certain factor of luminosity evolution has a more dramatic effect on the counts than the corresponding factor of density evolution. If the evolution functions are represented by $(1+z)^p$, then the value of p required to fit count data assuming density evolution is approximately twice that for luminosity evolution. The effects of including luminosity evolution of the *IRAS* luminosity function for explaining the counts of galaxies detected using *IRAS* and *ISO* are compared in Fig. 6. $p \simeq 3.9$ is needed to explain the data. Note that the assumed dust temperature affects the results (Fig. 5). A value of $p \simeq 3$ was often quoted for the form of evolution of dusty galaxies at low redshift, a value which made sense because it matches the rate of evolution of AGN. This value provides a good fit if $T \simeq 50\,\mathrm{K}$ is assumed. The value $p \simeq 4$ provides a better description in connection with a temperature $T \simeq 40\,\mathrm{K}$. Both p and T values are determined by the joint requirement to explain the observed counts of the more distant $175-\mu\mathrm{m}$ *ISO* galaxies (Fig. 14)[7].

Even a simple Schechter luminosity function representing the real Universe is sure to involve variations in both Φ_0 and L^* with redshift as the Universe evolves. However, the observational consequences are such that luminosity evolution must dominate, if we are to account for the observed counts without either generating too much background light, consuming all the baryons into stars or filling the Universe with relic blackholes [56].

The count determined for *IRAS* galaxies at $60-\mu\mathrm{m}$, in a conventional, but complex, notation, is shown in Fig. 5. Although the results vary between the different deep samples in different areas of sky, analysed using different techniques, there is clear evidence for an excess of sources at flux densities around 1 Jy, indicating positive evolution. By linking the results at $175\,\mu\mathrm{m}$ from *ISO* and at $60\,\mu\mathrm{m}$ from *IRAS* assuming a form of luminosity evolution, the fitted model curves shown in Fig. 5 are obtained.

3.3 Galaxy Formation Scenarios

IRAS observations by 1990 hinted that substantial evolution of the population of reasonably ordinary, representative dusty galaxies was taken place. However, most of the galaxies detected by *IRAS* were already detected in optical Schmidt camera surveys. They map the current end points of the ongoing

[7] Of course, there may be some favoritism towards $p = 4$ today to match the inferred rate of low-redshift evolution in optically-selected galaxies [177].

process of galaxy formation, unable to probe the evolution over an extended timeline. The only galaxies then detected in representative numbers at high redshifts were AGN. Not until 1995 did the first spectroscopically confirmed samples of Lyman-break galaxies (LBGs) at $z \simeq 3$ appear [253].

In 2002, the details of the process by which galaxies formed is still uncertain. There is a standard model, the hierarchical growth of structure dominated by a cold (non-relativistic) dark (non electromagnetically interacting) matter (CDM). A single, but revolutionary observation could change our understanding; however, the CDM model has proved to be robust, and it has been modified to account for almost a decade of increasingly rapid discovery. Clear alternative paradigms, such as the generation of large-scale structure in explosions at $z > 10$, by sweeping up shells of originally uniform IGM material into a honeycomb network, or the early formation of structures on the largest scales from the free streaming of hot (relativistic) dark matter into caustic structures – 'Zeldovich pancakes' – on mass scales of order $10^{15}\,M_{\odot}$, followed by their fragmentation to form galaxies, are all at odds with current data. In most cases, abundant high-redshift activity would heat the IGM to a temperature sufficient to wash out the observed CMB anisotropies [94]. An excellent review of these galaxy formation processes, dating back to when they were largely unchallenged by observations was provided by Efstathiou [112]. The detailed large-scale structure mapped by galaxy redshift surveys [83,58], which now reach out to redshifts where structure has evolved significantly, and certainly include representative cosmic volumes and the measured scales of the most significant structure in the CMB, all point to a Universe dominated by CDM. There has been discussion about the expected number of small dark matter halos, predicted by CDM simulations, that are not apparently associated with luminous galaxies [138]; however, this concerns the exact nature of small-scale substructure, and not its presence or absence. Without a significant density of CDM, then the bound structures observed at $z \simeq 6$ would have difficulty forming from the density contrast in baryonic matter that is measured via the CMB to be only about 10^{-5} at $z \simeq 1100$. Unless we are missing something significant, then the standard hierarchical CDM model seems likely to remain as the underlying description of the process of galaxy formation.

The standard CDM model involves the gravitational growth of bound structures under the influence of an initially almost smooth density field in an expanding Universe. Inhomogeneities were generated by quantum fluctuations during inflation – a period of very rapid early expansion. Inflations makes these inhomogeneities amenable to producing a Universe like our own. The most natural form of inflation renders the Universe spatially flat, and dramatically reduces the density of exotic particles that could have been created at even earlier epochs, such as magnetic monopoles, consistent with their non-detection.

This picture of the early Universe is consistent with both the observed properties of the CMB, which maps the density of baryonic matter at 're-combination', when the Universe cooled below the ionization temperature of hydrogen, and became transparent to radiation for the first time, and with the abundances of light elements formed by thermal nucleosynthesis about one minute after the beginning of the Universe and inflation: primordial big-bang nucleosynthesis.

In studies of galaxy formation, these details can be taken as initial conditions. This is no longer just an assumption. Concrete images of the CMB from the BOOMERanG balloon-borne experiment [94,89] and the ground-based DASI [69], CBI [202,238] and VSA [257] interferometers, at the South Pole, Atacama and Tenerife respectively, provide solid predictions. The results of the all-sky 10-arcmin resolution CMB image from the NASA *Wilkinson Microwave Anisotropy Probe (WMAP)* satellite (launched in June 2001) provided an extremely accurate image of the CMB in February 2003.

The distribution of dark matter that is consistent with CMB observations ensures that gravitational instabilities cause the first collapse of structures on small rather than large scales. The subsequent merging of collapsed structures then builds up increasingly massive objects. This process has been modeled in great detail using N-body gravitational simulations [207], and can be represented reasonably accurately by an analytic function due to Press & Schechter [216], extended [20], updated [237] and made more widely applicable [61] by others. The density contrast of linear overdense regions grows as $(1+z)^{-1}$ at high redshifts (Sect. 8.2). At lower redshifts, where a cosmological constant dominates the dynamics, the growth of structure is slowed. Hence, as recombination took place at $z \simeq 1100$, and the density contrast in baryonic matter revealed by CMB observations is only about 10^{-5}, at first sight this indicates that galaxies with non-linear overdensities $\gg 1$ required for the formation of galaxies could not have formed by the present epoch. Note, however, that the baryonic matter is prevented from collapsing prior to recombination by radiation pressure on the electrons, and the electrostatic transfer of this pressure to the baryons. CDM does not interact electromagnetically, and so couples to the baryons only through gravity. This means that overdensities in the dark matter field are unaffected by radiation pressure, and can start to grow as soon as the Universe becomes matter dominated [193], significantly before re-ionization, and provides an important head start for galaxy formation, leading to deeper potential wells for the newly neutral baryonic matter to fall into after recombination. In light of this argument it seems difficult to doubt the existence of non-baryonic dark matter.

The pre-recombination Universe is smooth and simple. The CMB provides a backdrop to the stage of galaxy formation; however, after recombination and the formation of the first objects a rich range of physical processes start to contribute and interact. The mix of relatively simple processes, such as the growth of gravitational instabilities in a matter field dominated by an unknown type of non-interacting gravitating matter, the hydrodynamics of collapsing gas and the radiation processes that cool the gas conspire to generate a wide range of possible outcomes. That is even before the complex details of the geometry of the cores of the dense gas clouds in which the first metal-free stars form, their stellar astrophysics, and the feedback effects from their perhaps exotic supernovae combine to make galaxy formation a complex process.

At present it is possible to model the process of structure formation in CDM through the history of the Universe on scales that include both the sub-structure in the densest clusters of galaxies [250] and the largest scale structures that can be observed at the present epoch [207]. If the processes of gas dynamics and cooling are included, then the resolution of the best current N-body simulations is close to matching the scales of the dense clumps of gas in which the first stars formed [1,65]. The feedback effects from supernova from these stars is, however, a quite different story. The messy astrophysics after the first stars form[8] may never be simulated in sufficient detail to capture the richness of the observed Universe on sub-Galactic scales [266]. This view may be unduly pessimistic, but it is hope that cosmic N-body simulations will be able to trace to the scales of planet formation and the origins of life.

3.4 Gas Cooling and Star Formation

At the present epoch, the regions in galaxies where stars form contain richly structured distributions of atomic and molecular gas and dust grains. These regions are dense (up to $> 10^6\,\mathrm{cm}^{-3}$) and cool (10–100 K), and there are a wide variety of ways for gas, dust and radiation to interact. They contain significant magnetic fields, which can have large effects on the dynamical processes.

These conditions are likely to be similar in recognizable high-redshift galaxies. While we do not understand all the details of star formation even in the Milky Way, it remains vital to study them, and the allied events taking place in nearby and distant galaxies. It is also important to bear in mind that the first stars generated the heavy elements whose properties now strongly influence the astrophysics of star formation in giant molecular clouds. The formation of this first generation of stars may have been very different. The regions where stars form in the Milky Way that can be studied relatively easily are not the same as the high-redshift clouds in which the first generations of stars form. The metallicities in the Milky Way should all be significantly greater, while there should be more dust, stronger magnetic fields and an ISRF [182,1].

The properties of the first generation of stars are critically important for determining the nature of subsequent galaxy evolution, and could synthesize

[8] Described by Dick Bond with the double entendre 'gastrophysics'.

the elements required for life, and for the study of the cold Universe. They are the targets for the next generation of large astronomical facilities: 30-m ground-based telescopes such as the California Extremely Large Telescope (CELT) concept, *JWST*, ALMA and others being designed and proposed, including a mid-IR 10-m-class single-antenna telescope, developed from the *SAFIR* 'sapphire' concept and a 1-km baseline interferometer operating at the same wavelength like the *SPECS* concept [191]. Ultimate understanding of the galaxy formation process requires that we understand the process of high-redshift star formation in detail.

In primordial gas, magnetic fields are expected to be small, prior to the action of dynamos in stars, rotating partially-ionized galaxy disks, or from electrons accelerated in AGN-powered radio galaxies, whose magnetized lobes of radio emission could ultimately permeate intergalactic space [131]. As a result, magnetic effects should not hold up the gravitational collapse of cool gas on stellar scales. The only addition to radiation from hydrogen that could be observable is from polar DH and LiH molecules. However, their abundance is low, and the only source of energy available to them will be the gravitational potential energy of contracting low-mass objects, and so it is unlikely that the effects will be dramatic.

Cooling by atomic hydrogen transitions is only effective if temperatures of $\sim 10^4\,\mathrm{K}$ are reached, when collisions can excite atoms into the ionized state required for radiative cooling. This is the virial temperature reached by collapsing gas clouds with masses $\sim 10^8\,\mathrm{M}_\odot$ at high redshifts. In hierarchical models of structure formation, such objects do not form until $z \simeq 8$–10. When ALMA and *JWST* probe this redshift range it may be possible to observe the first stars in these proto-galaxies directly.

Molecular hydrogen vibrational-rotational transitions offer a route to cooling gas that has been heated by shocks to much cooler temperatures of several hundred degrees. Clouds of dark matter and baryonic gas in virial equilibrium with this temperature have masses $\sim 10^5\,\mathrm{M}_\odot$ at high redshifts, which, being small, could form in reasonable numbers at redshifts as high as 20. The background abundance of molecular hydrogen H_2 at these redshifts compared to atomic hydrogen HI is expected to be of order 10^{-6} by number [84], and to form by the rare interaction of free electrons and HI to form H^-, and then H_2 in collisions with H. Direct formation of H_2 in H–H collisions is unlikely, owing to the very low ratio of binding energy to kinetic energy. The standard low-redshift mode of H_2 formation by surface chemistry on dust grains is impossible prior to metal enrichment. Once high enough densities are reached in cooling primordial clouds, 3-body collisions allow H_2 to form much more easily. However, in the absence of dust, the H_2 abundance is very fragile. As soon as the first stars form and UV light bathes the surroundings, molecular hydrogen will be destroyed throughout the birth halo of the star, preventing further cooling and star formation, even in the absence of supernovae to disrupt the cloud and disperse metals in stellar winds to modify the IGM chemistry. The UV light is also likely to inhibit cooling in neighbouring halos by reducing the H_2 abundance.

Once some metals have been formed, and some UV radiation has been released, the special simple conditions that apply to star formation in the early Universe are greatly modified. The process of star formation should then quickly become more akin to the complex form recognizable locally.

The outcome of these high-redshift processes are not yet constrained by observations: the study of the earliest stars will surely remain a key area of interest for both observational and theoretical astrophysics. One of the exciting recent developments is the association of high-mass stars with enigmatic long-duration GRBs [98]. If this association is genuine, then the extreme luminosity and unobscured γ-ray luminosity of GRBs could reveal the deaths of some of the earliest high-mass stars. These stars could have exotic properties, and masses up to $\simeq 300\,M_\odot$ [147]. Finding GRBs at very high redshifts may be the best way to investigate early star formation, and to constrain the consequences for metal enrichment and subsequent generations of cosmic structures.

4 Cosmology

We have already mentioned several features of the study of galaxy evolution that rely on the background cosmology, the properties of the Universe on the largest scales. The details are widely available [82,209,205,184]; however, it is useful to have them alongside the following discussion of the high-redshift Universe. In previous years it was important to discuss the range of possible values of cosmological parameters, but it now seems that the measure of the Universe has been taken successfully by combining the Hubble diagram of Type Ia supernovae [212] and the angular power spectrum of CMB anisotropies [94,173,69]. Other measures, more purely probing the density of the Universe, like the abundance of high-redshift clusters and the peculiar velocities of low-redshift galaxies are in general agreement with the results [64].

4.1 The Shape and Size of the Universe

The first descriptions of the Universe under General Relativity (GR) assumed homogeneity and isotropy on large scales, notwithstanding that its total extent was still sometimes assumed to be little bigger than the very anisotropic Milky Way. Ever since, observations have shown that these 'cosmological principle' assumptions seem to be reasonable. The deepest, largest redshift surveys [83,58] show that the Universe is accurately homogeneous on scales greater than $100h^{-1}$ Mpc[9]. Observations of the CMB show that it was homogeneous and isotropic at very early times [190,94]; both the *COBE*-DMR

[9] Hubble's constant $H_0 = 100\,h\,\mathrm{km\,s^{-1}\,Mpc^{-1}}$ measures the current expansion rate of the Universe.

and BOOMERanG CMB images include measurements of order 1000 causally disconnected regions. For the currently favoured expansion rate and density of the Universe, with $h = 0.65$ and a density parameter $\Omega_{\mathrm{m}} = 0.3$, this scale corresponds to a mass of order $3 \times 10^{16}\,\mathrm{M}_{\odot}$, which is reassuringly several times greater than the mass of the richest cluster of galaxies. Structures may form in the future on larger scales. On smaller scales the topology of the large-scale structure traced by galaxies is of overlapping sheets intersecting in filaments, which in turn radiate from clusters. This structure can be described accurately by a self-similar power-law power spectrum as a function of scale. This is reasonable, probably reflecting the complexity of the process of feedback and non-linear clustering that takes place on small scales during the assembly of galaxies, and does not necessarily reflect anything special about the initial conditions for the Universe. The connection between the amplitude of structures in the Universe on different scales, as traced from the small scales of galaxies up to clusters, filaments and supercluster sheets, up to the largest scales traced by observations of the CMB, can now be described and tested.

Describing and explaining an isotropic, homogeneous Universe is easier than any alternative. It was done early in the development of cosmology, almost coincident with the discovery of the expansion by all those listed in the title of the Friedmann–Robertson–Walker (FRW) metric. A distance element $\mathrm{d}s$ in this metric, the most general isotropic, homogeneous space that can evolve with time, is described by,

$$\mathrm{d}s^2 = c^2\,\mathrm{d}t^2 - R^2(t)\left[\frac{\mathrm{d}r^2}{1 - kr^2} + r^2\left(\mathrm{d}\theta^2 + \sin^2\theta\mathrm{d}\Phi^2\right)\right] . \tag{7}$$

The scalefactor R describes the evolution of the size of the Universe as a function of cosmic time t. The spatial coordinate $\boldsymbol{r} = (r, \theta, \phi)$ is comoving; that is particles moving with the Hubble expansion maintain the same values of $\boldsymbol{r}$. The evolution of the scalefactor is determined by the energy densities of different components of the Universe. $k = 1$, 0 or -1, and determines the geometry/curvature of the Universe. If $k = 1$, then the Universe is closed, with finite volume and a spherical geometry: distant sections at constant radius have a decreasing area, and so it can be possible for light rays to take laps around the Universe. If $k = 0$, as observations of the CMB strongly imply [94], then the Universe has a flat/Euclidean geometry, a steadily increasing area of spatial sections and an infinite volume. If $k = -1$, then the Universe is open, with infinite volume, and distant sections at constant distance have exponentially increasing areas. There is no reason why the Universe, if infinite, could not have a much more complex geometry on very large scales: observations only confirm that over the region we can see, the FRW description is an excellent one. Of course, the FRW description fails on small scales in the Universe, as the spacetime around the Sun and Milky Way is certainly not curved uniformly. Amongst the diverse benefits of having structure on small scales is the existence of life and the availability of gravitational lensing as a cosmological and cosmographic tool.

The effects of different curvatures k can be seen by finding the distance along the radial direction ($\mathrm{d}\theta = \mathrm{d}\phi = 0$) out to a comoving coordinate distance r at constant time. This 'proper distance' $\int \mathrm{d}s = d_{\mathrm{P}}$ is given by the product of the value of R at the epoch when it is evaluated (so $\mathrm{d}t = 0$) and either $\arcsin r$, r or $\operatorname{arsinh} r$, depending on whether $k = 1$, 0, or -1. In the $k = 1$ Universe the proper distance reaches a maximum and then returns to zero, and it increases ever more rapidly with r if $k = -1$. Note that the proper distance is difficult to determine in practice over a large distance; as it is defined at a fixed cosmic time, a pre-ordained conspiracy between observers is required to evaluate it.

Without finding the form of R, it is still possible to see how the properties of light traveling through the Universe are affected by the geometry. Light travels along geodesics with $\mathrm{d}s = 0$. Along a radial geodesic ($\mathrm{d}\theta = \mathrm{d}\phi = 0$) from a fixed comoving radial distance at which a light ray was emitted, r_{e}, at epoch t_{e} to the present time t_0, $r = 0$, the integral along a geodesic must be the same when travelling with adjacent wavecrests, as r is a comoving coordinate. As long as R does not evolve rapidly during the time interval that is the inverse of the frequency of the signal, then the integrals of R^{-1} over time, with limits that differ by the time separation of these wavecrests, from the metric, must also be the same. Hence, $(R\nu)^{-1}$ must be constant. The frequency of radiation received ν_0 thus differs from that emitted ν_{e} by the ratio of $R(t_0)$ and $R(t_{\mathrm{e}})$, and so the redshift, historically defined as $z = (\lambda_{\mathrm{e}} - \lambda_0)/\lambda_{\mathrm{e}}$ in terms of emitted and received wavelengths λ_{e} and λ_0 is given by $1 + z = R^{-1}$, if R is defined to be unity at the present epoch. The link between redshift, distance and time depends on the evolution of the scalefactor R.

4.2 Dynamics of the Universe

To find out how the scalefactor evolves, and thus determine all the time-dependent properties of a FRW Universe, it is necessary to solve the Friedmann equations,

$$\frac{\dot{R}^2}{R^2} + \frac{kc^2}{R^2} = \frac{8\pi G}{3}\rho + \frac{\Lambda}{3} , \tag{8}$$

and

$$2\frac{\ddot{R}}{R} + \frac{\dot{R}^2}{R^2} + \frac{kc^2}{R^2} = -\frac{8\pi G}{c^2}p + \Lambda . \tag{9}$$

These are the remains of the Einstein GR field equations after imposing the symmetry of the FRW metric. ρ represents the total mass density of the Universe, equivalent to the energy density in both matter and radiation, while p is the pressure due to both components. Λ represents the cosmological constant, a constant energy density that produces an effective negative pressure. This term is free to take any value in GR. It is also possible to make it a more general time-dependent function, as in 'quintessence' models, that effectively add a varying fifth force to the Universe on the very largest scales.

Quintessence, or Λ both affect the Universe on the largest scales, and have no observable effects on either solar system or Galactic dynamics. The same is true for any curvature term – it is only noticeable on very large scales. These factors do however affect the way in which overdensities grow in the Universe, a key process in galaxy formation (Sect. 8.2). Note that in an expanding, flat Universe on the largest scales, or at late times, the equations are always dominated by Λ.

The equations can be expressed in many different forms, but can be condensed into a single substantive equation, plus the conservation condition $\mathrm{d}(\rho c^2 R^3) = -p\,\mathrm{d}(R^3)$. p and ρ can be connected by an equation of state $p = w\rho c^2$. Pressure-free matter (often referred to as dust) has $w = 0$, while radiation has $w = 1/3$.

Note that with the exception of the cosmological constant, the Friedmann equations can be written down from the equations of conservation of energy and motion for a uniform spherical region in Newtonian physics. Since a zero-Λ model has no stable static solution, Newton or any successor could have predicted an expanding Universe, rather than just Einstein when he included Λ to stop the inevitable, unexpected expansion[10].

A very useful form in which to determine $R(t)$ and thus $z(t)$ is to combine the Friedmann equations so that all four terms can be expressed as equivalent densities that sum to unity, expressed as a time-dependent Hubble parameter,

$$H(t) = \frac{\dot{R}}{R} = H_0 \left[\Omega_\mathrm{m} + \Omega_\mathrm{r} + \Omega_\Lambda + \Omega_k\right]^{1/2} . \qquad (10)$$

$\Omega_\mathrm{m} = 8\pi G\rho_\mathrm{m}/3H^2$ is the density parameter in ordinary matter, with space density ρ_m, $\Omega_\mathrm{r} = 8\pi G\rho_\mathrm{r}/3H^2$ is derived from the equivalent matter density in radiation ρ_r, $\Omega_\Lambda = \Lambda/3H^2$ is the density equivalent to the cosmological constant and $\Omega_k = -kc^2/R^2H^2$ is the density equivalent to the observed (small) curvature of the Universe. These different density terms evolve as R^{-3}, R^{-4}, R^0 and R^{-2} respectively, and so their relative importance changes with epoch.

This expression can be integrated to obtain $R(t)$ in any specified model. Popular solutions over the last eighty years have included Einstein's unstable static solution with $\Lambda = 4\pi G\rho$, $k = 1$ and $rR = 3\,\mathrm{Gpc}$, de Sitter's exponentially expanding Λ-dominated model with $k = 0$ and $R \propto \exp t\sqrt{\Lambda/3}$, and the Einstein–de Sitter matter-dominated pressureless model (the hot theorists' favourite until 1999), with $\Omega_\mathrm{m} = 1$, $\Lambda = 0$ and $R \propto t^{2/3}$. In the radiation-dominated version of this model with $\Omega_\mathrm{r} = 1$, as expected at early times in all monotonically expanding Universes that contain radiation, $R \propto t^{1/2}$, while the oddly-popular zero-density Universe with $k = -1$ and $\Omega_\mathrm{k} = 1$ had $R \propto t$.

[10] That no-one did is perhaps strange – is that well over two hundred years of bland scientific conformity or cowed reverence? Maybe the ideas were at large, but not voiced clearly, or the idea of discussing the motion of a non-static Universe was really too much a leap to make.

It is possible for the sum of the density parameters to be greater than unity, but $k = -1$ is then required to restore order.

Current observations [212,94] suggest that $\Omega_\Lambda \simeq 2\Omega_{\rm m}$, Ω_k, $\Omega_{\rm r} \simeq 0$ and $h \simeq 0.65$. Because of the dependence of the terms on R, it is inevitable (barring a truly cataclysmic phase change when the Universe suddenly ends up in a lower energy state than at present) that Λ will dominate the dynamics of the expansion at late times. While a non-zero cosmological constant may not be the simplest, neatest form for the Universe, it does ensure that there is exponential expansion at late times, and thus a sort of satisfying time-independence of the ultimate evolution. The energy density in radiation $\Omega_{\rm r}$ is already smaller than $\Omega_{\rm m}$ by a factor of 17000, but dominated the energy density when $z > 17000$. An equivalent to Λ is also thought to have dominated the energy density in the early inflationary epoch, and then converted to conventional energy in the phase change that ended inflation and reheated the Universe.

4.3 Horizons

Whether the Universe has a horizon at early or late times, beyond which causal contact cannot be established or maintained depends on whether the integral of comoving distance along a geodesic $\int R^{-1}\,{\rm d}t$ converges. If it does, then there is a finite comoving distance beyond which we cannot see, either from $t = 0$ or to $t = \infty$. We seem to live in a Universe with $R \propto \exp(t)$ at both early and late times, and so the integral should converge when taken both from $0 \rightarrow t_0$ and from $t_0 \rightarrow \infty$. Hence, there are places in the Universe that are currently casually disconnected from us, but which may subsequently come into contact, places that are currently in contact but will subsequently be moved out of contact, and places that never have and never will be in causal contact. At very late times the visible Universe will be made up of only the bound local group of galaxies, with all other structures having been redshifted beyond sight and influence. It is likely that we would never be able to communicate with and receive a signal back from a galaxy at $z \simeq 2$ as the Universe is expanding too rapidly for causal contact to be maintained before there is time for a reply to be sent[11] [180].

4.4 Distance Measures

The comoving distance D ((3)) to a redshift z is just the comoving radial coordinate r in (9), $D = r$. This is the comoving or effective distance [184,205]. This measure of distance has the advantage that it does not change with time, and that the comoving volume element associated with a shell at radius D has the form ${\rm d}V = 4\pi D^2\,{\rm d}D/\sqrt{1-kD^2}$, read straight from the FRW metric with

[11] Alison Farmer's advice on reviewing this paper at the Journal Club in Caltech in November 2001 was "don't date (in the American sense) at $z > 2$".

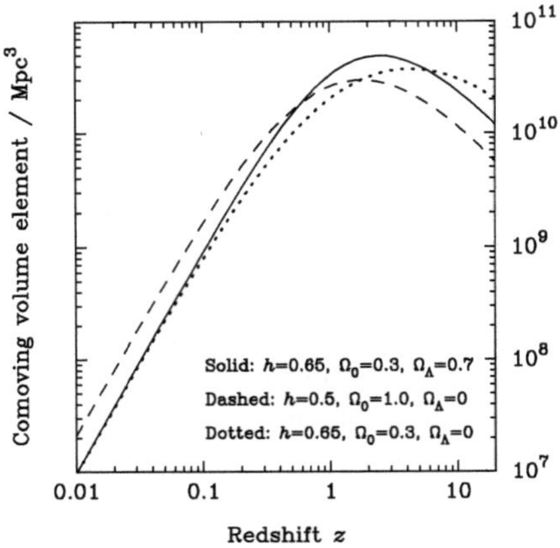

Fig. 7. The cosmological volume element $\mathrm{d}V/\mathrm{d}z$ as a function of redshift in three different cosmologies. The relative differences between the three volume elements is more significant than the differences in the distances shown in Fig. 8.

no redshift-dependent terms. The factor of $(1+z)$ in the denominator of (3) makes sense in this case, reflecting the competition between the increasing photon arrival interval, decreasing photon energy and increased fractional bandwidth to receive a signal in unit frequency range in the observer's frame as compared with its emission in the frame of the observed galaxy, all of which depend on $(1+z)$. The volume element associated with unit solid angle $\mathrm{d}V/\mathrm{d}z$ is shown as a function of redshift in Fig. 7.

Two other measures of distance are often used – the angular diameter distance d_A and the luminosity distance d_L. The angular diameter distance is the distance that describes the angle subtended by a rod of proper length l observed at cosmic time t_e by an observer at cosmic time t_0. If the rod is aligned in the θ direction on the sky (with $\mathrm{d}r = \mathrm{d}\phi = 0$), then the rod is defined to subtend an angle $\mathrm{d}\theta = l/d_\mathrm{A}$. From the FRW metric, $\mathrm{d}s = l = R(t_\mathrm{e})\, r_\mathrm{e} \mathrm{d}\theta$, and so $d_\mathrm{A} = R(t_\mathrm{e})\, r_\mathrm{e}$. Sensibly, this equals the proper distance to the rod observed at t_e in a flat Universe. If $k = -1$, then the light paths to each end of the rod diverge, increasingly quickly with increasing distance, and the proper distance exceeds d_A. If $k = 1$ then light rays to the end of the rod diverge decreasingly quickly with increasing distance, and the proper distance is less than d_A. The redshift dependence in the expression for d_A is solely due to the evolution of R since the radiation was emitted.

The luminosity distance d_L is defined in terms of the total energy received from a distant object over all frequencies or in a spectral line. In order to calculate the total flux density ($\mathrm{W\,m^{-2}}$) received from a source with a luminosity L (W), light emitted at time t_e from coordinate r_e should be considered at by the time it arrives at the observer at time t_0. The light has then spread out over a spherical shell with area $4\pi R^2(t_0)\, r_\mathrm{e}^2$. As a result, the flux density $S = L/[4\pi R^2(t_0)\, r_\mathrm{e}^2 (1+z)^2]$, where the two additional factors of

$(1+z)^{-1}$ reflect the reduction in both the rate of photon arrival and energy of each photon due to redshift $1+z = R(t_0)/R(t_e)$. d_L can then be identified in $S = L/(4\pi d_L^2)$, and so $d_L = R(t_0)\, r_e(1+z) = (1+z)^2 d_A$, differing from and exceeding d_A, once a further $1+z$ factor is included from scaling $R(t_0)$ from $R(t_e)$.

There are three practical definitions of distance, the effective distance D, which transforms spectral luminosity and the volume element most naturally, the angular diameter distance d_A, which transforms the angular size of objects on the sky, and the luminosity distance d_L, which transforms either monochromatic or bolometric luminosity. The relative redshift dependences of d_A and d_L are compared in Fig. 8. Angular diameter distances are always less than luminosity distances. This has the consequence that high-redshift objects become too faint to detect more rapidly than they become impossible to resolve. Intuitively, this can be thought of in terms of the ever increasing redshift sapping the energy of the source, while the finite distance to the edge of the observable Universe means that once a source is very far away, it is not easy to make it appear appreciably more distant by moving it farther into the background.

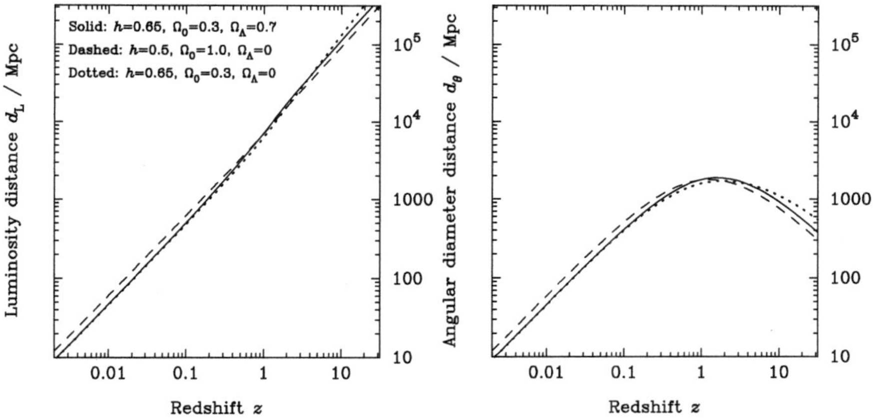

Fig. 8. The differences between luminosity and angular diameter distances for three historically reasonable values of the cosmological parameters. The flat, non-zero cosmological constant model represented by the solid line is now much favoured by observations. Note that the luminosity distance always increases, whereas the angular diameter distance increases and then declines. This is loosely due to the emission of photons at early time by objects that were then closer, and so 'looming over' us on the sky. This explains why CMB signals that started their journey towards us from points separated by only 300,000 lightyears on the sky 1.5×10^{10} years ago and have therefore traveled about 5×10^4 times further than their original separation are separated by the relatively large angle of 1 deg at the present epoch.

Surface Brightness Dimming: A consequence of the different dependences of angular diameter and luminosity distance on redshift is that the surface brightness of a source at redshift z, which depends on $d_{\rm A}^2/d_{\rm L}^2$, decreases rapidly with increasing redshift as $(1+z)^{-4}$: 'cosmological surface brightness dimming'. This can be especially significant when observing galaxies against a sky background due to the atmosphere or to blended emission from more nearby sources, and can possibly account for significant misinterpretations of the appearance of distant galaxies. If only the brightest parts of the galaxy can be detected, then apparently fragmented, multiple structure in a galaxy could simply be a biased view of the brightest regions of a larger underlying galaxy. Surface brightness selection effects are not currently important at mid-IR and longer wavelengths, as the resolution of telescopes operating at these wavelengths are insufficient to resolve distant galaxies: their surface brightness distribution is diluted by a point spread function much larger than the intrinsic angular size of the galaxy. This confers no advantage over a properly resolved image! In future continuum observations of resolved galaxies using ALMA, the peak surface brightness in an image could perhaps be used as a distance indicator: a detectable galaxy at redshift $z = 10$ would have to have an exceptionally bright region, with a surface brightness over 600 times greater, in order to equal the observed surface brightness of a galaxy at $z = 2$. Any galaxies with bright knots of emission would then have to be at relatively low redshifts.

4.5 Growth of Perturbations

Once the form of evolution of the scalefactor R is determined, then the backdrop of the evolving Universe is set, and it is possible to investigate the evolution of its contents in more detail. In particular, a description of the growth of overdensities is required to explain the growth of structure from initial smoothness towards the overdensities by factors of 10^{4-5} that exist now between the scales of galaxies and the IGM. At recombination, the CMB reveals density fluctuations in baryonic matter at the few 10^{-5} level, although the overdensities in CDM, being immune to radiation pressure, should be greater.

This can be done by a parallel analysis to the Jeans' analysis for star-forming clouds in the ISM, by linearizing the Poisson equation for the gravitational potential, the continuity equation for conservation of mass, and the Euler equation for fluid motion, in comoving coordinates locked within the expanding Universe. The analysis [82] leads to an equation for the evolution of the overdensity of a patch of the Universe with a mean density $\bar{\rho}$, $\delta = (\rho - \bar{\rho})/\bar{\rho}$ of the form,

$$\ddot{\delta} + 2H(t)\dot{\delta} - 4\pi G\bar{\rho}(t)\delta = 0 \ . \tag{11}$$

The equation involves the Hubble parameter, which includes the dynamics of the Universe. Note also that $\rho(t)$ has a strong R^{-3} dependence in a matter-dominated model.

In a static situation, $H(t) = 0$ and the $\dot{\delta}$ term vanishes, returning the familiar solution of an exponentially growing mode, from the star-formation Jeans' analysis. If this was the case, and there were no complications from magnetic fields, as arise in Galactic situations, then there would not be a problem to develop the rich structures we observe at the present epoch from almost smooth initial conditions at $z \simeq 1100$. However, in an expanding Universe, the first-order term does not vanish, and acts to reduce the speed of growth. In the case of a matter-dominated Universe, which is a good approximation at $z \sim 10$–1000 where the growth of perturbations starts, then $H \propto t^{-1}$, and so the results are a growing mode with $\delta \propto t^{2/3}$ (or $[1+z]^{-1}$) and a decaying mode with $\delta \propto t^{-1}$. A power-law growing mode is quite a challenge to structure formation – from recombination at $z \simeq 1100$, even a very rare 10^{-3} fluctuation will reach only $\delta \sim 0.1$ by the present epoch. As non-linear growth is expected to set in at $\delta \sim 1$, a recombination overdensity of order 10^{-2} would thus be required to meet this condition by the present epoch. Such overdensities are extremely rare, from the CMB fluctuation spectrum with an amplitude of several 10^{-5} [94]. The situation can be rescued if there is a dominant dark-matter field, in which fluctuations can grow without needing to wait for photon pressure to vanish at recombination. They can in fact start growing as soon as the Universe becomes matter dominated at $z \simeq 17000$ [193]. This allows a very typical 10^{-5} dark-matter perturbation to reach a significant overdensity $\delta \simeq 0.17$ at the present epoch, and for a not unusual 10^{-4} fluctuation to reach $\delta \sim 2$, assuming linear growth.

The cosmological constant does not yet completely dominate the dynamics of our Universe, but will do several Hubble times into the future. When the cosmological constant dominates the dynamics, then the growth of structure is powerfully suppressed and perturbations do not grow: H is then constant, not falling, and so the $\dot{\delta}$ term remains large, while the driving density falls rapidly due to the exponential growth of R. Hence, the Universe we see today is unlikely to develop a much richer array of structure as it enters a Λ-dominated phase. In an unfavoured open, low-density model Universe the growth of perturbations can continue at a low level, but structure formation is retarded at $z < \Omega_{\mathrm{m}}^{-1}$, leading to the formation of massive structures at earlier times than in denser models. The abundance of clusters at $z > 0.5$ supports a low-density model: if $\Omega_{\mathrm{m}} = 1$, then it is difficult to generate sufficient early growth of fluctuations to lead to the observed abundance of clusters, without too much structure in σ_8, the width of the distribution of the density contrast on $8h^{-1}$ Mpc scales at the present epoch.

Carroll et al. [71] provided a very useful and accurate analytic fitting function to compare the growth of fluctuations in different cosmologies,

$$\frac{\delta(z=0)}{\delta_{\rm EdS}(z=0)} = \frac{5}{2}\Omega_{\rm m}\left[\Omega_{\rm m}^{4/7} - \Omega_{\Lambda} + \left(1+\frac{\Omega_{\rm m}}{2}\right)\left(1+\frac{\Omega_{\Lambda}}{70}\right)\right] , \tag{12}$$

in which $\delta_{\rm EdS} \propto t^{2/3}$ is the overdensity in an Einstein–de Sitter model. The overdensity at the present epoch is expected to be about 65% of the value in an Einstein–de Sitter model for $\Omega_{\Lambda} = 0.7$ and $\Omega_{\rm m} = 0.3$ and the same value of h. The results of some exact calculations for $\delta(z)$ are shown in Fig. 9.

The growth of small CDM perturbations in the early Universe generates potential wells into which baryonic matter can collapse and cool, beginning the process of galaxy and star formation. The scale and rate at which it takes place are controlled by the details of the astrophysics in each halo, and by the properties of the initial density field. For example, if there is a preferred scale on which fluctuations are unusually large, then this could be represented in the observed distribution of galaxies. Note that the power-law faint-end slope of the Schechter luminosity function implies that either this is not the case, or the structure was subsequently washed out as galaxy formation continued. No trace of a strongly preferred scale appears to remain in the galaxy correlation

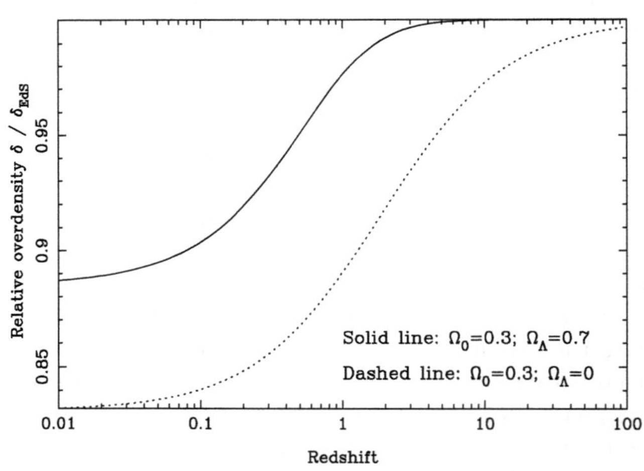

Fig. 9. The form of growth of perturbations in popular Universal models as compared with the Einstein–de Sitter model ($\Omega_0 = 1$; $\Omega_{\Lambda} = 0$). The value of δ is important as it controls the linear growth of overdense structures, and the redshifts at which structures greater than a certain mass appear. In the Einstein–de Sitter model $\delta \propto (1+z)^{-1} = R$. In lower density models the grow of perturbations is retarded at later times, but only by a small factor. Identical normalisation at early times is reasonable, as all models are matter-dominated at early times. The flattening of the curves at low redshifts reflects the small interval of cosmic time available there in the plot. Note that the growth rates are not converging to the same value at the present epoch.

function [83]. The angular scale corresponding to the first peak in the CMB power spectrum is on supercluster scales, and so does not appear to have had a significant influence on galaxy properties.

4.6 The Growth and Collapse of Individual Halos

The Jeans' analysis discussed above provides a way to assess the growth of small statistical fluctuations. However, it provides no insight into how to track the growth of an individual overdense halo. For this study it is essential to track density contrasts as they grow non-linearly to $\delta \sim 1$ and greater, corresponding to the formation of bound galactic halos. This can be done in an Einstein–de Sitter model (a reasonable approximation at early times) by tracking the evolution of individual overdense regions in the Universe, described as closed ($k = 1$) sub-Universes with their own solutions to the Friedmann equations. The radial evolution of 'successful' perturbations that collapse from the Hubble flow and form bound objects has a cycloid solution by analogy. Expansion occurs to a maximum extent – 'turnaround' – before recollapse towards a point, following the solution for radius r at time t:

$$r = A\,(1 - \cos\theta), \quad t = B\,(\theta - \sin\theta)\ . \tag{13}$$

In order to ensure that a mass M is enclosed, and the acceleration $\ddot{r} = -GM/r^2$ is correct at the edge of the sub-region, $A^3 = GMB^2$. By expanding both r and t as functions of θ, for small t, the θ dependence can be eliminated, and

$$r \propto \frac{A}{2}\left(\frac{6t}{B}\right)^{2/3}\left[1 - \frac{1}{20}\left(\frac{6t}{B}\right)^{2/3} + \mathrm{O}(t^{4/3})\right] \tag{14}$$

is found. Note that to first order the linear size of the region grows in the same way as the scalefactor in an Einstein–de Sitter model, $t^{2/3}$, while the density contrast increases as 3 times the first-order term in the bracket, and so also depends on $t^{2/3}$ for small t, as expected from the Jeans' analysis. Differences arise when the density contrast becomes greater. Turnaround occurs when $\theta = \pi$ (and $t = B\pi$). At that point, $r = 2A$, and by comparing the volume enclosed within this radius with the size to which the region would have expanded if $\delta = 0$, the overdensity at turnaround

$$\delta_{\mathrm{TA}} \simeq \frac{1}{(2A)^3}\left(\frac{A}{2}\right)^3\left(\frac{6t}{B}\right)^2 = \frac{9\pi^2}{16} \simeq 5.6\ . \tag{15}$$

That is, in an Einstein–de Sitter model, the density of a halo is about 5.6 times greater than the average density of the Universe when turnaround takes place. If the overdensity had grown linearly, then the density contrast at turnaround would be $(3/20)(6t/B)^{2/3}$, only $\simeq 1.06$ at the corresponding epoch. Hence, non-linear growth is important by the time of turnaround, as $\delta\rho/\rho > 1$. Note that without considering non-linear growth, the overdensity of the collapsing region would be underestimated severely. Note, however, that this enhanced growth factor is not sufficient to remove the need for dark matter to drive early fluctuation growth.

What happens as overdensities collapse? At some point shortly before or after turnaround the linear solution will fail, and an overdensity will break away from the cycloid solution. Being too dense to be affected by the surrounding Universe, the collapse will then proceed more rapidly. The gas and dark matter within will orbit and scatter, and should settle down into virial equilibrium – distributing the potential energy liberated during collapse, $V_{\rm TA} - V_{\rm V}$ into kinetic energy T. As the condition $T = -V_{\rm V}/2$ holds in virial equilibrium, and $V_{\rm TA} - V_{\rm V} = T$, this requires $V_{\rm TA} = V_{\rm V}/2$, and so a contraction by a factor of 2 from turnaround to the final state. In this way, the object becomes self-supporting at half of the turnaround radius, and never contracts to a point. In the cycloid solution, the value of r reaches half of its value at turnaround when $\theta = 3\pi/2$, which could be associated with the time of virialization. The corresponding fractional overdensity is 146.8. If θ is taken all the way to 2π, at $t = 2\pi B$, then the linear growth solution predicts $\delta \simeq (3/20)(12\pi)^{2/3} \simeq 1.69$. This condition $\delta = 1.69$ is often used as a threshold to define the collapse of structures in linear theory.

Hence, bound galactic dark-matter haloes would be expected to be about 150 times denser than the background Universe at the time they became bound/formed. At later times, the background Universe continues to expand and become less dense, but the size/density of the bound object remains fixed. There may be some minor reduction in size due to losses of energy in radiation and the ejection of angular momentum, but the density of dark-matter objects could thus provide an indication of the redshift of formation $z_{\rm f}$. For example, the present density of the Universe is about $3.5 \times 10^{10}\,{\rm M_\odot\,Mpc^{-3}}$, but within a radius of 1 Mpc, rich clusters of galaxies are approximately 10^4 times denser. Hence, the Universe would be expected to have expanded by a factor $(1 + z_{\rm f})^3 \sim 10^4/150$ since clusters virialized, implying $z_{\rm f} \sim 3$. That $z_{\rm f}$ is remotely similar to a plausible formation redshift for clusters suggests this is plausible. A value of $z_{\rm f} = 3$ actually makes more sense as the redshift at which the cluster member galaxies formed. In N-body simulations, it is conventional for bound objects resembling rich clusters in mass and extent to form at $z < 1$.

This treatment refers to the progress of the collapse of an overdense region of the Universe composed predominantly of dark matter. As described earlier, the complexity of describing the cooling baryonic material within and the onset of star formation and feedback is considerably greater. Simulations of these processes cannot currently claim predictive power. It is possible that we may always have to rely on observations to understand these early phases of galaxy evolution.

4.7 Press–Schechter Formalism and Extensions

One approach to studying structure formation and galaxy formation is to make N-body simulations of the gravitational interaction of dark-matter particles, which if convergence is obtained, should provide an excellent guide to the gravitational development of structure. This allows the properties of individual galaxy dark-matter haloes to be related to the initial conditions in the Universe set in the simulation.

A very neat way to combine the form of growth of perturbations and the initial fluctuation spectrum to generate a mass spectrum of bound objects without numerical methods was developed in 1974 by Press & Schechter [216]. The result is analytic and includes in a relatively transparent way the length- and time-scales that are important in determining the properties of the galaxy population. The formalism has been extended [61,59] to account for the abundance of collapsed bound halos that are part of larger structures, which it is necessary to treat when considering galaxies within clusters. The formalism has also been adapted [237] to make a better fit to the results of high-resolution numerical simulations [159], by taking into account that halos are typically non-spherical, and so are expected to collapse along their shortest axis first. This factor was neglected above: at the level of our understanding of galaxy formation is probably not the worst assumption made.

There are two parts to the Press–Schechter description. First, tracking the growth of the perturbations that eventually collapse to form galaxies, which determines when an overdense region turns around and starts to collapse. Second, the way in which the amplitude of fluctuations depends on the initial conditions imposed on the density field.

We investigate the consequences of the Press–Schechter description for the rate of mergers between galaxies, which are likely to be the biggest factor controlling the evolution of galaxies at far-IR wavelengths, as there is a clear association between merging objects and the most luminous dust-enshrouded galaxies [226].

Scales of Overdensities: An overdensity field $\delta(\boldsymbol{x})$ can be expressed as a sum of Fourier modes in a large volume with periodic boundary conditions as $\sum \delta_{\boldsymbol{k}} \exp(-i\boldsymbol{k}.\boldsymbol{x})$. When integrated over a representative volume V,

$$\int_V \delta(\boldsymbol{x}) \exp(-i\boldsymbol{k'}.\boldsymbol{x})\, \mathrm{d}^3\boldsymbol{x} = \int_V \sum_{\boldsymbol{k}} \delta_{\boldsymbol{k}} \exp[-i(\boldsymbol{k'} - \boldsymbol{k}).\boldsymbol{x}]\, \mathrm{d}^3\boldsymbol{x} = V\delta_{\boldsymbol{k'}} \,. \quad (16)$$

If the variance in the density field is defined as σ, then

$$\sigma^2 = \sum_{\boldsymbol{k}} \langle |\delta_{\boldsymbol{k}}|^2 \rangle = \frac{1}{V} \sum_{\boldsymbol{k}} \delta_k^2 \,, \quad (17)$$

where δ_k is a scalar, whose properties can be described by a one-dimensional power spectrum in Fourier space $P(k) = \delta_k^2$, as $V \to \infty$ and the universe becomes fully homogeneous and isotropic. The variance in the density field can be represented by,

$$\sigma^2 V = \int \sigma^2 4\pi l^2 \, \mathrm{d}l = \frac{1}{(2\pi)^3} \int P(k) \, 4\pi k^2 \, \mathrm{d}k \ . \tag{18}$$

It is desirable to exclude contributions from fluctuations on scales greater than a cutoff radius, which defines the mass associated with a fluctuation of interest, and so it is better to define the spread in the overdensity spectrum by,

$$\sigma^2 V = \frac{1}{2\pi^2} \int_0^\infty P(k) \, W(k) \, k^2 \, \mathrm{d}k \ , \tag{19}$$

including a window function $W(k)$ to sample the fluctuation spectrum on a chosen range of scales. A Gaussian window function $W_r \propto \exp(-(r/R)^2)$ transforming to $W_k \propto \exp(-(kR)^2)$ has the advantage of no sharp features in either real or Fourier space. Note that σ^2 will be convergent if $P(k) \propto k^n$ with $n > -3$ on large scales ($k \to 0$) and $n < -3$ on small scales ($k \to \infty$) without a window function. Hence, the fluctuation power spectrum $P(k)$ must curve to avoid inconsistency.

$n = 1$ is predicted by inflationary theories of seed structures that grow from quantum fluctuations as the Universe expands exponentially on all scales. This is a 'scale-independent' or Harrison–Zeldovich spectrum, for which the typical amplitude of fluctuations entering the horizon is constant in time, or equivalently, for which the depth of potential wells associated with fluctuations on different spatial scales is always the same. It is sufficient to illustrate using an Einstein–de Sitter model, because at early times after recombination the Universe is close to being matter-dominated. On a chosen scale, $\sigma^2 \propto M^{-(3+n)/3}$, where M is the average mass enclosed within that scale. Fluctuations, and thus σ, grow as $t^{2/3}$, and so $\sigma \propto M^{-(3+n)/6} \, t^{2/3}$. On the horizon scale, the enclosed mass is the product of the density, which depends on $R^{-3} (\propto t^{-2})$ and the cube of the horizon scale ($\propto t^3$). Hence, the mass enclosed within the horizon $M \propto t$, and within the horizon $\sigma \propto t^{(2/3)-(3+n)/6}$. Thus, for σ to be independent of epoch, $n = 1$. Causal processes can only operate on scales smaller than the horizon, and so the amplitude of fluctuations that can start to grow and begin to affect the galaxy formation process is constant in this situation.

Once the process of galaxy formation gets underway, non-linear collapse modifies the spectrum of fluctuations [204]. The two-point correlation function of galaxy positions on the sky can be transformed to yield the dependence $P(k) \propto k^{\sim -2.5}$ on small scales. Only on the largest scales, corresponding to both CMB fluctuations over 10-deg in extent on the sky, and spatial scales greater than the largest superclusters is the spectrum close to scale-invariant. All-sky and wide-area CMB observations confirm that n is within about 5% of unity on these scales. The all-sky CMB results from *WMAP* provide a very accurate measurement of n on the largest scales.

Components in the Press–Schechter Formalism: If the expected spread in the magnitude of the fluctuations on a scale R, involving a typical mass M, are described by σ, and if the fluctuation spectrum is Gaussian, then the probability of finding a region of mass M with overdensity δ exceeding a certain value δ_c is

$$F = \int_{\delta_c}^{\infty} \frac{1}{\sqrt{2\pi}\sigma(M,t)} \exp\left[-\frac{\delta^2}{2\sigma^2(M,t)}\right] \mathrm{d}\delta \, , \tag{20}$$

which can be re-written as a mass spectrum,

$$N_{\mathrm{PS}}(M)\,\mathrm{d}M = -\frac{\bar{\rho}}{M}\frac{\partial F}{\partial M}\,\mathrm{d}M. \tag{21}$$

σ depends on M via the links to the power spectrum $P(k) \propto k^n$, with $\sigma \propto M^{\gamma/2}$ where $\gamma = -(3+n)/3$, and its time dependence is determined by the evolution of δ.

The criterion usually applied for the formation of a bound object is that an overdensity turns around from the Hubble flow and goes on to collapse to a point in a time-symmetric way by epoch t if the overdensity exceeds the critical value $\delta_c \simeq 1.69$ for linear growth that matches the whole collapse cycle. As the mean size of fluctuations grows steadily with cosmic time, this corresponds to a greater number of bound objects exceeding the δ_c barrier at later times. More sophisticated descriptions of the process in careful, rigorous statistical terms, involving the trajectories of random walks through the density field, can be found elsewhere [20,237]. Care is necessary to normalise the Press–Schechter function correctly – simple treatments only recover half of the total density in the final function. The reason for this is still rather nebulous, but is linked loosely to the lack of treatment of underdense voids in the description, and tends to be explained away by the claim that a fraction of material accretes gradually into halos after their initial collapse. However, the agreement between the predictions and the results of high-resolution N-body simulations [159] is good, especially after modifications have been made to the empirical/collapse condition [237]. The Press–Schechter function can be used in variables δ_c and $\sigma(M,t)$, in which case

$$N_{\mathrm{PS}}(M,t) = -\frac{\bar{\rho}}{\sqrt{2\pi}}\frac{1}{M}\frac{\delta_c}{\sigma^2}\frac{\mathrm{d}\sigma}{\mathrm{d}M}\exp\left(-\frac{\delta_c^2}{2\sigma^2}\right) \; ; \tag{22}$$

however, a more useful description [112] is to convert to the mass of bound objects as the main variable and define $M^*(t)$ to represent the typical mass of a bound object as a a function of time. In the hierarchical picture that the formalism represents, M^* increases monotonically with time. When the mass description is used,

$$N_{\mathrm{PS}}(M) = \gamma\frac{\bar{\rho}}{\sqrt{\pi}}\frac{1}{M^2}\left(\frac{M}{M^*}\right)^{\gamma/2} \exp\left[-\left(\frac{M}{M^*}\right)^{\gamma}\right] \, . \tag{23}$$

This form of this function is related directly to the mass of bound objects, and can thus be converted into a luminosity function via a mass-to-light ratio, or moments can be determined to study the cross section to gravitational lensing, for example.

The empirical modifications to the Press–Schechter formalism to account for the non-spherical collapse of haloes [237] are described by,

$$N_{\rm ST} = A\sqrt{a}\left\{1 + \left[2a\left(\frac{M}{M^*}\right)^{\gamma}\right]^{-p}\right\} N_{\rm PS}\ , \tag{24}$$

where $A \simeq 0.32$, $a \simeq 0.7$ and $p \simeq 0.28$ [237,211]. For the standard Press–Schechter function $A = 1$ and $p = 0$. The difference from the Press–Schechter function is most pronounced at M^* and below.

It is necessary that both $\int N\,\mathrm{d}M = \bar{\rho}$ and $\int M\dot{N}\,\mathrm{d}M = 0$ to ensure that the total mass is normalized correctly, and that mass is conserved as the mass function evolves. It is possible to restrict the lower limit of these integrals to bound the possible values of dark-matter halo mass, and there could be gradual growth of bound objects from an almost homogeneous IGM component with time rather than all being derived from the merger of pre-existing bound objects.

A simple merger rate of galaxies can be derived, to provide a neat account of the rate and properties of the major mergers between galaxies with similar masses that likely account for most of the activity seen in dusty galaxies. The merger rate can be determined from $\dot{N}_{\rm PS}$. At low masses, galaxies continually merge from a chosen mass to form bound objects at greater masses, and while less massive objects continually combine to form new haloes at the chosen mass. The slope of $N_{\rm PS}$ at low masses should not change with epoch, as on sub-cluster scales that are safely above any minimum halo mass, and below the evolving value of M^* the power-law distribution of masses of nearby halos should lead to self-similar, complex, random merger behaviour. To conserve mass, the value of $\dot{N}_{\rm PS}$ must be negative at low masses, and positive at high. This reflects the hierarchical progression of structure formation, sweeping up from low to high masses with cosmic epoch.

Including the Sheth–Tormen modification, there is no requirement for a low-mass cutoff X in the density normalisation integral, as

$$\int_X^\infty M N_{\rm ST}\,\mathrm{dM} = \frac{A}{2}\frac{\bar{\rho}}{\sqrt{\pi}}\int_X^\infty U^{-1/2}\left(1 + \frac{1}{2}U^{-p}\right)\exp(-U)\,\mathrm{d}U\ . \tag{25}$$

For both the Sheth–Tormen description (unless $p > 0.5$), and for $A = 1$ and $p = 0$ (standard Press–Schechter), this integral converges to $\bar{\rho}$ for $X = 0$. The value of $\dot{N}_{\rm PS}$ or $\dot{N}_{\rm ST}$ at small masses ($M \ll M^*$) must also be negative: this requires the same constraint on p.

Some form of empirical correction is then required at low masses to obtain a 'formation' rate, which must at least be zero for all masses. The slope of

$N_{\rm ST}$ and $\dot{N}_{\rm ST}$ at low masses is the same, and so

$$\dot{N}_{\rm form} = \dot{N}_{\rm ST} + \phi N_{\rm ST} \tag{26}$$

is a good choice at these low masses. At large $M \gg M^*$, $\dot{N} \gg N$, and the straight derivative $\dot{N}_{\rm ST}$ provides a good description of the formation rate [48,49]. An additional term or two to ensure good behaviour at $M \simeq M^*$, where most activity occurs is also necessary. However, the details are not important here. The most significant part of the treatment is to overcome the negative $\dot{N}_{\rm PS}$ at small masses, as mergers must still be taking place on those scales, at least within a few orders of magnitude of M^*.

To find the total amount of material involved in mergers as a function of time, integrate $M\dot{N}_{\rm form}$ over mass. The result depends only on $\dot{M}^*/M^*$, and thus on the rate of fluctuation growth, and explicitly not on the value of p or A. Hence, the rate at which mass is processed through merging galaxies depends only on $\delta(t)$. It is then a question of determining the efficiency of each merger in forming stars or fueling an AGN, which could depend on both mass and cosmic epoch, and have a complex internal time evolution, in order to relate the results of observations.

Building an evolving mass spectrum for galaxies based on both the gravitational growth of dark matter fluctuations in specific cosmological models, and the initial conditions describing these fluctuations provides a neat, analytic way to investigate the formation of galaxies. It is in the transformation from mass to luminosity that most of the complexity and uncertain astrophysics lurks.

4.8 Angular Momentum of Galaxies

Most stars in the Universe are in pressure-supported Galactic bulges [130], but rotationally supported spiral galaxies are most numerous at the present epoch. Their angular momentum was picked up somewhere along their formation path, and has not been erased by merging activity. As well as the mass distribution of the dark-matter haloes that comprise the underlying building blocks for the galaxy-formation process, the issue of the distribution of galaxy spins is important, especially as many of the most luminous low-redshift dusty galaxies appear to be merging.

The key process that generates angular momentum is tidal torquing of non-spherical halos against each other during the linear phase of growth of fluctuations. Analysing the numerical halos formed in N-body simulations tends to indicate a log-normal distribution of the dimensionless angular momentum parameter $\lambda = J|E|^{1/2}G^{-1}M^{-5/2}$, where J is angular momentum, M is mass, and E is the binding energy of a virialized halo. For a body with size r and typical internal velocity $\bar{v}$, $J \simeq Mr\bar{v}$, $E \sim M\bar{v}^2$, and so $\lambda \sim r\bar{v}^2/GM$. If it is rotationally supported, then $\bar{v}^2 \sim GM/r$, and so $\lambda \simeq 1$. For the Milky Way, $r \simeq 8\,\rm kpc$ and $\bar{v} \simeq 220\,\rm km\,s^{-1}$ enclosing $M \sim 10^{11}\,\rm M_\odot$, and so $\lambda \simeq 0.7$.

Tidal torquing takes place as overdensities grow. If two adjacent overdensities with density contrast δ and mass M are separated by a comoving distance r, then their mutual acceleration $a \sim GM\delta/r^2R^2$, where R is the scalefactor. Only the overdensity unbalanced by other coordinating clouds leads to an acceleration. If the overdensities are asymmetric, then a torque can act with a lever arm $l < rR/2$, less than half the separation of the clouds. The torque $T \simeq aMl = GM^2\delta/2Rr$. Note that as fluctuations grow as $\delta \propto R$, matching the growth of the Universe, the torque is constant. After turnaround at $t_{\rm TA}$, torque from the asymmetry becomes less effective, due to the lever arm vanishing as the halo collapses. Angular momentum should thus build up uniformly from very early times until turnaround, when it reaches $J = Tt_{\rm TA}$, and $E \simeq GM^2/rR_{\rm TA}$. Recall that the overdensity at turnaround is expected to be $9\pi^2/16$. By expressing the time at turnaround $t_{\rm TA} \simeq t_0(R_{\rm TA}/R_0)^{3/2}$ and $M \simeq \bar{\rho}_0(4\pi/3)(rR_0)^3$, the value of $\lambda \simeq t_0\sqrt{G\bar{\rho}_0}$, or about 0.3 [205].

This means that galaxies are not too far from being rotationally supported, whatever their formation epoch, and motivates the idea that the earliest objects are likely to be disks rather than monolithically collapsing slabs or low-angular momentum spheres of gas, which could exist but are probably infrequent. Thus perhaps a picture of the formation of elliptical galaxies as mergers of ensembles of small rotationally supported disks is the most likely scenario for galaxy formation, as included in the recipes for most semi-analytic models. However, these models tend to fail to reproduce a sufficiently large number of very luminous galaxies, which may form stars in a more efficient mode that consumes more of their gas content more rapidly. Direct observational untangling of these issues is a key goal for future observations of high-redshift galaxies, and these questions have certainly not yet been settled. These considerations also apply only to dark matter. Once collapse begins, the angular momentum of the baryons and dark matter can evolve differently.

4.9 Correlations and Peculiar Velocities

Another fossil record of the galaxy formation process is the clustering strength of galaxies, which seems to be a Universal power law, with the probability of finding a galaxy a distance r from a chosen object varying as $\xi \simeq (r/r_0)^{-\gamma}$ with $\gamma \simeq 1.8$ and with $r_0 \simeq 5 - 12h^{-1}$ Mpc depending on the type of galaxy. The 2-point correlation function, the Fourier transform of the power spectrum of fluctuations $P(k)$, is being measured in unprecedented detail by the 2dF and SDSS projects [210,256]. Higher-order correlations between galaxy positions carry information about non-Gaussian structure present either at the beginning of (not expected in simple inflationary models) or imposed during the process of galaxy formation. Peculiar velocities induced by overdensities also offer a complementary approach to probing large scale structure.

4.10 Cosmology and Studies of Galaxy Evolution

The details of the galaxy formation process are likely to be dominated by the internal astrophysics of galaxies, strongly affected by environment, and properties like metallicity that are determined by the previous history of the material in the galaxy. The easiest effect of the background cosmology to quantify is on the appearance of a galaxy as a function of redshift, via the interaction of the shape of the redshifted SED of a galaxy and the shape of the Universe. This provides a distinct advantage when investigating galaxy evolution via emission for cool galaxies at submillimetre and far-IR wavelengths.

The counts and background radiation intensity are affected by the values of the cosmological parameters, via $r(z)$ for the background ((4)), and also by the distance parameter $D(z)$ for the counts ((2)). These parameters are probably now reasonably well-determined, and over the next few years, they should be fixed to better than 1% accuracy by precise measurements of the CMB.

4.11 K Corrections

The detectability of distant galaxies depends on several factors. The sensitivity of a particular instrument often imposes a limit to how many distant galaxies can be studied, and only if the instrument can search a wide enough area to detect a high-redshift galaxy in a practical integration time can cosmological observations be carried out. For example, in the 1960's only the very brightest of distant QSOs could be detected using photographic techniques, even on the 5-m Hale telescope. The introduction of electronic detectors since the 1970's has allowed quantitative high-redshift astronomy to develop.

The brightness of foreground emission from the telescope, atmosphere, Zodiacal light and low-redshift objects can also be difficult to handle. The relative importance of these effects differs from waveband to waveband. In general, the finer the angular resolution, the less important the foreground signal. Cosmological surface brightness dimming could easily render all or part of more diffuse high-redshift objects undetectable, and could account for part of the unusual morphologies suggested for galaxies in the deepest *HST* observations. The overlapping signals from faint galaxies clustered on the scale of the observing beam [52,150,206], source confusion noise, now limits the depth of submillimetre-wave surveys using 10-m class telescopes, as it once limited deep radio and X-ray observations. Once all these issues are understood, the intrinsic brightness of the galaxies can be considered.

Depending on the form of the SED, galaxies at different redshifts can be relatively easier to detect at a particular wavelength. Investigations of the distant Universe at radio and optical wavelengths probe relatively flat SEDs, with $f_\nu \propto \nu^\alpha$ with $\alpha \simeq -1$, and so distant galaxies become steadily harder to detect, both due to inverse square law dimming, and to redshift reducing the fraction of the total energy that arrives in the observing band. This continues to affect the deepest optical and radio observations [222]. In Fig. 10 we show the steadily reducing flux density expected from a standard, but increasingly distant galaxy in these bands. Surveys at near-IR wavelengths could potentially probe the stellar population of more distant galaxies out to high redshifts, at which absorption in the Lyman-α transition in intervening neutral matter is redshifted into the observing band. They also detect very red galaxies, with steeper spectra, $f_\nu \propto \nu^{\sim -1.6}$ (the Extremely Red Objects 'EROs' with $R - K > 5.5$). When redshifted, this steep spectrum makes a galaxy rapidly more difficult to detect.

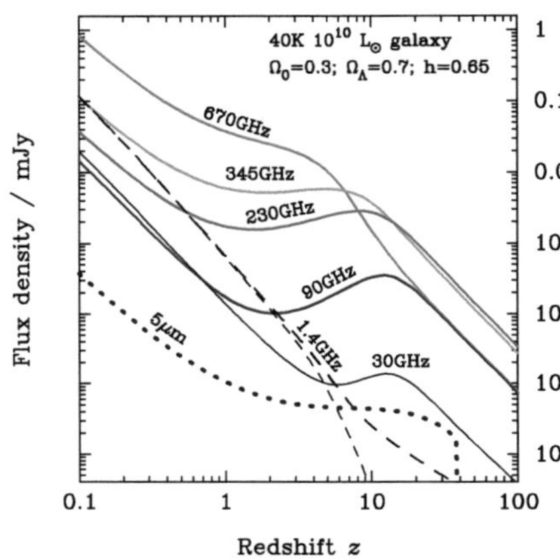

Fig. 10. An illustration of the K corrections expected for a dusty star-forming galaxy from realistic to very high redshifts. The low luminosity chosen perhaps corresponds to a distant 'first-light' object. There are several key features. First, the dramatic inverted (sub)millimeter K correction due to redshifting the steep dust spectrum, in comparison with the undramatic monotonic decline of the 1.4-GHz radio flux density. Secondly, the peak of the flux density–redshift curves occurs at a constant (high) redshift for frequencies between about 30 & 300 GHz. This is because a minimum dust temperature is imposed by the CMB, $2.73(1+z)$ K; and this controls the redshift at which the peak of the SED appears in the observing band. Thirdly, the detectability of the near-IR (5 μm) and radio emission from very distant objects. As the SED of starlight in the mid-IR is almost a blackbody, the $5-\mu$m emission experiences a significant K correction for $z \simeq 3-30$, and is cutoff at the redshift above which Lyman-α absorption is redshifted into the observing band: about $z \simeq 38$ at 5 μm. The thin dashed line representing radio emission includes synchrotron emission, but not free-free emission, which has an almost flat spectrum and dominates at the highest redshifts. The spectrum including free-free emission is shown by the thick dashed line. As the Universe is optically thin to radio emission, the free-free emission from ionized gas in the very first galaxies at $z \simeq 20$ could be detected using a future Square Kilometer Array (SKA) radio telescope, if the regions are sufficiently compact and the resolution is sufficiently great to avoid source confusion.

For historical reasons the effect of redshifting the spectrum is described as a K correction. At small redshifts, the correction from redshifting the SED of the target population could be added as a minor correction term to a magnitude–redshift diagram, in addition to the small 'E-correction' from galaxy evolution. There are three especially interesting spectral regions for the K-correction. One is in the (sub)millimetre waveband, where redshifting the steep $f_\nu \propto \nu^{\sim 3.5}$ SED can lead to distant galaxies with the same bolometric luminosity becoming brighter with increasing redshift (Fig. 10). Another is the rest-frame mid-IR waveband for galaxies that contain little hot dust, but have an evolved stellar population. The Rayleigh–Jeans thermal tail of stellar photospheric emission with $f_\nu \propto \nu^{\simeq 2}$ can then be probed. The third is at soft X-ray energies, of order $1-5$ keV, where Compton-thick AGN surrounded by column densities of about $10^{24}\,\mathrm{cm}^{-2}$ of gas can have the X-ray emission from their innermost regions absorbed, leading to an extremely blue X-ray spectrum at frequencies less than the peak of the X-ray background radiation at about 30 keV. This is closely analogous to the submillimetre situation, with the caveat that there are a large number of much redder foreground X-ray sources to weed out before the most obscured distant objects can be detected.

Optical and Radio *K* Corrections: The effects of different K corrections are illustrated in Fig. 10 at a range of frequencies, extending out to potentially extremely high redshifts. The appearance of an O-star photospheric SED in the near-IR waveband is shown for comparison. The K corrections for this source at $z \leq 1$, and for a 1.4-GHz radio source are also shown. Both display a steep monotonic decline with increasing redshift. Similar declines in the detectability of galaxies are typical in the mid- and far-IR observing bands. The effects are illustrated for a very low luminosity galaxy, that might be typical of the very earliest luminous objects, and is a factor 10^{3-4} times less luminous than the most exotic QSOs and ultraluminous dusty galaxies known at high redshifts [226].

Selection Effects at Long Wavelengths and Photometric Redshifts: The Universe is transparent to radiation at mid-IR and longer wavelengths. Beyond 300 μm, the K-correction effect of redshifting the spectrum renders very distant galaxies almost uniquely easy to detect (Fig. 10). However, these observations are not immune to selection effects. In particular, it is difficult to distinguish a high-redshift galaxy from one with a cooler SED at lower redshift [109,45,57]. The effect is illustrated in Fig. 11 at wavelengths of 1.1 mm, where it is very significant, and at 450 μm and 175 μm where it is much less dramatic. It is important to be aware of this effect when interpreting the results of far-IR and submillimetre-wave galaxy surveys and high-redshift observations of known galaxies, which can be very sensitive to their assumed SEDs. Based on long-wavelength continuum observations alone it seems to be impossible to avoid this temperature–redshift degeneracy [53].

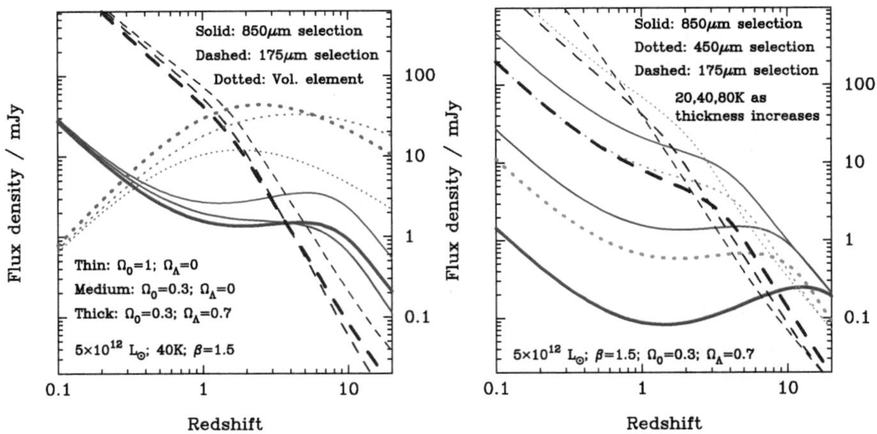

Fig. 11. Some of the uncertainties that affect the interpretation of the results of submillimetre and far-IR surveys. The left panel shows the effects of changing the cosmological parameters on the K correction and the volume of a unit-redshift shell in the Universe. While the (0.3,0.7) model almost certainly represents the real Universe, the other models can be seen to produce rather similar effects. The right panel shows the much more significant effect of different dust temperatures/SEDs. Low-temperature galaxies with a fixed total luminosity are significantly easier to detect at all redshifts at submillimetre wavelengths, and at redshifts $z \leq 1$ at far-IR wavelengths (175 μm). At shorter wavelengths this effect is much less significant, as the SED is observed close to the peak restframe wavelength. The effects of modifying α and β, the other SED parameters are much less significant [57]. The temperature dependence is sufficiently strong that the quantity $T/(1+z)$ is constrained much better by colour data than T or z independently [45,57,53].

In the light of *SIRTF* observations, it may turn out that there is a clear luminosity–temperature relation which will allow redshifts to be estimated using a colour–magnitude diagram. However, the existence of warm, low-luminosity M82 and cool, high-luminosity *ISO* galaxies [76] tends to argue against a simple reliable application of such a technique to individual galaxies. The range of possible SEDs, as a function of luminosity for galaxies with well-defined redshifts, multi-colour far-IR and radio data, assuming that the far-infrared–radio correlation holds is shown in Fig. 12. The significant minimum 25% scatter sets a limit to the maximum possible accuracy of a redshift determined using a multi-colour photometric technique [151].

5 The Cosmic Microwave Background (CMB)

The CMB dominates the energy density in radiation in the Universe, exceeding the cumulative background radiation intensity from galaxies integrated over all wavelengths by a factor of order 100. Its intensity decays as R^{-4}. In an accelerating Universe, this will lead to its eventual demise as the dominant

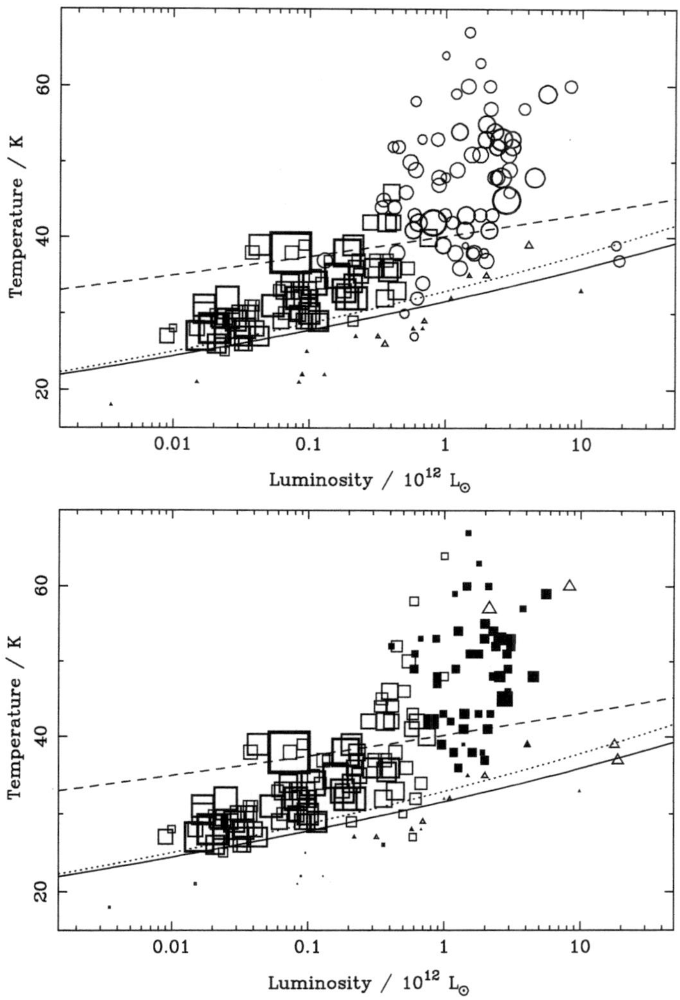

Fig. 12. Luminosity–temperature relations for dusty galaxies with known redshifts [53]. In the upper panel the galaxies are denoted by their sample origin: squares low-redshift *IRAS* galaxies with submillimetre data [107]; moderate-redshift faint radio galaxies in the HDF with *ISO* detections [133]; and very luminous *IRAS* galaxies with positions from the FIRST-VLA survey [29], and spectroscopic redshifts [252]. In the lower panel, sources from all samples are described according to their redshifts: empty squares at $z < 0.2$; filled squares at $0.2 \leq z < 0.5$; empty triangles at $0.5 \leq z < 1.0$; filled triangles at $1 \leq z < 2$; empty circles at $2 \leq z < 3$; and filled circles at $z \geq 3$. Sources represented by larger symbols have more accurate data. The lines in both diagrams represent other derived relations. The solid line shows results from low-redshift *IRAS* galaxy colours [79,91], while the dashed and dotted lines represent merging and quiescent populations of dusty galaxies respectively [53].

radiation field, as long as galaxies continue to generate significant amounts of energy in the future. It is an extremely valuable tool for investigating the early Universe, and its properties can be modified by scattering along the line of sight to the last-scattering surface of its release from equilibrium with electrons at recombination ($z \simeq 1100$), allowing physical conditions at lower redshifts to be probed.

The key features of the CMB are its extremely accurate blackbody temperature, and its isotropy at the 10^{-5} level. The detailed departures of the CMB from isotropy are very close to those predicted by the interaction of radiation and matter in the pre-recombination Universe. The low level of density fluctuations, simple physics of Compton scattering and simple chemistry act to produce predictable peaks in the power spectrum of initial density perturbations on angular scales close to 1, 0.5 and 0.25 deg ($l = 213^{+10}_{-13}$, 541^{+20}_{-32} and 845^{+12}_{-25}, where l is the Legendre polynomial index of the power spectrum [94]). These peaks represent the imprint of stationary phases in the gravity-driven, photon-electron pressure supported oscillations in the radiation–electron–baryon–CDM fluid prior to recombination. On scales corresponding to oscillations that are moving at recombination, the phases of the signals do not add constructively and so the amplitude is lower.

The time available for oscillations to occur is governed by the time interval between the epoch a particular scale becomes smaller than the horizon and recombination. Smaller scales came through the horizon earlier, and thus can go through a greater number of cycles before the pressure drops dramatically at recombination. The largest-scale peak corresponds to clouds that had time to make only a single infall to maximum density before recombination. Other odd-numbered peaks correspond to maximum density at recombination after additional oscillation cycles. Even-numbered peaks correspond to minimum density states: stationary rebounded clouds. There is a natural reduction in the intensity of CMB fluctuations on smaller scales due to the diffusion of photons out of overdense regions: Silk damping. Scales spaced equally between the peaks are associated with maximum velocities at recombination. Because moving electrons polarize scattered photons, the initially unpolarized thermal CMB should become polarized most strongly on these scales, with a polarization anisotropy amplitude about two orders of magnitude less than the ($\sim 10^{-5}$) amplitude anisotropies.

The first hints of peaks in the CMB angular power spectrum was established by a wide range of experiments probing degree angular scales throughout the 1990's. The clearest signal was provided by the ground-based Saskatoon experiment [199], which detected significant power on scales sandwiched between the well-defined anisotropy signal determined by the DMR instrument aboard the *COBE* satellite on scales between the dipole and 7 deg [34], and limits to the fluctuation amplitude matching the DMR level on smaller 10-arcmin angular scales. The smaller-scale limits using SuZIE [132] and the OVRO 5-m antenna [174], and the detection of a DMR-level signal on intermediate 10-arcmin scales using the first-generation CMB interferometer CAT [18][12]. The existence of detectable coherent peaks in the CMB itself testifies that the Universe remained neutral from $z \simeq 1100$ until $z \sim 20$, otherwise the angular structures observed on the CMB would have been blurred out substantially by line-of-sight scattering. The definitive detection of the first peak was provided by the BOOMERanG and MAXIMA balloon experiments in 1998/1999 [94,173], and subsequent detailed mapping of the structure in 10-deg^2 fields using the DASI [69], CBI [238,208] and VSA [257] interferometers.

The power spectrum provides a direct measurement of the angular size of an object with the physical scale of the horizon at recombination, allowing the angular diameter distance to be measured, and so providing an accurate constraint on the value of $\Omega_{\mathrm{m}} + \Omega_{\Lambda}$ [94]. The power spectrum has been measured using a variety of independent techniques, and is now one of the most solid observations in cosmology. The astrophysics of the interaction of thermal radiation with primordial ionized gas is also one of the cleanest processes in the Universe, and so the results put the initial conditions of galaxy formation on a very firm footing. In January 2003 *WMAP* reported a high-quality detection of the power spectrum all the way from the dipole to 10-arcmin scales, including the crucial region at $l \simeq 50$, where the rise of the power-spectrum from large scales to the first peak is sensitive to the speed at which recombination occurred. The detection of significant non-Gaussian anisotropies, or the failure to reproduce the existing power spectrum, would provide startling news. Measurements of CMB polarization in the next decade should provide an absolute test of the paradigm of initial conditions for galaxy formation building from CMB observations.

5.1 Secondary Anisotropies

Scattering of CMB photons from ionized structures since recombination also provides a useful information about galaxy evolution. The Universe was still very smooth when it became neutral at recombination, and there are no expectations that baryonic galaxy-formation processes could occur in almost uniform gas at over 3000 K. However, the reionization of the Universe began prior to $z \simeq 6$[30,99], due to either the first stars forming or the first AGN beginning to accrete. Once reionization began, scattering of the CMB could occur.

The Sunyaev–Zeldovich (SZ) effect [255] is the most significant mechanism for imprinting structures on the CMB. There are two components of the effect: a thermal and a kinematic part. The thermal effect is produced by Compton scattering of CMB photons off hot electrons. It imparts a characteristic increase to the net energy of CMB photons at the expense of the electrons, leading to a reduction in the intensity of the CMB at frequencies

[12] For a summary of about 20 first-generation CMB experiments see [172].

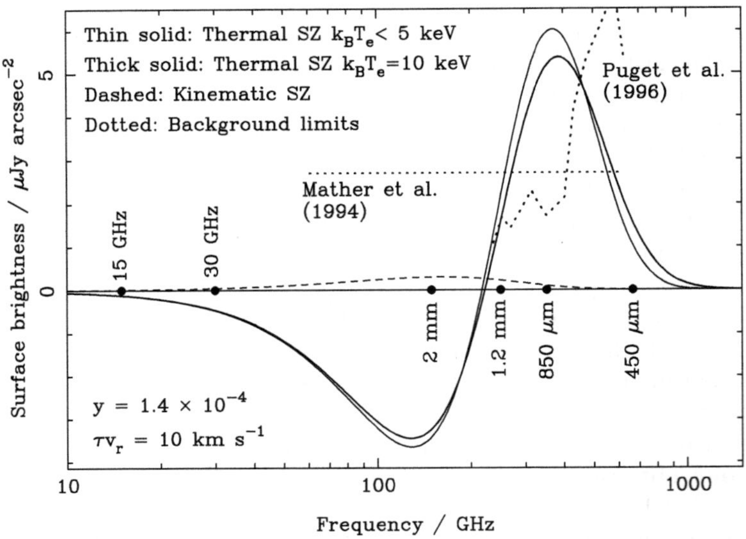

Fig. 13. A comparison of the SZ spectrum in the centre of a reasonably massive cluster of galaxies, with an early limit to the background radiation intensity in the submillimetre waveband [190] and the first detection [217]. The relative sizes of the thermal and kinematic effect and the effects of including relativistic electrons can be seen. Note that at wavelengths longer than about 600 μm the SZ effect from clusters is likely to be larger than the typical integrated background signal from field galaxies.

below 217 GHz, and to an increase at higher frequencies, that peaks at about 370 GHz. The form of the intensity change,

$$\frac{\Delta I}{I} = \frac{\Delta T}{T} = \frac{x \exp x}{\exp x - 1} \left[x \frac{2x \exp x}{\exp x - 1} - x - 4 \right] \frac{y}{2} , \tag{27}$$

where $x = h\nu / k_{\rm B} T_{\rm CMB}$ (see Fig. 13). The term before the square brackets converts from a fractional temperature change to a fractional flux density change. The size of the effect is set by the parameter $y = \sigma_{\rm T} k_{\rm B} / (m_{\rm e} c^2) \int n T_{\rm e} \, {\rm d}l$, which includes the integral of pressure through an ionized structure, via the electron space density n and temperature $T_{\rm e}$; $\sigma_{\rm T}$ is the Thompson cross-section of the electron. At long wavelengths, where most measurements of the signal have been made, $\Delta I / I = -2y$, independent of frequency.

From *COBE*-FIRAS measurements of the CMB spectrum averaged over the whole sky $y < 1.5 \times 10^{-5}$ [121]. In rich clusters of galaxies $y > 10^{-4}$ and $T_{\rm e} \simeq 10$ keV, leading to a probability of scattering $\tau = \int n \sigma_{\rm T} \, {\rm d}l \simeq 10^{-2}$. The detailed SZ signal from clusters has been detected by several experiments, including cm-wave detectors of the SZ decrement at both the OVRO/BIMA millimetre-wave interferometers [162] and the cm-wave Ryle Telescope in Cambridge [161], and at short, SZ increment wavelengths using the SuZIE and

SEST experiments. The well-sampled baseline structure of the OVRO/BIMA observations provide the finest images. With increased sensitivity, less dense structures such as intercluster filaments, and even individual galaxy halos may ultimately be detectable. There are no clear detections of the SZ effect from an ionized IGM or in smaller groups. This could change after reports of excess power on small angular scales detected by CBI announced at the AAS in January 2002 [238]. The excess power is greater than predicted previously for the SZ effect, and if confirmed, there seems to be a significant contribution to the background radiation intensity from the formation of clusters, with more energy being injected into the intracluster medium (ICM) than expected, leading to a hotter/denser gas. This type of signature should be detected clearly by dedicated cm-wave, arcmin-scale SZ interferometers under development. With close-packed arrays of tens of 4–5-m sized antennas to provide faithful images of the relatively low surface brightness of the SZ effect [149,167], these instruments will make accurate measurements of the resolved profile of the SZ effect in known rich clusters, search for the signature of connecting filaments between clusters, and seek high-redshift clusters.

If the scattering electrons are relativistic, or if there is a significant probability of multiple scatterings, then the SZ spectrum can be subtly modified. For relativistic scatterers, the null frequency increases slightly, and the increment signal at high frequencies changes shape [221] (Fig. 13). In the standard model of structure formation by hierarchical clustering, the most massive objects with the highest virial temperatures ($T_V \propto M^{2/3}$) are expected at low redshifts, and it seems unlikely that numerous high-redshift clouds of relativistic electrons are lurking to be detected this way. In principle, the relativistic SZ signal could allow the measurement of the temperature of electrons in the ICM; however, this can probably be done more reliably using X-ray observations.

The kinematic SZ effect is produced by scattering from electrons with a net peculiar velocity, and has the spectral dependence of the CMB. If the scatterers are approaching, then the intensity of the CMB is boosted, if receding, then the CMB intensity is reduced. The signal depends on only the peculiar velocity of the cluster. For reasonable cluster parameters, the peak of the kinematic signal is never likely to exceed about 10% of the thermal signal, except near the 217-GHz null frequency of the thermal effect, where it is the only contribution. It may be possible to make a peculiar velocity survey of clusters to high redshifts using this technique [39], especially using the SZ null frequency channel of the all-sky *Planck Surveyor* CMB mapping mission.

The surface brightness of the SZ effect is independent of the redshift of the scattering object. While cosmological surface brightness dimming reduces the intensity of the signal as $(1+z)^{-4}$, this is counteracted by the increase in the energy density of the CMB by the same factor at the redshift to the SZ scatterer. By comparison with the intensity of X-ray emission from the same electrons that scatter the CMB, which depends on the different integral of $n^2T^{1/2}$ along the line of sight, the absolute size of the cluster can be determined, if T is known and uniform, and the cluster is spherical. By comparing with its angular size on the sky, H_0 can then be determined. This is becoming more realistic, but more complicated than first thought, as high-resolution X-ray images of clusters from *Chandra* showing substantial substructure are combined with resolved centimeter/millimetre-wave SZ images.

There are other less-significant secondary effects. The Ostriker–Vishniac (OV) effect, is the integrated equivalent to the SZ effect due to scattering from diffuse, ionized gas not associated with deep cluster potential wells. The Rees–Sciama (RS) effect is due to the evolution of cosmic structures as CMB photons cross them, leading to CMB photons being reddened as they climb out of a potential well that deepens during the crossing time. These effects have the same spectral form as the CMB, and are also independent of redshift, but are not expected to be greater than about 10% of the SZ signal [233].

The ionized gas that leads to the detectability of the SZ effect could have been generated by collisions as clusters formed, or by radiation released during the formation of the cluster member galaxies or their progenitors. The SZ effect provides a way to probe both the existence of bound virialized structures at high redshifts, and the way in which diffuse gas in these structures was ionized. Within about twice the arcmin-scale gravitational lensing Einstein-ring radius of a foreground cluster, the lensed background SZ signal from more distant groups and clusters superimposed on the cluster's own SZ effect should be smoother than that in a general field. Independent of redshift, this could provide a way to enhance fainter OV, RS or SZ effect signals.

Non-Gaussian Features: Foregrounds, and especially sharp-edged linear and two-dimensional features like cosmic strings and textures, could introduce non-Gaussian features in the CMB. There are currently no hints of such structure, either via individual detections of unexpectedly hot or cold regions or of a non-zero higher than second-order correlation function of the CMB brightness. Any detection of a different type of fluctuation beyond the simple Gaussian field expected from quantum fluctuations during inflation could reveal dramatic new physics[13]. *WMAP* offered the first opportunity to detect such oddities, both by providing a large number of pixels for determining correlation functions, and covering more sky to increase the chance of detecting a rare, rogue feature; however, no clear signs of anything untoward emerged [168].

---

[13] The two-point correlation function $C_2(|\mathbf{r_0}|) = \langle T(\mathbf{r})\,T(\mathbf{r}-\mathbf{r_0})\rangle$ fully describes the statistical properties of a Gaussian field: all higher-order correlation functions should be zero.

5.2 Foregrounds and Signatures of Galaxy Formation

Both primordial and secondary features in the CMB are prone to being masked by and confused with foreground emission, both diffuse radio and infrared synchrotron, free-free and dust emission in the Milky Way, and the contributions of radio and submillimetre galaxies at high redshifts. In some regimes, it is not possible to detect CMB anisotropies, as the foreground signals dominate. This has been investigated using synthesis models of known populations [258]. It has now been verified directly that the effects of galaxy foregrounds are not significant for primary, non-polarized anisotropy measurements by BOOMERanG, DASI, CBI, VSA and *WMAP*. Polarisation has also been detected [169,203].

The determination of the angular power spectrum of CMB fluctuations on scales from the whole sky to several tens of arcmin, at Legendre polynomial indices $l < 4000$, is a striking recent highlight in observational cosmology, and is reaching maturity with a wide variety of ground-based, and balloon- and space-borne surveys.

Future surveys may allow foreground galaxies to be detected. For example the *Planck Surveyor* space mission will detect up to 10^5 dusty galaxies at 350 to 850 μm. Almost all of the issues concerning foreground contamination of the CMB signal should be settled when *Planck* flies in 2008. Pointing away from Earth, Moon and Sun at the Earth–Sun L2 point, *Planck* will map the whole sky 4 times over 2 years, imaging the CMB in 10 channels from 30 to 857 GHz, with a resolution of 5 arcmin at frequencies greater than 200 GHz, and diffraction-limited resolution for its 1.2-m off-axis aperture at lower frequencies. While having limited polarization sensitivity, *Planck* will provide a definitive map of the whole submillimetre-wave sky on super cluster scales.

Increasingly accurate CMB measurements will continue to make this cold component of emission from the Universe a vital one for both measuring the size and shape of the Universe and investigating its contents.

6 The Evolution of High-Redshift Galaxies

At some wavelengths, especially those of interest here, a simple census of the contents of the Universe is still being taken (see Fig. 14). At others, especially optical and radio wavelengths, details of the properties of most of the galaxies detected (masses, metallicities and physical sizes) [235] have already been determined. It is easy to lose track of the fact that resolved, multicolour images of typical high-redshift galaxies have only been available in significant numbers since the observations of the HDF-N in 1995 [269]. There are sure to be more surprises lurking as the Universe is surveyed to greater depths, even at the well-studied wavelengths. We are still far from observing a representative sample of the first objects that re-ionized the Universe at $z \simeq 6$, and their progenitors.

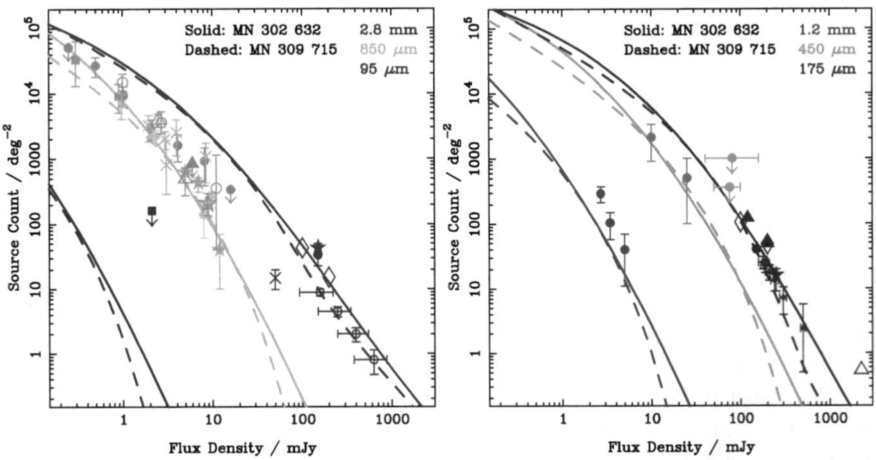

Fig. 14. Counts of galaxies detected at a range of far-IR, submillimetre and millimetre wavelengths, by a wide range of groups and techniques (a full summary can be found in [57]). This represents the full knowledge of the counts of dusty distant galaxies from published 175 − μm *ISO*, 450/850 − μm SCUBA and 1.2-mm MAMBO surveys to August 2002. Note that the range of flux densities probed is not very great, about a factor of 10. There is a need for both deeper surveys to inform us about the evolution of galaxies less luminous than L^* – equivalent to about 0.1 mJy at 850 μm (see Fig. 10), and for wider surveys of the rarest, most luminous galaxies at brighter flux densities. It is possible that the bright counts fall very rapidly as a physical limit to the luminosity/mass of a galaxy is reached [142], before bright low-redshift galaxies or high-redshift gravitationally lensed galaxies [43] dominate the counts at the brightest flux densities. The overplotted lines correspond to luminosity evolution and hierarchical models of galaxy formation [54,56].

6.1 Optically Selected Galaxies

The best-studied population of distant galaxies are the Lyman-break galaxies (LBGs) [253,254], with luminosities ranging from about 10 times greater than the Milky Way. These are selected using multicolour optical images from 4-m class telescopes, by searching for a sharply-defined, red colour break between adjacent filters, due both to neutral hydrogen absorption both internally and in intervening gas at the redshifted 91.2-nm Lyman-break frequency. Candidate LBGs at $2.5 < z < 4.5$ can be isolated in observations extending to the *I*-band. Spectroscopy of these photometric candidates using 10-m class telescopes confirms that about 70% are at these redshifts. The LBGs have a well-determined luminosity function [3], and detailed astrophysical information is generally available from optical and near-IR spectral properties [235,214]. At lower-redshifts, a comparable Balmer-break sample can be selected, while higher-redshift samples could be detected, at lower efficiency, by extending the range of filters used out to near-IR wavelengths. The LBGs are a very important, well-defined baseline population of distant galaxies for comparison with the other galaxy populations in the distant Universe.

6.2 Red Galaxies and Extremely Red Objects (EROs)

By the nature of the SED break signature sought in the rest-frame UV waveband, the LBGs have emergent rest-frame UV continuum flux, and are necessarily either rather blue in colour or very luminous if they are red and emit a trace of their total luminosity there. They display a wide variety of well-studied spectral features at longer wavelengths, but it is certainly very likely that the LBG selection process excludes galaxies with very red SEDs, or too faint, as insufficient UV flux escapes to yield a detectable signature.

Adelberger & Steidel [3] have discussed the luminosity function of the LBGs and their relationship to other classes of high-redshift galaxies. A class of galaxies definitely eludes Lyman-break selection: the mixed bag of objects defined by extremely red colour, $R-K > 5.5-6$ – the Extremely Red Objects (EROs). This is amongst the least satisfactory of astronomical classifications. The condition $R-K > 5.5$ applies to less than 1% of galaxies in deep survey fields, however it is typical of both evolved stellar populations and strongly dust-extinguished galaxies with intrinsically blue stellar/AGN SEDs at moderate redshifts. Almost any red low-redshift galaxy would become extremely red and faint at $z \simeq 1$, and invisible by $z \simeq 2.5$. Hence, the classification fails either to refer to a unique intrinsic physical property or to select a homogeneous population. Imposing a more severe constraint on $R-K$ to ensure more extreme redness assists the detection of unusual objects, but even then careful follow up is required to sift out the luminous, obscured galaxies. Historically, the interest in EROs reflects the rare detection of interesting, anomalously red galaxies in narrow-field near-IR images taken in the mid-1990's [140]. The advent of wide-field near-IR imaging has led to a flood of ERO samples, including relatively shallow surveys with $K < 20$ [90] and $H < 21$ [81], in which at least 80% of the detected galaxies are simply evolved ellipticals at moderate redshift $z \simeq 1$, with follow-up observations [197] failing to detect radio and infrared emission to diagnose significant ongoing activity. Deeper surveys [245], most notably the Subaru Deep Field [259] may make a different selection. Unfortunately, no deep radio imaging is available in this field to select out the vigorous star-forming EROs. The $K \simeq 20$ EROs are strongly clustered [90], but as they are amongst the oldest populations of galaxies at moderate redshifts, they would be expected to have formed in the most overdense regions at $z \simeq 3$ [20], and so this is not surprising.

A link between EROs and distant submillimetre galaxies is expected: if the dust in these galaxies has the geometry of a screen, then their emission should be strongly reddened at optical wavelengths. That may be the case, although several of the best studied submillimetre galaxies are expressly not EROs [154,155], even if parts of these galaxies do have the required red colour [157]. Most of the submillimetre population is optically faint with $I > 26$

[243]. Verification that it is also extremely red requires imaging observations to $K \simeq 23$, a feat that requires significant investment of observing time even with 8–10-m class telescopes [127]. At present, the association between distant ultraluminous infrared/submillimetre-selected galaxies and a subset of EROs is an open question.

There is a direct link between the surface density of faint blue galaxies and the amount of energy they release. This is because the effects of redshifting a relatively flat blue spectrum $f_\nu \propto \nu^0$ expected from a young stellar population counteracts the $1+z$ factor in the equation for background energy density ((4)) [249]. There is no such analogous conversion between the flux density of EROs and their cosmological significance. The hope is that the study of EROs develops in a similar way to how LBG studies grew from the faint, blue galaxy surveys in the 1980's. The most important role they are likely to play is providing information about populations of evolved galaxies at moderate redshifts, revealing details of the process of elliptical galaxy formation at earlier times. As there is much speculation about the connection between submillimetre-selected galaxies and the formation of elliptical galaxies and spheroids, this may have a direct bearing on studies of the formation of dust-enshrouded high-redshift galaxies: EROs at $z \simeq 1$ may be the fossil remains of all the submillimetre galaxies.

A further motivation for seeking ERO counterparts to submillimetre galaxies is the need to determine accurate positions. As a fraction of EROs are expected to be associated with dusty galaxies, then the detection of a relatively rare ERO within the positional error of a submillimetre galaxy is likely to provide a strong indication of a possible correct counterpart [127].

There is no published case of an EROs selected in a systematic near-IR survey being detected in the submillimetre, despite some searches, but many submillimetre galaxies have ERO counterparts [241,134,185], or are associated with slightly bluer, but still very red galaxies [126]. Only one has so far been verified as an undeniable counterpart with a secure redshifts [128]. In two cases, millimetre-wave interferometer observations have reduced the positional uncertainty of the initial submillimetre detections, co-locating the millimetre-wave emission and ERO, and reducing the probability of a false association [134,126].

6.3 Active Galaxy Selection

Both optically- and radio-selected AGN have been detected at great distances since the beginnings of observational cosmology. Only relatively recent developments in large telescope design have allowed more representative high-redshift galaxies (the LBGs) to be detected in significant numbers. High-redshift AGN were amongst the first distant objects to be detected at submillimetre wavelengths in the mid 1990s, and dedicated searches continue [153]. A recent development is the first sign that QSOs at redshifts $z \simeq 6$ are being viewed through an increasingly neutral IGM [119,140], so their UV light is absorbed by Lyman-α transitions in the intervening IGM – the Gunn–Peterson effect [144]. The re-ionization that makes the Universe transparent at lower redshifts is a key part of the history of galaxy formation, and the recent detection of the epoch at which it ceases to be effective represents real progress. Note, however, that the detection of re-ionization is not going to affect progress in deep far-IR/submillimetre observations, where the opacity along the line of sight is negligible to even the highest redshifts. A more pressing concern for probing further at these wavelengths is that the metals necessary to allow the detection of dust and gas may not be available at higher redshifts, if re-ionization takes place at an epoch just prior to $z = 6$.

6.4 Faint Radio Galaxies

Up to three orders of magnitude less luminous than radio-loud AGN at radio wavelengths are the faintest population of galaxies detectable using current radio telescopes, with 1.4-GHz flux densities down to about 10–20 μJy. This is an interesting population with a surface density similar to the dusty galaxies detected using SCUBA and MAMBO (Fig. 14). If they lie on the tight far-IR–radio correlation observed at low redshifts [85,275], then they should sample the same population of dusty, luminous, high-redshift galaxies (Fig. 10). These studies [24,72] have targeted faint radio galaxies with the faintest optical counterparts, expected to indicate high redshifts, for searches for thermal dust emission with the SCUBA submillimetre camera. The result is a sample of well over 100 submillimetre galaxies with accurate sub-arcsec radio positions. This sample is likely to be biased to lower redshifts than the full submillimetre population, due to the less favourable radio K correction, and to include an enhanced AGN fraction due to excess AGN-powered non-thermal radio emission. This radio population does, however, offer much easier follow up observations than galaxies selected only at submillimetre wavelengths [77].

6.5 Submillimetre Galaxies and the Optical and Radio Populations

The selection effects at work in surveys for optical, radio and reprocessed thermal dust emission were discussed extensively by Adelberger & Steidel [3] and are illustrated in Fig. 10. They sought to exploit their large, well-studied, spectroscopically-confirmed sample of several 1000 optically-selected LBGs (at $2.5 < z < 4.5$) to seek any less numerous, more luminous submillimetre-selected galaxies within, using the correlation between the rest-frame UV spectral slope of galaxies [194] which holds to luminosities $L \simeq 10^{11}\,L_\odot$ for low-redshift galaxies, based on UV observations using the *International UV Explorer (IUE)* satellite. If the correlation holds all the way to the reddest LBGs, and to the luminosities of the submillimetre-selected galaxies, then this is plausible. It appears to work for at least one identified

SMG, SMM J14011+0252 (Fig. 22) [155]; however, this galaxy is exceptionally bright at optical wavelengths; typical SMGs are up to 6 magnitudes fainter [243]. The most luminous *IRAS* galaxies at low-redshifts also have far-IR luminosities underestimated by the UV–far-IR relation [139]. This is easy to understand, as there are densely-enshrouded regions in the most luminous galaxies from which almost no UV light escapes, whereas more diffuse low-luminosity systems contain many fewer totally opaque regions. Thus the mean opacity (reflected in the UV–submillimetre ratio) and luminosity (from the sum of the emission in the optical/UV and submillimetre wavebands) appear to scale together.

Submillimetre observations of optically-selected LBGs have tended to detect no emission [73], with the occasional, dramatic exception [75,155]. The best limits were obtained for cB58 [17], which is strongly magnified by a gravitational lens. cB58 appears to be 40 times brighter than a typical LBG, and yet can only just be detected at (sub)millimetre wavelengths: a typical LBG thus probably has a very low flux $850-\mu$m flux density of order 0.1 mJy (see Fig. 14). This is consistent with the detection of positive fluctuations in the noise from the positions of faint LBGs in the very deep SCUBA $850-\mu$m image of the HDF-N [206], and the statistical detection of net positive $850-\mu$m flux density from the directions of LBGs in the CUDSS SCUBA survey [265]. The deepest reliable $850-\mu$m photometry at present reaches down to about 2 mJy for individual objects, but stacking large numbers of images to obtain deeper limits for populations seems reliable.

There is certainly a Universal multi-wavelength luminosity function describing the galaxy population at high-redshifts: one of the key goals of observational cosmology is to determine it. However, existing submillimetre instruments are not sufficiently sensitive to detect the bulk of the optically-selected population, while current optical observations are not deep enough to detect submillimetre galaxies in large numbers. This is probably due to the submillimetre galaxies being deeply extinguished rather than to them being at very great distances, based on the non-detection of three example to deep $3-\sigma$ limit of $K_{\rm s} \simeq 21.9$ using VLT [92]. Other evidence is provided by the possible detection of a very red counterpart to the brightest submillimetre source in the HDF-N at $K_{\rm s} \simeq 23.5$ [106] using the Subaru telescope, and the detection of red K-band counterparts to most submillimetre galaxies observed at depths as deep as $K \simeq 24$ at $2-\sigma$ using the Keck-NIRC camera [127].

Deeper surveys from each direction, exploiting both *JWST* to image submillimetre galaxies and ALMA to image the optical population should resolve this question directly by 2010. The deepest images from both instruments could extend current sensitivity limits by at least 2 orders of magnitude. Note, however, that it is possible that many LBGs contain little dust, and so ALMA may be hard pressed to detect them – many LBGs may just not be dusty. The same is unlikely to be true for the the detection of the optical emission from the reddest and most distant submillimetre galaxies for *JWST*; some of the SCUBA galaxies could have I-band magnitudes in the mid 30's if they are intrinsically red and located at $z > 3$ [95].

6.6 Gravitational Lensing

Gravitational lensing provides powerful evidence for the presence of dark matter and for the accuracy of General Relativity, especially in well-constrained sets of images, where the image geometries are securely determined and easy to interpret, such as an Einstein ring of emission from a large source behind an almost circular lens. It is also a useful tool to detect galaxies that would otherwise be too faint to observe: a gravitational telescope. A description of the extensive formalism developed to treat gravitational lensing can be found in [230,27].

General properties of gravitational lensing are the existence of a weak all-pervading cosmic shear field, induced by the small deflections of light from all mass inhomogeneities along the line of sight to distant objects. Cosmic shear distorts the images of background galaxies, without magnification, modifying their axis ratios by several percent at high redshifts. The deflection is cumulative, but only by of order 10 milliarcsec. In the last two years this signal been detected in large, deep, high-resolution optical images, for example [16,189]. It is currently well beyond the capability of submillimetre and far-IR instruments to detect cosmic shear, as this requires high-resolution and large samples of galaxies. Weak shear cannot be measured unambiguously on an object by object basis; however, there are hopes for the future: ALMA will have the sensitivity and angular resolution to measure the shapes of high-redshift galaxies very accurately, and by measuring molecular rotation line emission, could trace out their rotation curves. Cosmic shear distorts galaxies in a unique way that can be distinguished from a rotation and change of inclination, and so deep ALMA observations could perhaps measure shear accurately from a small sample of galaxies[47].

On top of the sea of weak cosmic shear distortions, individual mass concentrations produce greater distortions, and in some cases, large magnifications and multiple images. Light is deflected by a lens by an angle α that depends on the mass enclosed within the impact parameter b of the deflected ray $M(< b)$ divided by b, or equivalently on the circular velocity or velocity dispersion of the lens at a radius b. α thus generally depends on the angle between lens and image on the sky $\theta_{\rm I}$. The general link between image and undeflected source positions as compared with the lens, $\theta_{\rm I}$ and $\theta_{\rm S}$ respectively, is established by the lens equation, which for the often reasonable approximation of a circular lens is,

$$\theta_{\rm I} = \theta_{\rm S} + \frac{D_{\rm LS}}{D_{\rm S}}\,\alpha(\theta_{\rm I}) \ . \qquad (28)$$

The angular diameter distances between the observer and source and between lens and source are D_S and D_{LS} respectively. The magnification obtained can be approximated accurately by $\mu = (\theta_I/\theta_S)(d\theta_I/d\theta_S)$. For singular isothermal spheres with a density profile $\rho \propto r^{-2}$, α is independent of the impact parameter. A galaxy with an isothermal halo and 300 km s^{-1} velocity dispersion deflects light by 2.2 arcsec [230]. This leads to a characteristic Einstein radius θ_E for the formation of a ring image around a circular lens when lens and source are exactly aligned ($\theta_S = 0$), with $\theta_E = \alpha D_{LS}/D_S$. Einstein radii for galaxies are of the order of an arcsecond. For clusters, which have circular velocities about 5 times greater, the corresponding angle is of order $0.5 - 1$ arcmin.

As the angular separation between lens and source is reduced, into the regime where lensing magnification becomes significant, the image is first sheared into a magnified arclet with an increasing elongation. Somewhere close to a magnification $\mu = 2$, multiple images form, with separations comparable to the Einstein radius, if the projected surface density of the lens on the sky exceeds a distance-dependent critical value. For a distant lens this is comparable to all the mass in an outward-pointing section through an $\Omega = 1$ Universe projected onto a single plane at $z = 1$. Most moderate redshift galaxies and clusters exceed the critical surface density, whereas most lower density groups of galaxies do not.

Here we will discuss gravitational lensing as a tool to learn more about distant galaxies at long wavelengths. The spectacular internal structure visible in lensed images of galaxies from *HST* [166] cannot yet be reproduced at millimetre–far-IR wavelengths. The best that can be achieved is arcsec resolution using millimetre-wave interferometers, sufficient to resolve the multiple images of strongly lensed QSOs [4,276].

The exploitation of gravitational lenses for the study of the evolution of dusty, gas-rich galaxies has so far been confined to a small sub-region of all lensing studies, in which distant strongly-magnified galaxies have been detected when their unmagnified counterparts would not have been at the sensitivity limits of existing surveys. This includes the brilliant *IRAS* galaxies F10214+4726 [223,234] and APM 08279+5255 [152], at redshifts $z = 2.3$ and 3.8 respectively, much greater than the bulk of the *IRAS* population, the LBG cB58 [115], and a triply-imaged background galaxy in the field of the cluster A 2218 [170], all parts of which are readily detected in a deep SCUBA image.

As angular resolution and multi-waveband coverage is at a premium at long wavelengths, the ability to examine lensed sources with large magnifications is very useful. Many of the galaxies with SEDs plotted in Fig. 1 have multiple datapoints only because of gravitational lensing. An interesting feature is that some of these galaxies appear to have unusually flat mid-IR SEDs. This may be due in part to selection effects, such as bright optical emission being essential for their detection, identification or recognition [152]; however, it could also reflect the spatial structure of the dust emission within the objects. Hotter dust emission might be expected to be more compact, and thus potentially lead to a greater magnification [113,44].

The effects of source confusion noise can also be reduced by lensing, which both separates galaxies on the sky, making them easier to resolve apart, and makes them brighter, and thus easier to detect. *ISO* and SCUBA images in the lensed fields of rich clusters of galaxies [7,42,240,88] are effectively deeper than those in the field, taking advantage of this reduction of confusion noise. This is especially useful at these wavelengths, where potentially confusing emission from passively evolving cluster galaxies is relatively weak, and the available angular resolution is relatively coarse. The Einstein radii of clusters subtend angular scales of order 1 arcmin at $z \simeq 0.2 - 0.4$, well matched to the fields of view of the SCUBA, MAMBO and SHARC-II (sub)millimetre-wave cameras.

The effects of lensing on the detectable population of galaxies depends on the from of the galaxy counts. Lensing conserves surface brightness: it neither creates nor destroys photons, and does not change their frequency. Hence, the total intensity of background radiation from the sky is not modified by lensing. This condition leads to a modification of the counts. A magnification μ changes a differential count

$$N_{\mathrm{d}} = \frac{\mathrm{d}N(> S)}{\mathrm{d}S} \tag{29}$$

to

$$N'_{\mathrm{d}} = \left(\frac{1}{\mu^2}\right) \frac{\mathrm{d}N[> (S/\mu)]}{\mathrm{d}S} \tag{30}$$

[60]. The ratio of N'_{d} to N_{d} is the magnification bias B: if $N_{\mathrm{d}} \propto S^{\alpha}$, then

$$B = \frac{N'_{\mathrm{d}}}{N_{\mathrm{d}}} = \frac{1}{\mu^{2+\alpha}} \; . \tag{31}$$

Hence, if the slope of a differential count α is steeper than -2, corresponding to an index steeper than -1 for the integral counts shown in Fig. 14, then $B > 1$, and the magnification bias is positive for positive magnifications $\mu > 1$, and the number of detectable sources is increased. At submillimetre wavelengths the counts can be very steep, and have slopes that change rapidly with increasing depth (Fig. 15), and so the magnification bias can be very significant, especially at the brightest flux densities.

Magnification bias at these wavelengths is likely to remain an important effect in the future, helping to overcome limitations of angular resolution to study deep fields and the internal structure of sources. Relatively shallow wide-field surveys in these wavebands, that probe the steep bright counts should reveal many very luminous lenses. The first examples should come from large-format ground-based (sub)millimetre-wave cameras including BOLOCAM, 100-deg^2 surveys from *Herschel* and the *Planck Surveyor* all-sky survey [43].

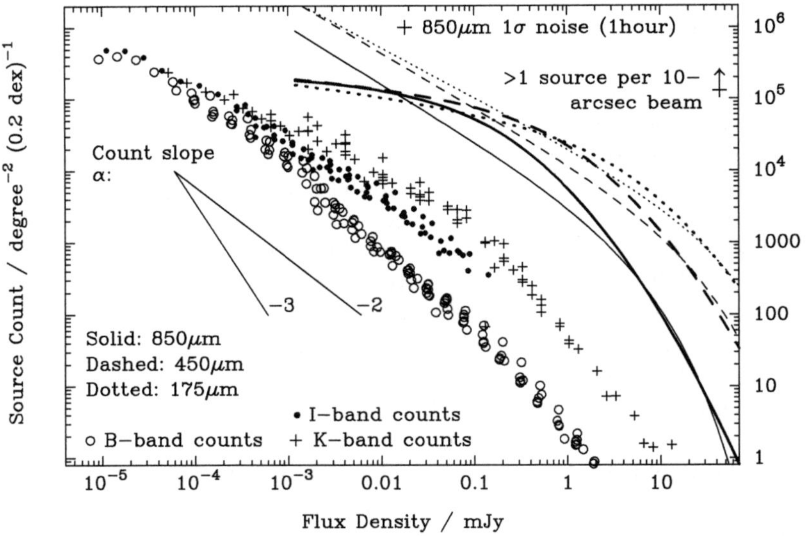

Fig. 15. A comparison of the counts of galaxies at optical, near-IR, far-IR and submillimetre wavelengths. The thick and thin lines correspond to the different models shown by the heavy lines in Fig. 14. The slopes of the lines are overplotted to show where magnification bias is likely to be large. The submillimetre counts almost certainly change slope through the significant $\alpha = -2$ value for which the magnification bias changes sign at the depths that can be probed by ALMA in a reasonable time, reflecting that mJy sources generate a significant fraction of the background radiation intensity and the faintest counts flatten. The difference between the slopes of the deep optical and submillimetre counts further emphasizes that they provide different information about galaxy evolution.

Perhaps very close to the centres of clusters of galaxies ALMA can exploit the lensing effect to increase the effective field of view of the telescope [46]. If the counts are flat, with $\alpha > -2$ fainter than 100 μJy, and the cluster has a sharply-peaked central density profile, then the swarm of demagnified images expected to lie close to the core of a cluster could be detected using ALMA, which has the sensitivity to detect these faint images, and the resolution to separate them on the sky.

6.7 The Process of Constraining Galaxy Evolution from Submillimetre Observations

The key non-intuitive flat flux-density–redshift relation (Fig. 10) should always be remembered when discussing galaxy evolution from submillimetre-wave data. Galaxies at moderate and high redshifts detected in the submillimetre are selected by luminosity not distance, subject to their SED (Fig. 11). The surface density of SCUBA galaxies brighter than about 2 mJy is about 0.5 arcmin^{-2}. Each of these, with a 40-K dust SED, has a luminosity of about

$5 \times 10^{12}\,L_\odot$ – a very luminous galaxy. If this assumption is wrong, and the galaxies had SEDs like the Milky Way, this would still correspond to a luminosity of about $5 \times 10^{11}\,L_\odot$, an order of magnitude greater than the dust luminosity of the Milky Way. If the temperature was in fact greater than 40 K, then the intrinsic luminosity would be increased. The volume in which these submillimetre-selected galaxies lie is a very long, thin pencil beam, with a great extent in the redshift direction, extending from $z \simeq 1$ to $z \sim 5$. The volume beyond $z \simeq 5$ is relatively small, and so the comoving volume occupied by these galaxies is thus of order $10^{-4}\,\mathrm{Mpc}^{-3}$ (Fig. 7, whereas *IRAS* galaxies with this luminosity have a space density $< 10^{-6}\,\mathrm{Mpc}^{-3}$ at the present day (Fig. 4). Hence, regardless of the lifetime or duty cycle of the submillimetre galaxies, they evolve very dramatically as compared with their ultraluminous counterparts at low redshifts, being at least about 200 times more numerous in terms of space density at $z > 1$.

Without redshifts for a large number of submillimetre galaxies, it is difficult to be certain of the role they play in the process of galaxy formation [192,240,178]. Much has been discussed about their potential identification with the epoch of formation of the spheroidal components of galaxies. Since most of the stars in the Universe are in elliptical galaxies [130], these arguments implicitly accept the cosmological significance of the submillimetre population. However, it is possible to make several simple arguments which leave this question open, and have a significant impact on our understanding of galaxy evolution.

Elliptical Galaxies: The space density of high-redshift SCUBA galaxies is at least 3 orders of magnitude less than that of L^* galaxies at the present epoch, and similar to the space density of giant elliptical galaxies [105,109]. This could indicate that every submillimetre galaxy is destined to be a giant elliptical galaxy at the present epoch, and that only giant ellipticals formed in this way. However, less dramatically, a much greater fraction of galaxies could go through a submillimetre luminous phase, perhaps during the most significant merger event in their history. If this phase lasts only about 0.1% of the Hubble time, then this would be compatible without generating too many submillimetre sources. It is thus possible that all galaxies could be visible in a brief submillimetre-luminous phase, and then quietly slip back onto the ordinary luminosity function. These numerous brief bursts of activity also do not exceed the measured background radiation intensity. If brief enough, then the number of galaxies undergoing this phase would by chance be consistent with the observed space density of giant ellipticals. The necessary duration of every galaxy's submillimetre phase would have to be 10^7 years, which is not an unreasonable duration for an intense, centrally condensed burst of activity in a massive galaxy, corresponding approximately to the Galactic rotation period. It is reasonable for at least a significant minority of all galaxies to undergo a submillimetre luminous phase at some stage in their evolution, or for the galaxies to be extremely rare.

The correspondence of the apparent space density of submillimetre galaxies and low-redshift giant elliptical galaxies is interesting, but it could simply be coincidence. The stars in giant ellipticals are very old, but could have formed in a series of mergers through a relatively short interval at high redshifts, each of which may not have been luminous enough to be detected as a submillimetre galaxy. These galaxy sub-units would subsequently congregate together at lower redshifts to make the elliptical galaxy [248]. The star-formation process would have to deplete enough gas to prevent substantial low-redshift star-formation activity, while not disturbing the colours of the galaxies. ERO samples with $K \simeq 20$ present plausible evidence that many elliptical galaxies are old and in place at $z \simeq 1$, but cannot reveal whether their formation involved a submillimetre bright phase.

Elliptical Galaxies and AGN: Perhaps the best argument in favour of a rapid formation timescale for ellipticals comes from the bulge–blackhole mass correlation, which is much easier to imagine coming from a common single event than a drawn-out series of events in the history of all galaxies. However, in that case where are all the typical L^* galaxies in submillimetre surveys? The blackholes in spiral galaxy bulges and in smaller ellipticals seem to trace the stellar mass in the same way as in the giant ellipticals and yet they are not represented in existing submillimetre surveys. Is there a conspiracy to generate about 50% of the background radiation in galaxies with a very sparse spatial distribution that are much brighter than L^* at high redshifts, and yet for almost all galaxies to have a scaled-down bulge component with a blackhole whose properties are linked in the same way? This is quite acceptable: only a small fraction ($< 1\%$) of galaxies are in the luminous phase at any time, and yet most could have gone through such a phase, generating most of their total emitted energy during it. If they turn out to be strongly clustered, with a correlation length greater than $10h^{-1}$ Mpc, then it perhaps indicates that they are associated with the densest, earliest collapsing, most active regions in the early Universe. However, if their luminosity is short-lived or merger-driven, as in all low-redshift galaxies of comparable luminosity [226], then the associated brief flashes of activity are likely to be less correlated. The submillimetre galaxies may only be located in the most overdense regions, but with a short duty cycle, with only one ever being observed per cluster. The test will be a survey sufficiently large to detect a representative number of the largest overdensities of submillimetre galaxies on the sky, several square degrees in extent. So far the total area surveyed to significant depths by submillimetre telescopes is less than $0.3\,\mathrm{deg}^2$, and no striking overdensities have been found. Useful upper limits have been imposed, but no convincing 2-point correlation function has been detected [232,265,68]. Some possible examples have been found – the surface density of submillimetre galaxies around some high-redshift radio galaxies [156] and clusters [41], and associated with one dense field of LBGs [74] is about an order of magnitude greater than in the field. The nature of the ultimate remnants of the submillimetre galaxies are intractable at present, despite some interesting insights [127]. At present, it is much easier to address the evolution of the submillimetre galaxy population as a whole.

6.8 The Evolution of the Submillimetre Galaxy Population

As with other populations, it is necessary to determine possible forms of dusty galaxy evolution that can explain the observed counts and background radiation. As discussed above, the submillimetre galaxies certainly evolve very strongly, when compared with galaxies of similar luminosity at the present epoch. However, from their contribution to the background radiation intensity it is clear that this evolution has to be much closer to pure luminosity evolution than pure density evolution – to boost the space density of detectable galaxies by the factor of 200 required to explain the counts, evolution in number density by this factor would lead to the measured background radiation intensity being greatly exceeded.

6.9 Simple Forms of Evolution

The most direct way to fit the results of observations is to take a luminosity function and choose evolution functions to match. It is important to be sure that the model is not too complex, that is the number of parameters describing it can be adequately constrained by the available data.

When looking at the evolution of dusty galaxies, the natural luminosity function is that determined from observations of low-redshift *IRAS* galaxies at 60 μm [228,247]. In order to explain the shape of the *IRAS* galaxy counts (Fig. 5) strong evolution is necessary out to the $z \simeq 0.1$ limit of the *IRAS* survey. Without this evolution, the predicted faint counts would be too few. The form of evolution required is similar to the evolution of optically-selected galaxies [177] and QSOs [62] at $z < 1$. The strength of evolution required is only weakly dependent on the assumed SEDs of *IRAS* galaxies, as they sample a small redshift range over which the K correction is low, regardless of the SED.

The next piece of information is provided by deep *ISO* surveys at 175 μm, which probe similar rest-frame wavelengths, reaching out to $z \simeq 1$. At 15 μm the SEDs of *IRAS* galaxies include spectral lines and small-grain PAH emission/absorption features. The counts at 15 μm probe out to similar redshifts, and also show strong evolution. As 15 and 60 μm are on the same side of the peak of the SED, the relative counts provide little information about the temperature of the dust. However, by combining an evolution function with a simple single-temperature SED, the *IRAS* and $175 - \mu$m *ISO* results can be seen to be compatible with a 40-K temperature and luminosity evolution of the form $(1+z)^{\simeq 4}$ (Fig. 6). This temperature is significantly higher than that in quiescent galaxies [219], but agrees with values measured from submillimetre observations of low-redshift *IRAS* galaxies [107], from some high-redshift dusty galaxies, for which SED measurements exist on both sides of the rest-frame peak [154,155] (see Fig. 1), and from other multiwavelength investigations of evolution [261]. It is also possible to explain the form of low-redshift evolution by pure density evolution; however, the redshift dependence must then be steeper, as $(1+z)^{\simeq 7}$, and must not continue to the redshifts of the *ISO* galaxies, or else the radiation intensity would exceed that observed by at least an order of magnitude. Whatever form of evolution applies, it must appear similar to pure luminosity evolution.

The existence of more distant galaxies was not probed by the *ISO* surveys, which reached the confusion limit of a 0.6-m antenna with a detector that undersampled the telescope point-spread function, and covered only small areas of the sky. To extend the available information to higher redshifts, observations at longer wavelengths are required. Both the spectrum of background radiation (Fig. 16) and the counts of submillimetre galaxies (Fig. 14) provide this information. The K correction allows submillimetre galaxies to contribute to the counts out to the highest redshifts; however, while the background spectrum incorporates the emission from all redshifts, but the effect of redshifting the spectrum and diluting it by a factor $(1+z)$ ((4)), ensures that it is much more sensitive to activity at $z \simeq 1$ than at $z \simeq 5$. The constraints on a form of high-redshift evolution imposed by these data are illustrated in Fig. 17. The 'anvil' model used includes luminosity evolution of the form $g(z) = (1+z)^{\simeq 3.9}$ out to redshift $z_{\rm max}$, followed by a constant luminosity function modified by the function $g(z) = (1+z_{\rm max})^{3.9}$ for $z_{\rm max} < z < z_0$, with no galaxies beyond z_0. The requirement to match both the $850-\mu$m galaxy counts and the background spectrum at wavelengths longer than 200 μm constrain a track through the $(z_{\rm max}, z_0)$ parameter space in similar directions. There is no unique best-fit value. This is a generic feature of all evolution functions, including models without either a sharp change of slope or a sharp cutoff.

This degeneracy remains even for very accurate submillimetre counts. The width of the permitted swath in parameter space is reduced if the error bars are tightened, but no unique clear maximum emerges. Points along the best-fitting track correspond to models in which strong evolution takes place out to an increasing value of $z_{\rm max}$, but which is traded off against a reduced upper limit for the population z_0. At all points on the track, the total amount of energy generated, the evidence of which now resides in low-mass stars, stellar remnants, metals or supermassive blackholes, agrees with observations, subject to a stellar initial mass function (IMF) that has a low-mass cutoff at $\simeq 1\,{\rm M}_\odot$, and a metallicity in the Universe that is raised to close to solar values globally, in the absence of a contribution from AGN. Based on the bulge–blackhole mass ratio, star-formation activity is likely to dominate AGN power

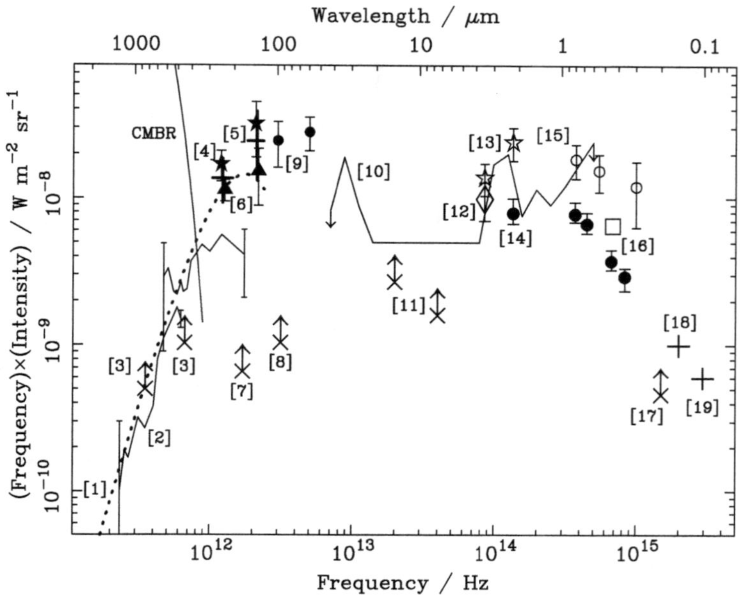

Fig. 16. Various measurements of and limits to the background radiation intensity from galaxies, and the spectrum of the CMB, from the radio to UV wavebands, plotted in units of equal energy per logarithmic frequency interval. The numbers refer to different observations: a detailed key can be found in [57]. Over most of the range where dust emission contributes strongly to the background spectrum, from 3 mm–10 μm, the intensity of the background has been measured to an accuracy within a factor better than about 3, generally in the last 5 years. *COBE*, *ISO* and SCUBA observations in the 1990's were essential for providing these measurements. Note that the energy density in the CMB is about 100 times greater than the total included in all the measurements shown. The background intensity in the X-ray waveband is the only significant component that is not shown, making up a few percent of the total non-CMB background intensity.

by a factor of about 5 [51]. Further details can be found in [56], while similar models have been discussed elsewhere [224,80].

Any reasonable current form for the evolution of dusty galaxies is shown in Fig. 24, in which the evolution function

$$g(z) = (1+z)^{3/2}\,\mathrm{sech}^2\,[b\,\ln(1+z) - c]\cosh^2 c\,, \tag{32}$$

with $b \simeq 1.9$ and $c \simeq 1.8$. This functional form accounts for all the most current data shown in Figs. 14 & 16 and what is known about the redshift distribution of the submillimetre sources [243]. By using this more adaptable evolution function, a less degenerate result can be obtained: see Fig. 18, in which the values of parameters b and c that can account for the data are constrained to lie within a more compact region of parameter space.

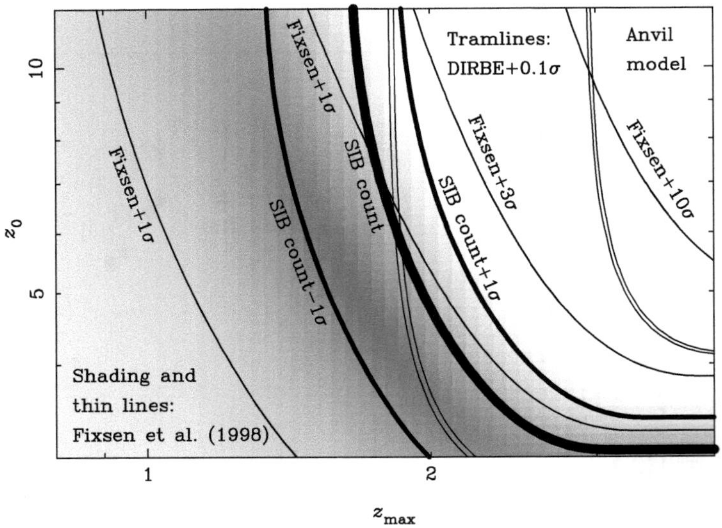

Fig. 17. The relative importance of different high-redshift information for constraining a model of galaxy evolution [56]. The anvil model is used for illustration. The data is for the far-IR/submillimetre-wave cosmic background radiation measured from *COBE*-DIRBE and FIRAS [122], and from the first measurement of the counts of submillimetre-selected galaxies [240]. The parameters are the redshifts at which $(1 + z)^{3.9}$ luminosity evolution ceases, $z_{\rm max}$, and at which the population ceases to be present beyond z_0. Between $z_{\rm max}$ and z_0 the evolution function is $(1 + z_0)^{3.9}$. The constraints from the submillimetre-wave counts are tighter than from the background determination; however, they provide a constrained swath of parameter space rather than a peak probability. This is a generic feature of the models, and the reason that breaking the counts into a redshift distribution is a crucial step for making progress.

Note that none of this discussion includes any details of the physical processes that generate the evolution or luminosity. This is necessary for a full understanding of the situation, but it is important to remember that any such model must predict a luminosity function that evolves in a similar way to that derived here.

6.10 Caveats

The process described here assumes a constant SED for the submillimetre galaxies, and that their luminosity function at high redshifts can be described by the form observed for *IRAS* at low redshifts, subject to luminosity evolution. These assumptions appear to be reasonable based on existing data, but need further testing. In particular, as discussed above in the context of observational selection effects, systematic changes of either the typical dust temperature or the width of the range of temperatures with increasing red-

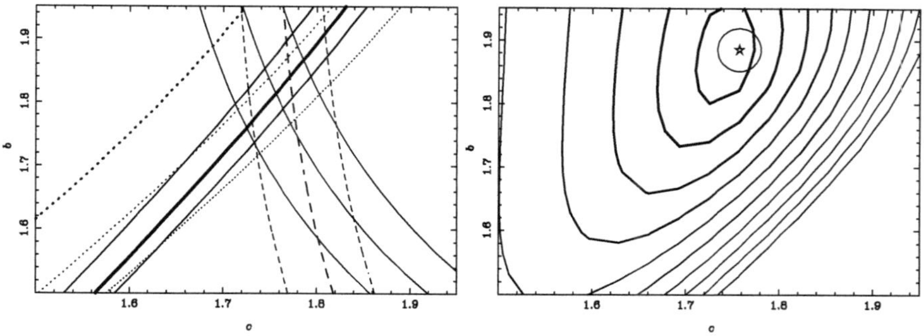

Fig. 18. The restrictions on parameters b, c (Equ. (32)) required to account for different types of data. In the left panel the $\pm\sigma$ errors on counts are shown by the dashed and solid lines: thick solid - submillimetre at 850 μm; thin solid far-infrared at 175 μm and dashed radio at 4.8 GHz. The dotted lines show the 1-, 2- and $3-\sigma$ quality of fit to the background spectrum (Fig. 16) in order of decreasing line thickness. The right panel shows the combined probability contours spaced in units of σ away from the best-fit value shown by the encircled star. The minimum value of $\chi^2 = 4.1$.

shift could easily lead to errors of interpretation. These issues have been discussed and simulated by Eales et al. [109]. If high-redshift galaxies had systematically cooler dust temperatures, then less dramatic evolution would be required to account for their properties. This would be a real concern if the value of $T \simeq 40$ K that accounts for the low-redshift data were greater. However, at present it seems reasonable that this temperature remains reasonable for the cores of luminous dust-enshrouded galaxies out to high redshifts. No very luminous galaxies containing only cold dust are known – the coolest *ISO* selected galaxies at moderate redshifts have temperatures $T \simeq 30$ K [76]. The mass of dust in a cold, very luminous galaxy would also have to be much greater than in a hotter galaxy with the same luminosity. At some point the available metallicity imposes a lower limit to the dust temperature for a given luminosity [56].

6.11 Redshift Distributions

The way to break the degeneracies between different forms of evolution is to determine the redshift distribution of a sample of submillimetre galaxies [56]. By comparison, improving error bars on the counts and backgrounds, or finding samples at other submillimetre wavelengths does not narrow the track through the parameter space shown in Fig. 17 dramatically. The redshift distributions predicted by some plausible models for the evolution of galaxies based on counts and background as described above are compared in Fig. 19. This includes all the models described in [56]. Note that the anvil models tend to provide better fits to current data. A revised version of the plot, including the best fitting 'Anvil-5' model, and the new best-fitting sech^2

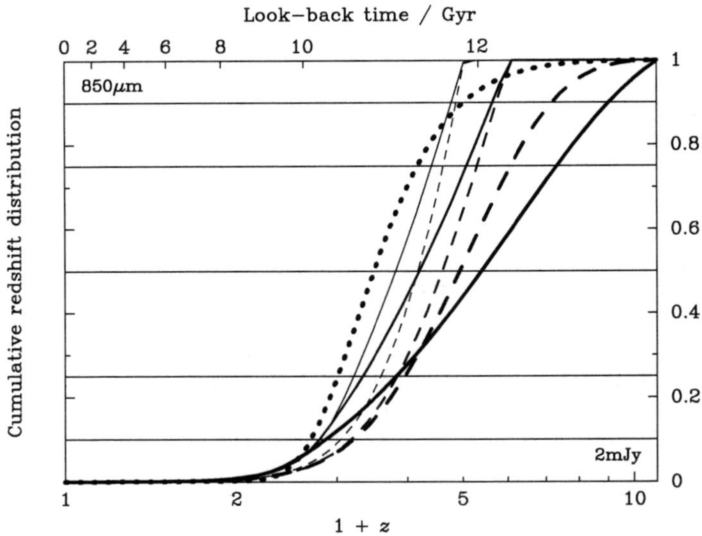

Fig. 19. The redshift distribution predicted for galaxies selected using SCUBA to a depth of 2 mJy at 850 μm in a variety of models that provide reasonable account of the far-IR/submillimetre-wave counts and background. The model predicts these quantities to within about a factor of 2 of each other [56], with the exception of the two thinnest dashed lines, which predict counts that are higher than the cloud of current results shown in Fig. 14. The dotted line corresponds to a model with a burst of activity with a Gaussian profile 0.8-Gyr wide in time centred at $z = 2.6$: see the thin dotted line in Fig. 24. The solid lines correspond to 'anvil models', with a sharp change of slope in the evolution function, and the dashed lines to smooth 'peak models'. The significant spread in redshift distributions emphasizes the importance of measuring the distribution to discriminate between the models. The observational determination of a complete redshift distribution for submillimetre galaxies is a vital goal. No data is plotted, as it is not yet possible to provide a reliable estimate of the redshift distribution. Radio-selected surveys detect up to 70% of the submillimetre galaxies detected in direct submillimetre surveys [24,72] should soon provide reasonable redshift statistics. Based on the flux density–redshift relations shown in Fig. 10, most submillimetre galaxies are unlikely to be at $z \gg 3$. Deep radio observations of submillimetre galaxies [242,77] indicate a median redshift in the range $2 < z < 3$. There is no evidence for a significant tail of submillimetre-selected galaxies at $z < 1$ [243] where the K correction is less favourable. The models represented by the two thinner solid lines and the dotted line provide the best description of the redshift distribution known at present.

model ((32)) are compared at selected interesting depths and wavelengths in Fig. 20. This shows both the range of possible redshift distributions that can be accommodated at 850 μm, especially concerning the number of galaxies at low (as defined in the submillimetre – $z < 1$) and very high ($z > 5$) redshifts. It is difficult to determine the redshift distribution [243]; however, to be sure of the form of high-redshift galaxy evolution, it must be found.

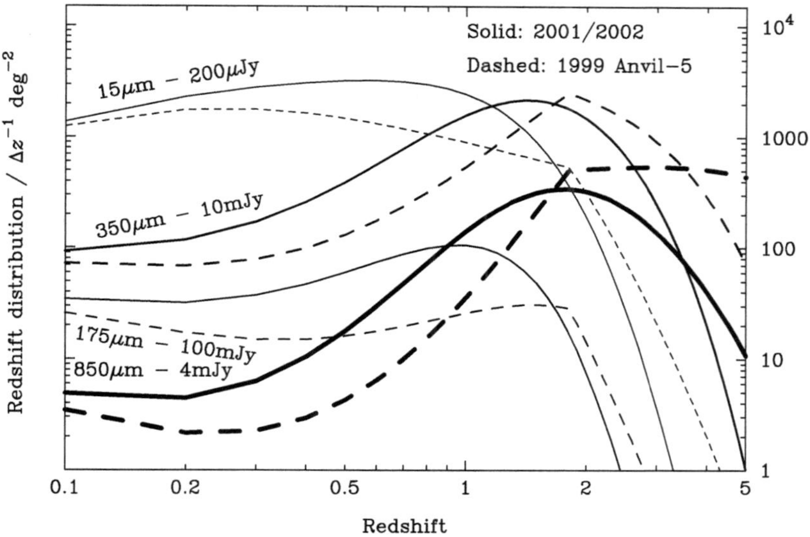

Fig. 20. The predicted redshift distributions of deep mid-IR, far-IR and submillimetre galaxies at sensitivities that correspond to the maximum achieved using *ISO* at 15 and 175 μm, SCUBA at 850 μm, and practical depths using the forthcoming 350 − μm SHARC-II camera at CSO. One of the models from Fig. 19 – 'Anvil-5', with strong evolution to $z = 1.8$ and then a constant luminosity function to a cut-off redshift $z = 5$ – still accounts well for the observations (*dashed lines*), and is compared with a more recent, smoothly varying model [57,243] (*solid lines*). The differences between the predictions are most striking at high redshifts and at long wavelengths – the numbers of 850 − μm detected galaxies at redshifts approaching 5 are expected to be much greater in the 1999 Anvil-5 model. The determination of the redshift distribution of submillimetre galaxies will readily distinguish between these models, despite their very similar counts and backgrounds.

Radio detection of counterparts to submillimetre galaxies has helped constrain redshifts [67,242,274], including samples selected first as optically-faint radio galaxies, and only then observed at submillimetre wavelengths [24,72], to allow more rapid sample compilations than in blank-field submillimetre surveys. Deep near-IR follow-up observations appear to identify a significant fraction of submillimetre galaxies at $K \simeq 22-23$ [14]. The problem is that current limits for detecting the continuum from these galaxies in spectroscopic observations generally require $K \simeq 20$ or less [92,106,127]. If high-redshift galaxy targets have strong emission lines, then fainter galaxies with $K \simeq 21.5$ could be observed, but the faintest of these objects will probably only yield redshifts using a future 30-m ground-based telescopes or *JWST*. Searches for

[14] Relevant observations were carried out (during European daylight while we skied) using the NIRC camera at the Keck telescope in Hawaii [127]. Thanks to collaborators Lee Armus, Dave Frayer, Naveen Reddy and Nick Scoville.

CO emission to determine and confirm redshifts absolutely are also possible [73].

In a remarkable successful run of observations also coincident with the Saas Fee school, a radio pre-selected submillimetre galaxy sample [72] was observed using the new blue-sensitive arm of the LRIS spectrograph at Keck, and a crop of over 20 redshifts for radio–submillimetre-selected galaxies were found [77]. The associated spectra provide evidence for AGN features in a minority of the sources, and evidence for outflowing absorbing gas [244] in a majority [78]. Optical multi-object spectroscopy targeting accurate radio positions appear to be the most efficient way of determining redshifts for submillimetre galaxies.

6.12 Semi-analytical Modelling

As there is no certain redshift distribution for submillimetre-selected galaxies, despite the progress on faint radio pre-selected samples, only so much can be determined about their evolution. There are a variety of models that describe their properties. Some of the most interesting attempts integrate semi-analytical galaxy formation models, that combine the results of N-body simulations of the gravitational growth of structure with generally simple recipes for the physics that describes the cooling of gas in the galaxies, and the subsequent star-formation and feedback processes. A look into the multiphase, very optically thick, dynamically-shredded ISM of even the relatively low-luminosity Antennae galaxies [271] should perhaps caution against treating these processes too simply. This empirical caution is supported by the results of gas dynamical simulations to trace the evolution of gas in merging galaxies, in which a rich array of complex, short-lived, rapidly variable behaviour can be seen [195,31]: the results depend strongly on the initial conditions of the galaxy mergers – the impact parameter, energies, relative inclination etc.

The first attempt to incorporate details of the astrophysics of forming galaxies to explain the submillimetre-wave counts and background was made by Guiderdoni et al. [143], as the first results of surveys for high-redshift dusty galaxies arrived in 1997. The standard recipes invoked did not predict enough very luminous systems. These were legitimately added as a separate population of galaxies (in their E model) to account for the data. Semi-analytical models [267,163,248,35] had traditionally accounted for the growth of stellar populations out of cooling gas that settles in virialized halos to form rotationally supported disks. If this additional population is due either to the collapse of low-angular-momentum haloes or the violent mergers of galaxies, then their properties lie outside this realm of gradual, regular activity. In particular, most recipes involve a close link between mass and luminosity, with the most luminous objects, especially giant elliptical galaxies forming late in the Universe's history when halo masses grow large enough. This does not necessarily mean that the stars within formed late, just that the galaxies assembled late.

More recently, another semi-analytic model, which includes a more efficient mode of cooling of gas in the deepest potential wells, thus generating more rapid evolution and more luminous, but shorter-lived galaxies, has been proposed by Granato et al. [142]. This approach explicitly includes the formation and fueling of AGN, which are not typically involved in semi-analytic models. The AGN luminosity increases exponentially through the luminous phase of its duty cycle, but the star-formation activity is more smoothly distributed in time. Hence, later in the duty cycle a massive galaxy would be more likely to be recognized as an optical QSO, but is more likely to be detected as a star-formation powered object at much earlier times. Feedback from winds generated by supernovae and by the AGN strip fuel to terminate star formation activity and AGN accretion. This feedback is expected to be less effective in suppressing activity in deeper potential wells, leading to a 'reversed hierarchical' picture in which the most massive galaxies form in single events, relatively early, in regions destined to be rich groups and clusters of galaxies at late times. By construction, this model accounts for the Magorrian relation between the mass of supermassive blackholes and the velocity dispersion of the surrounding stars [188]. A similar explanation of very powerful radio galaxies [12], in which dust-enshrouded star-formation can only take place in young systems that the AGN has not grown large enough to disrupt.

Other scenarios are also possible in which winds driven by high-mass stars rather than by AGN drive out gas from a galaxy to prevent further accretion and/or star formation [239], or consume all the available gas into stars, again preventing further activity. Perhaps dark-matter could also accrete onto a seed blackhole at a low rate in a manner determined by the halo potential, gradually and inconspicuously generating the observed bulge–blackhole correlation over a significant fraction of a Hubble time. For the most recent discussion of the links between star formation activity and AGN see [264].

Without a direct, detailed time-series of observations of a large number of elliptical galaxies in the process of formation, or a detailed time sequence of galaxy mergers, providing a time-lapse image of the velocity and constitution of the gaseous, stellar and dark matter components of these objects, then the issues of the power source of the luminous dusty high-redshift galaxies are unlikely to be settled soon. At present, the early formation of elliptical-like stellar systems in Granato et al.'s picture [142] seems to be more in keeping with the detection of quiescent EROs at $z \sim 1$, than more gradual formation pictures that tend to have problems generating both the required numbers of high-luminosity, high-redshift dusty galaxies, and the associated evolved remnant EROs at $z \simeq 1$. Whether it can account for the evolution of dusty galaxies awaits measurements of spectra and masses for submillimetre galaxies.

The results of different models all account for the published data, this usually being a necessary requirement for referees to accept papers and for editors to publish them! However, differences between the redshift distributions predicted by various models are very valuable for learning about galaxy formation once more data becomes available.

6.13 Background Radiation Constraints

The background radiation intensity at far-IR wavelengths is now determined reasonably accurately (Fig. 16). There should be relatively little freedom remaining to modify the form of evolution described dramatically and still reproduce the results. One important area for future development is the use of $\gamma\gamma \rightarrow e^+e^-$ opacity between CMB/infrared background photons and TeV energy cosmic-ray photons to probe the radiation field. The interaction cross section for photons peaks sharply just above the electron–positron pair creation energy of 1.02 MeV, and corresponds to the interaction of a low-energy far-IR background photon with a wavelength in μm that is equal to the numerical value of the energy of a high-energy photon in TeV. The optical depth is such that observations of downward curvature in the measured γ-ray SEDs of low-redshift blazars are interpreted as being due to γ-ray attenuation by this effect. So far, the limits imposed by this method [251] are subject to uncertainties based on lack of knowledge of the intrinsic γ-ray SEDs of blazars, and to small numbers of both detectable blazars (2 so far) and numbers of detected photons from each [146]. However, more capable ultrahigh energy γ-ray observatories are coming into service, and the results should allow an accurate spectrum of the mid- and far-IR background radiation intensity to be determined from higher-quality observations of a larger sample of blazars, less susceptible to concerns about rapid variability modifying the spectrum.

6.14 Constraints from Bright Counts

The counts of distant dusty galaxies predicted by a variety of models generally agree with the faint few-mJy counts of submillimetre galaxies determined by SCUBA and MAMBO surveys [243,57] and the 100-mJy counts of far-IR galaxies detected by *ISO* at 175 μm. However, one of the most interesting questions is what happens at the brighter flux densities. (Fainter probes are currently not possible due to the limitations of resolution, and hence source confusion.) As a result of the strong K correction at submillimetre wavelengths, the deep SCUBA counts mostly consist of high-redshift galaxies, with a relatively small contribution, perhaps 10%, from galaxies at redshifts $z < 1$. However, there are luminous low-redshift dusty galaxies with $60 - \mu$m flux densities up to about 100 Jy, corresponding to about 1 Jy at 850 μm. By moving to brighter flux densities than are probed by existing surveys, both more luminous distant galaxies and more proximate galaxies will begin to influence the counts. As high-redshift galaxies have submillimetre fluxes that are almost independent of distance, the slope of the submillimetre counts at flux densities greater than about 10 mJy should correspond quite accurately to the high-redshift submillimetre luminosity function. To ensure convergence to the measured background intensity, this must decline more rapidly than $\Phi \propto L^{\alpha}$, where $\alpha < -2$. Low redshift *IRAS* galaxies have a power-law luminosity function with $\alpha \simeq 2$ at the highest luminosities, and this could conceivably continue out to the high redshifts. This slope of the luminosity function reflects processes taking place in the most luminous, short-lived, phases of galaxy evolution, and so a direct determination of its form is sure to be important. However, mass functions of bound objects are predicted to have exponential forms at the highest masses. As a result, a cutoff mass/luminosity might be expected in the distribution of high-redshift galaxies. How does this interact with the observed more gradual decline of the luminosity function of the most luminous low-redshift galaxies? The decline of the high-redshift 10–100-mJy $850-\mu$m counts and the flux density above which the low-redshift Euclidean population of less exotic galaxies should start to dominate the counts is very uncertain. In some models [224] a large density of cool low-redshift galaxies are included, in addition to the extrapolations of the *IRAS* $60-\mu$m luminosity function to longer wavelengths using the SEDs determined for samples of $60-\mu$m-selected *IRAS* galaxies [107,108]. In others, a precipitous drop is predicted [142] over a narrow range of flux densities. These models should be easy to discriminate using the bright counts at flux densities of several 100 mJy that should be detected at three submillimetre wavelengths by the *Planck Surveyor* space mission [43].

To recap, the most sensitive indicator of the concordance between a model and data is the predicted redshift distribution of the sources: as we have seen, shifting activity to higher redshifts reduces the intensity of the background radiation associated with the generation of a certain amount of energy, by a factor of $1+z$, while having little effect on the counts. The redshift distribution of submillimetre galaxies can be affected in a radical way, however (see Fig. 19). Published predictions of redshift distributions are not available for all the models listed, as observational results concerning the distribution for submillimetre galaxies selected without the aid of radio information are sparse. A picture of a radio-selected sample with a median redshift $z \sim 2.5$ is starting to emerge [242,77]. Be aware that the predicted distributions can be biased to unrealistically high redshifts, especially in models where the evolution of luminosity density in the Universe remain flat over long intervals of redshift beyond $z \simeq 1$ (Fig. 24).

7 Properties of High-Redshift Dusty Galaxies

The submillimetre-selected galaxies are a key component of the distant Universe, and are likely to have contributed a significant fraction of the energy released by galaxies over the history of the Universe. They have only been known since 1997, when the SCUBA camera first detected them, and at present it is challenging to study large or complete samples at other wavelengths [243]; many are extremely faint with optical magnitudes $I > 26$, and it is difficult to identify more than 50% in even the deepest images. Part of this is perhaps due to their uncertain positions to about 5–10 arcsec, with candidates offset in position, but areas extending 15 arcsec have been searched for counterparts. A clear example is the 8-mJy SCUBA galaxy in the Hubble Deep Field [101,150], whose identification remains enigmatic, and could be associated with an extremely faint red galaxy [106].

The submillimetre galaxies can genuinely be called part of the cool Universe: even the most extreme example APM 08279+5255, an *IRAS*-selected $z = 3.8$ $R \simeq 15$ red QSO [152] has a dust temperature of about 110 K. Although there is a significant contribution to the total luminosity of this galaxy shortward of mid-IR wavelengths, from both very hot dust and an AGN accretion disk, it is cool dust that makes it shine so brightly at far-IR wavelengths. There is an excellent chance that this most extreme object, which is magnified by a factor of several tens by an intervening lensing galaxy, has hot inner regions that are magnified by a greater amount than the cooler outer regions, boosting its optical and near-IR emission [176]. However, this would lead to an overestimate of its dust temperature by only about 20% [44].

It is possible, but far from certain, that submillimetre galaxies correspond to forming elliptical galaxies [109,178]. In favour of the argument is the factor that only of order 0.1% of the galaxies in the Universe appear to be submillimetre luminous, comparable to the fractional density of giant elliptical galaxies at the present epoch. The inferred luminosities and gas masses of the submillimetre galaxies also correspond to the consumption of gas on a sub-Gyr timescale at high redshifts, sufficient to generate a homogeneous and uniformly old stellar population typical of elliptical galaxies. However, it is unclear that elliptical galaxies are as old as their stellar populations, and they could have assembled at lower redshifts. It is not at all clear that the submillimetre-selected galaxies that have been studied in detail are destined to become elliptical galaxies. They certainly look destined to have an appropriately large stellar mass, but at $z = 2.55$ and $z = 2.81$ they are not at very great redshifts (see Figs. 21 [154,124,263] and 22 [73,155,157]). The well-studied examples are also unusually bright at optical wavelengths, being over 6 magnitudes brighter than most submillimetre-selected galaxies [243]. They are relatively easy to identify and study at other wavelengths, but may not reflect the properties of the other 90% of the population.

Figure 23 shows a cautionary, and perhaps more realistic, example of an identification of an example of the submillimetre population. Without obtaining radio and near-IR images of the field this galaxy would not have been identified. It appears to be associated with an ERO, rather than with the edge-on spiral galaxy nearby that was considered as an early plausible

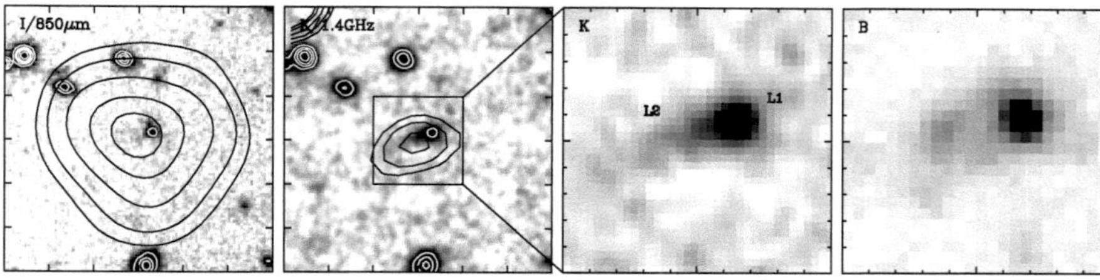

Fig. 21. The first identified submillimetre galaxy, SMM J02399–0136 behind the cluster Abell 370 [240]. The four images show a variety of multiwaveband data. White contours represent flux levels in the saturated parts of images. North is up and east to the left. The irregular pair of merging galaxies near the centroid of the submillimetre emission, represented by contours in the 30×30-arcsec far-left image, was targeted for optical spectroscopy using the 4-m Canada–France–Hawaii Telescope (CFHT) in late 1997, a few months after the submillimetre galaxy was detected [154], revealed a redshift $z = 2.80$ and a clear AGN spectral signature in the brightest westerly component. X-ray observations from the *Chandra* satellite [28] reveal that the galaxy definitely contains an accreting AGN. The detection of redshifted CO(3-2) emission at the same position and redshift using the OVRO Millimeter Array [124] confirmed absolutely that the counterpart was correct, and revealing a large mass of molecular gas. Subsequent spectroscopy from the ESO VLT [263] has revealed a halo of Lyman-α emission extending over 13 arcsec. Radio emission (black contours in the 30×30-arcsec second panel) from the galaxy provides a better location for the submillimetre emission. The Abell 370 field has some of the deepest radio data ever taken, and both components of the galaxy are clearly detected in the images [158]. The SED of the galaxy is close to that expected from the observed far-IR–radio correlation at low redshifts. 10×10 arcsec near-IR (K-band) and optical (R-band) images are shown in the right-hand frames. The new *HST* ACS camera should soon reveal exquisite resolution images of this galaxy.

identification, assuming that it contained extensive cool ($\simeq 20$ K) dust [241]. However, its optical spectrum shows no sign of the vigorous star-formation activity from the [OII]/[OIII] line ratio or a strong Hα recombination line, and so the ERO H5 is a much more compelling candidate.

7.1 Radio (and Optical) Pre-selection

We previously discussed the success and otherwise of making connections between the populations of distant galaxies detected at optical and submillimetre wavelengths. To first order, there appears to be little connection between the populations at the current mutual detection thresholds. If deeper images are ultimately available, then any object with the SED of a typical low-redshift *IRAS* galaxy should be detectable at both wavelengths. The mismatch between the angular resolution of the first generation of far-IR and submillimetre surveys and the exquisite sub-arcsec images flowing from *HST* and ground-based near-IR instruments remains a problem. The total number, surface density and limiting luminosity for detection of optically-selected

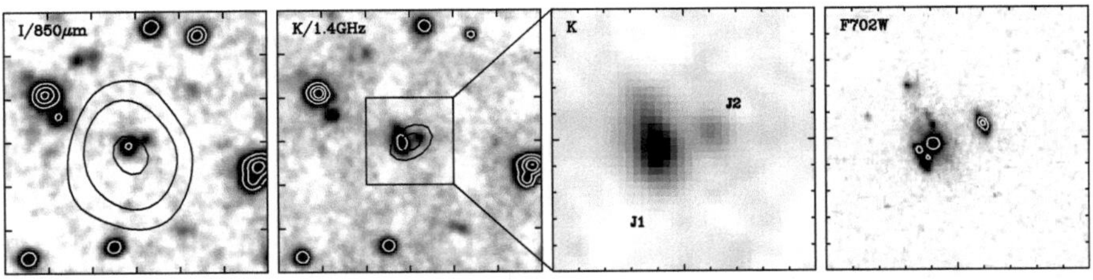

Fig. 22. Another high-redshift ($z = 2.56$) submillimetre-selected merging galaxy, SMM J14011+0252 behind the cluster Abell 1835 [157]. The multiwaveband data [243] is presented in the same format as Fig. 21. This time an *HST* image is available, showing a dramatic interacting and distorted red eastern component to the galaxy (J1) [155]. The galaxy was detected in the redshifted CO(3-2) line using the OVRO MMA, absolutely confirming the identification [73]. The northwesterly blue component at the same redshift could be selected in a LBG survey. However, radio (*black contours, second panel*) and millimetre-wave interferometer images show that the luminosity of this galaxy at all wavelengths is likely to be dominated by emission from the very red region to the north of the fragmented J1 [157].

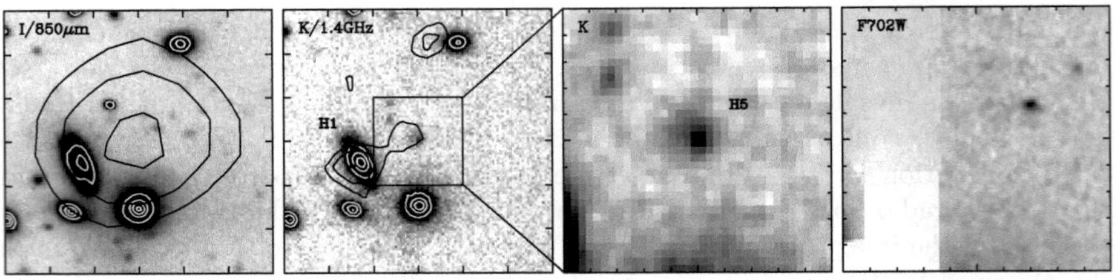

Fig. 23. A more typical submillimetre galaxy with no redshift and confirmed identification. The images are presented in the same format as the preceding figures [243]. The ERO close to the centroid of the submillimetre emission is the likely association. However, at $K \simeq 19.1$ it is several magnitudes brighter than the bulk of submillimetre galaxies. The near-IR K-band data and radio emission is a crucial link to the proposed identification: the presence of radio emission reveals that, even without considering the submillimetre emission, the ERO is unlikely to be a passively evolving elliptical galaxy with red colours due to a quiescent population of evolved stars, but is probably due to a strongly reddened young stellar population.

galaxies still exceeds the equivalent values in the submillimetre by at least an order of magnitude: well over a thousand high-redshift optically-selected LBGs have detailed spectra [254]. Nevertheless, observations of reprocessed dust emission are important to account for the large fraction of energy that can be missed in optical surveys, and to find galaxies that fall outside their selection criteria. It is important to link together the properties of all high-redshift galaxies if we are to understand galaxy formation in full. Some of the interesting populations of galaxies that have been defined at other wavelengths but investigated in the submillimetre waveband are listed below.

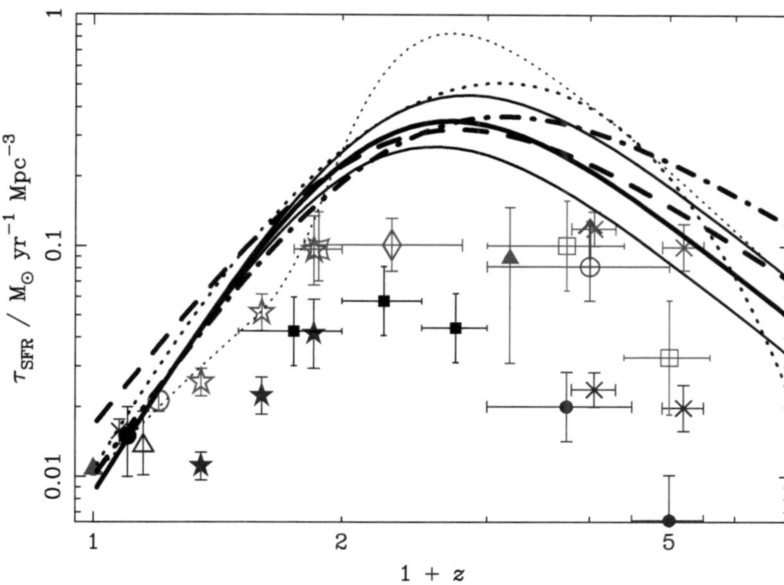

Fig. 24. The history of energy generation in the Universe, parameterized as a star-formation rate per unit comoving volume. The absolute normalization of the curves depends on the assumed stellar IMF and the fraction of the dust-enshrouded luminosity of galaxies that is generated by AGN. The points show results derived from a large number of optical and near-IR studies, for which detailed references can be found elsewhere [57,243]. The most important results are the filled stars [177] and high-redshift diagonal crosses [254]. The up-pointing arrow comes from a submillimetre-based estimate of star-formation rate from the HDF [150]. An important recent measurement of the extinction-free low-redshift star-formation rate from radio data from [275] is included. The thick solid and dashed lines represent current best fits to far-IR and submillimetre data in a simple luminosity evolution model and a hierarchical model of luminous merging galaxies respectively, as updated to reflect additional data and a currently favored non-zero-Λ cosmology. The thinner solid lines show the approximate envelope of 68% uncertainty in the results of the luminosity evolution model. The thin and thick dotted lines represent the best-fitting results obtained in the original derivations [56,54].

Optically-Selected Galaxies: LBG samples [254] provide thousands of spectroscopically confirmed galaxies with luminosities as low as L^* at redshifts $2.5 < z < 4.5$, by detecting the redshifted 91.2-nm hydrogen absorption edge between various standard optical observing bands. The selection effects that apply to LBGs, faint radio sources and submillimetre galaxies are quite different [3]. The key question is whether sufficiently deep LBG surveys can always detect faint traces of optical emission escaping from dusty galaxies and identify them. This is difficult to decide based on the limited ability of existing submillimetre-wave telescopes to probe the faint fluxes expected from LBGs. As discussed in Sect. 6.5, while low-luminosity galaxies can probably always be detected both from stellar and dust emission, at the highest luminosities dusty galaxies appear to be more difficult to detect optically, and it is certainly very difficult to assign accurate luminosities from optical observations alone [139].

Near-IR Selected Galaxies: EROs and even 'Highly ERO (HEROs)' [259], have been detected using large format near-IR cameras. These populations are likely to overlap with dust-enshrouded galaxies at some level. At $K \simeq 20$ most EROs seem to be evolved elliptical galaxies at moderate redshifts ($z < 1$); however, it is likely that the small energetic contribution of redshifted optical emission from dusty high-redshift galaxies will be detectable in deeper near-IR images [127].

7.2 Gamma-Ray Burst (GRB) Host Galaxies

There is a race to understand and identify the host galaxies of GRBs [98], which are typically associated with faint blue galaxies. In some cases, X-ray spectra with unusually high iron abundances have been inferred, that could be the debris of a high-mass star ejected in powerful winds or in a supernova. Red transient emission has also been detected, that would be consistent with a supernova light curve in the decaying power-law emission from the cooling GRB fireball. These are all potential signs that GRBs are associated with the deaths of at least some massive stars. This could provide an obscuration-independent route to investigating star-formation in the distant Universe, as γ-rays are immune to extinction by dust, and so neither the presence or absence of dust in the hosts should bias their detection at γ-ray energies, unlike far-IR or optical observations of the host itself. So far, several bursts appear to be associated with dusty galaxies [36,123], broadly consistent with a picture in which a significant fraction of high-redshift, high-mass star formation activity is obscured by dust. However, no systematic deep survey of GRB hosts has been made to test this view [246,26].[15] Much more extensive observations are necessary to confirm and quantify the connection. The resolution and sensitivity of ALMA should make this an easy task, with minute-long exposures being required to image each of the host galaxies of the several thousand GRBs that should have been catalogued with accurate positions by 2010.

8 Galaxy Evolution: A Global, Multiwavelength View

Since there are now many thousands of optically-selected galaxies at redshifts greater than 3, and at least 100 AGN at $z > 5$ and about 300 dusty

[15] A program at JCMT with Vicki Barnard and Nial Tanvir to search more systematically for host emission began in acceptable weather in Hawaii the week after the Saas Fee meeting [26]. See also Berger et al. [37].

ultraluminous galaxies likely to lie at $z > 1$, we are starting to build up a reasonable census of the contents of the high-redshift Universe (still subject to potential surprises!). The challenge to ensure that there are no classes of objects missing from the census, and to push deeper to try to see the first luminous objects is now a real one. A key question is whether any inferred form of the evolution of galaxies shown in Fig. 24 is correct, and to extend the results to higher redshifts. There are hints that the opacity of the Universe in Lyman-α is rising through unity at $z > 6$, and so rest-frame UV observations will provide a guide to the more distant Universe, unless the galaxies ionize large enough surrounding regions to redshift their emission out of the absorption range. Rest-frame optical emission can be sought in the difficult to observe near-IR waveband, while dusty objects can be observed beyond reionization at (sub)millimetre wavelengths, and the associated ionized gas can be detected from their radio free–free emission (Fig. 10). At even higher redshifts, the dust abundance is expected to decline, yet free-free emission from even the earliest galaxies could still be detected using a sufficiently sensitive telescope, requiring nothing more than the presence of ionized hydrogen.

It will be a great challenge to find sources sufficiently bright against which to probe the pre-reionization Universe in absorption at these wavelengths; however the radio, X-ray and IR continuum emission from GRBs associated with the very first stars may offer a powerful, if short-lived and unreproduceable source.

8.1 Observing First Light

The next ten years of observational cosmology is full of promise. New powerful multi-object spectrographs on 8–10-m class telescopes, including DEIMOS on Keck, VIMOS on VLT and GMOS on Gemini are poised to provide new very deep redshift surveys to extend the influential results from the 4-m-class CFRS redshift survey [177] out to $z > 1$. With surveys from the forthcoming *SIRTF* IR space mission, these should define the evolution of galaxies and their constituent stellar populations accurately out to redshifts $z \simeq 2$.

At longer wavelengths, ALMA will provide extremely sensitive resolved images of dust and molecular gas in metal-enriched galaxies at any redshift, by exploiting the favourable submillimetre K correction out to redshifts where dust and metals first formed. The time required to generate dust in a first generation of stars is very small, corresponding to the lifetime of the most massive stars ($\sim 10^6$ yr), combined with the comparable wind crossing time of a star-formation region (assuming 100 pc and 1000 km s^{-1}). If the epoch of reionization has now indeed been detected, then perhaps the next key target in understanding galaxy formation is to determine the maximum redshift at which dust and metal-enriched molecular gas can be detected by ALMA.

At higher redshifts, only the unobscured restframe visible and UV light from the first stars and free-free emission from the ionized gas could be detected, at mid-IR and radio wavelengths in the observer's frame. At redshifts in excess of 6 this is a key goal of deep surveys using the *JWST*. Ground-based telescopes are likely to be prevented from achieving the sensitivity required at near-IR wavelengths by atmospheric emission, while *SIRTF* does not have a large enough aperture to avoid confusion from the overlapping images of less distant galaxies and reach the necessary depths.

Inverse Compton scattering from CMB photons will limit the lifetime of electrons generating synchrotron emission from supernova remnants at very high redshifts. The luminosity of radio synchrotron radiation depends on the energy density in the magnetic field, while the energy lost from the electron population to inverse Compton scattering depends on the energy density of all sources of radiation. The tight correlation between the radio and far-IR emission from galaxies requires that the ratio of magnetic and radiation energy densities is almost constant over 4 orders of magnitude of galaxy luminosity at low redshifts [85]. Note that in the most vigorous star-forming regions, dust temperatures are of order 100 K, indicating ISRF intensities that exceed the CMB energy density at $z \simeq 0$ by a factor of $(100/2.73)^4 \simeq 1.7 \times 10^6$. The intensity of the CMB increases as $(1+z)^4$, although it is a coincidence that this matches the inferred strength of evolution of the luminosity density of galaxies at low redshifts. The intensity of the CMB would not match this ISRF until a very high redshift $\simeq 36$. In a more reasonable representative region of a luminous star-forming galaxy at $T = 40$ K this occurs at up to $z \simeq 14$. In a galaxy like the Milky Way with $T = 20$ K it occurs at $z \simeq 6$.

However, the relativistic electrons that generate the radio emission are not bound to, and do not necessarily remain near to star-forming regions, and so are likely to experience a lower ISRF where the synchrotron emission is generated. Given the volume and total luminosity of the Milky Way, the energy density in radiation averaged over the whole volume is only about 2.5 times greater than in the CMB. Hence, equality in the ISRF and CMB energy densities will occur at only $z \sim 1.5$, and beyond $z = 2$ the CMB will dominate the radiation field, effectively quenching synchrotron emission by providing much more rapid cooling of electrons. The resulting upscattered photons will be in the X-/γ-ray regime, but their total luminosity will be extremely small and difficult to observe, as the fraction of the bolometric luminosity of galaxies radiated in the radio waveband is only of order 10^{-3}, less than the standard X-ray emission from compact binaries in star-forming galaxies.

In hotter dusty galaxies, reflecting a more intense ISRF, the domination of the energy density by the CMB will not occur until higher redshifts. Since the intensity of thermal emission, determined by the intensity of the ISRF, scales as $T^{4+\beta}$ and the intensity of the CMB scales as $(1+z)^4$, equality would occur at $z \simeq 5.5$ for a 40-K dust source and $z \simeq 15$ for an 80-K dust source. There is a small secondary effect, that heating of dust by the CMB imposes a lower limit to the dust temperature, but this modification is much less significant than the fractional reduction in the radio emission. Note, that even if the

CMB quenches synchrotron emission, it will not affect the observability of free-free emission from ionized gas at even the highest redshifts. The chance of detecting almost flat-spectrum free-free emission from very high redshift galaxies using a future centimetre-wave radio telescope is excellent. A galaxy with a bolometric luminosity of $10^{10}\,L_{\odot}$ is expected to generate a free-free flux density of about $1\,\mu$Jy at any observed radio frequency in the GHz range out to $z \simeq 20$ (Fig. 10). If any such objects are compact, and sufficient angular resolution is available to avoid the effects of confusion, then the signs of the first ionization by the youngest galaxies in the Universe could be imaged in this way.

9 Future Observations of the Distant Cold Universe

The opportunities to study distant galaxies in detail at submillimetre wavelengths depends on the availability of instruments with sufficient sensitivity and angular resolution to provide detailed astrophysical diagnostics for the detected sources.

The availability of the SCUBA and MAMBO cameras, with fields of view of order 5 arcmin made the first surveys of the distant dusty Universe possible, while at shorter wavelengths and lower redshifts, *ISO* provided the first glimpse of distant galaxies (Fig. 14). From these galaxy populations, and the background radiation intensity derived using the *COBE* satellite, the surface densities of faint dusty galaxies are known within a factor of about 30%, and the intensity of the background radiation has been constrained to within a similar factor from 2 to 200 μm. At shorter wavelengths, the background intensity is constrained to within a decade-wide strip throughout the far- to near-IR wavebands. A reasonable picture of the significance and nature of the emission of dust radiation from galaxies has thus been developed.

However, our exploration of the distant Universe at these long wavelengths has just begun. Only about 0.3 deg^2 of sky has been mapped in submillimetre-wave surveys, while the extent of *ISO* images that are sufficiently deep to provide cosmological information is only about 50 arcmin^2. None of these images has the resolution to resolve distant galaxies. We will now address the issues of sensitivity, resolution and diagnostic power, and instruments that are being developed to study the cold Universe in more detail over the years ahead. It is important to remember that all these observations will require multiwavelength support from radio, optical and X-ray observations to maximize their scientific return.

9.1 Sensitivity and Field of View

There is no substitute for a telescope with a large total collecting area, to improve signal-to-noise ratios, even without the advantage of improved diffraction limited performance. It is important, however, to limit the thermal emission from the telescope, as this can easily be the dominant source of noise at mid-IR and longer wavelengths. This thermal restriction has so far limited space-borne apertures to under 1-m. *IRAS*, *ISO*, *SIRTF* and *ASTRO-F* all lie in this range. The first larger aperture to fly in space will be the 3.5-m *Herschel* observatory, at about 70 K. The SOFIA airborne observatory and the BLAST balloon-borne instrument will field comparably sized warm, more emissive antennas, as they are in contact with the 230-K stratosphere. The remnant atmosphere about 15000 m for SOFIA and 40000 m for BLAST will also contribute noise.

In 2–3 years, SOFIA is expected to reach the confusion limit imposed by the beamsize and the surface density of galaxies after about 1 week of observations staring at the same field, a comparable time to SCUBA at present. As a result, it will be suitable for making very narrow-field (arcmin^2) deep surveys, to probe between the much larger number of galaxies that will be detected in wider fields at coarser resolution using *SIRTF*, and to provide a first look at the sub-mJy mid-/far-IR sky. At $z \simeq 1$ the SOFIA confusion limit corresponds to the flux density of a 10^{11}-L$_\odot$ L^* spiral galaxy. To probe large numbers of galaxies to these depths will require the greater sensitivity of *Herschel* in space, which should reach the confusion limit in a 200 − μm staring observation in about 1 hour. With a field of view of order 40 arcmin^2, a *Herschel* survey of 1 deg^2 would take approximately 4 days of constant observing. The all-sky (sub)millimetre-wave survey that will be carried out by the *Planck Surveyor* space mission, primarily to image the CMB on scales larger than 5 arcmin will provide a sensitive comparison image for these future surveys. *Planck* is currently scheduled to fly alongside *Herschel*. The ultimate sensitivity of *Planck* at the shortest wavelengths will be limited by the unresolved structure from confused galaxies [52], and it will detect a large number (10^{4-5}) of submillimetre galaxies at least about 10 times brighter than those that are being reported from SCUBA surveys.

The largest existing ground-based submillimetre-wave telescope is the 15-m JCMT, and the comparably sized 8 × 6-m SMA interferometer. At millimetre wavelengths the 30-m aperture IRAM telescope at Pico Veleta, and the 6 × 15-m aperture IRAM PDB interferometer provide the greatest collecting areas. The forthcoming CARMA composite array of 8 × 6-m and 6 × 10.4-m antennas in the Inyo mountains of California will provide a collecting area approximately 70% as large, with the potential for excellent image fidelity from a greater number of longer baselines. The 100-m clear-aperture Greenbank Telescope (GBT) should be capable of operating at 3 mm during the winter in West Virginia, while a 50-m millimetre-wave telescope – the LMT/GTM is being constructed near Puebla in Mexico to operate at 1.1 mm. The 64 × 12-m ALMA will provide a collecting area comparable to the GBT, in a variety of interferometer configurations, with baselines up to 10 km.

For interferometers the collecting area is the key guide to sensitivity, assuming a reasonable number of several tens of baseline pairs and sensitive receivers. However, for single-antenna telescopes the maximum size of the field of view, and the fraction filled with detectors is usually more important. For example, current bolometer arrays, SCUBA and MAMBO, are only a small factor more sensitive per detector than the single-pixel bolometer detectors that they superseded. Their advantage comes from imaging several tens of beams on the sky simultaneously, and from being able to track and subtract fluctuations in atmospheric emission across the whole array.

The next generation of bolometer cameras will be several times more sensitive, and field much larger numbers of detectors. The BOLOCAM and SHARC-II cameras are being commissioned at the CSO, using monolithic arrays of 151 and 384 detectors respectively in the 2.1/1.4/1.1-mm and 350$-\mu$m atmospheric transmission windows. BOLOCAM can image the Sunyaev–Zeldovich effect from clusters out to any redshift, and detect luminous distant dusty galaxies. SHARC-II provides wide-field images of star-forming regions in the Milky Way, make follow-up observations and obtain accurate photometry for distant galaxies detected at other wavelengths, and make dedicated pencil-beam deep surveys to determine the deep counts of distant galaxies [7]. It is now possible to begin the design process for cameras with many thousands of pixels, potentially able to image fields almost 1 deg^2 in size on a 10-m telescope. A large-format camera such as SCUBA-II being designed at the UK-ATC in Edinburgh would be able to reach the confusion limit on a 15-m telescope in about 1 hr per 64-arcmin^2 field in good weather from the 4000-m Mauna Kea observing site. A 64 $\times$ 64 element array has been discussed as a possible detector in this camera. This is possible because of the development of very stable, superconducting transition edge sensor (TES) bolometers, as demonstrated for the first time in May 2001 with the FIBRE spectrograph [33]. There are several other highly innovative bolometer detector developments under way. Developments in technology could make a 256 $\times$ 256 array practicable at the time that *Herschel* flies and ALMA is in service in about 2010, to complement these facilities. With a deep mapping speed of order 1 deg^2 per night, submillimetre-wave surveys using such a camera could eclipse the existing capabilities of optical and radio telescopes to detect raw numbers of galaxies at $z > 1$. Large-format arrays may even allow accurate subtraction of varying atmospheric emission, allowing surveys to continue in less-stable daylight conditions. Remember that in the almost 5 years of submillimetre observations using SCUBA, only about 0.3 deg^2 of sky has been imaged to a cosmologically useful depth, less than a third of the minimum area required to survey a representative region of the high-redshift Universe.

9.2 Resolution

Future ground-based and air- and space-borne instruments will have the capability to survey representative volumes of the distant Universe at long wavelengths. However, *SIRTF*, SOFIA, SCUBA-II and *Planck* will not have the resolving power required to detect substructure in the galaxies they find.

Studies of the nature of the detected objects will require observations either at shorter wavelengths, or using interferometers. To keep pace with the detection rates, the sensitivity of ALMA will be required. ALMA will be able to image a 2-mJy galaxy detected at 850 μm in about 1 min, allowing a systematic high-resolution programme of follow-up observations. With a 1-mm resolution of 20 milliarcsec on a 10-km baseline, ALMA will easily resolve 200-pc scale structures in high-redshift galaxies.

At shorter wavelengths, there are several routes to higher-resolution images of distant galaxies. A filled-aperture 10-m class space telescope has been discussed (*SAFIR*), which would provide a very sensitive deep sub-arcsec resolution survey at a wavelength of order 10 μm. For more detailed studies of the astrophysics of distant galaxies, resolving individual examples in detail, a prototype 10–20-m baseline mid-IR interferometer *SPIRIT* has been proposed, as a vehicle for technology development leading to 1-km-scale baseline free-flying interferometers like *SPECS* [191].

These instruments are all matched to the complementary high-resolution capabilities of the near-IR *JWST*, and to the extended VLA and SKA radio telescopes. For spectroscopic follow-up of the most distant galaxies detected, a 30-m class optical telescope, such as the California Extremely Large Telescope (CELT) concept that is under study will be required. This type of facility may already be required to follow up a significant fraction of existing submillimetre-selected galaxies in detail.

9.3 Other Techniques

A first glimpse of the high-redshift dust-enshrouded Universe has been provided by imaging continuum detectors. As described elsewhere in the volume [84], vital astrophysical information is provided by observations of submillimetre-wave spectral lines.

One way to make additional progress, is to combine spectral line information for galaxies selected by their continuum emission. If the output is dispersed in frequency, then there is additional protection against the effects of confusion noise [63]. At shorter mid- and far-IR wavelengths, fine-structure emission lines from low-ionization gas in excited regions of distant galaxies, and from the molecular hydrogen rotational and vibrational line emission expected to be associated with gas cooling in the first galaxy halos will be targets for detection, if wide-band instruments are available. Spectral data would allow sources close enough on the sky to have ambiguous positions and continuum flux densities to be distinguished, unless they were physically associated the confused galaxies are unlikely to lie at the same redshift. This will be important if existing samples of submillimetre-selected galaxies are observed at the resolution of *Herschel*.

Ground-based and space-borne line surveys will provide a valuable complement to future very deep continuum images. ALMA [272] and *SPECS* [191] already include the possibility of making combined line and continuum surveys, although their potential 10-milliarcsec and sub-milliarcsec resolution respectively means that neither has to rely on spectral line observations to avoid confusion. It is likely that both innovative spectrographs on even relatively small, cold space-borne telescopes, and wide-band CO line survey instruments on the ground will allow the distant cold Universe to be explored at a drastically increasing rate over the next decade.

10 Summary

The goal of this contribution has been to point out some of the special opportunities for studying the formation and evolution of galaxies in the wavelength range from 10 μm to about 3 mm, especially using dust continuum radiation.

This represents one of the last wavebands to be investigated at the relevant depths for cosmological surveys, and has yielded a haul of unexpectedly luminous galaxies. The most luminous QSOs detected at optical wavelengths since the early 1960s are matched in luminosity by galaxies that emit the great majority of their luminosity at far-IR wavelengths, and at submillimetre wavelengths where they could be detected out to extremely high redshifts. As some of the most interesting galaxies in the Universe, they must be explained by any complete model of galaxy formation.

At present only about 0.3 deg^2 of sky has been imaged at submillimetre wavelengths, detecting several hundred galaxies. These vital first steps were possible by the great commitment of time and effort of the teams centred at Edinburgh and Bonn that designed and built the SCUBA and MAMBO (sub)millimetre-wave cameras for the JCMT and IRAM 30-m telescope. This field is only beginning; it should mature rapidly now that larger submillimetre cameras are available at CSO (SHARC-II and BOLOCAM), and work is underway on LABOCA, a development of MAMBO for the APEX telescope[16] at the ALMA site in Chile, and the SCUBA-2 camera for the JCMT. At shorter wavelengths the *SIRTF* and *ASTRO-F* space missions will provide much more information about mid- and far-IR galaxies; *ASTRO-F* is planned to produce a much more sensitive revised version of the *IRAS* all-sky survey. The 3.5-m *Herschel* satellite will subsequently provide better resolution, and its fellow traveler *Planck Surveyor* will provide a sensitive 5-arcmin-resolution map of the submillimetre sky, a comparable resource at long wavelengths to the *IRAS* survey that spurred most of the interest in the cold Universe. The 2.5-m aperture SOFIA airborne observatory and ultralong duration balloon payloads, such as BLAST, with comparable warmer apertures, will provide opportunities for projects with shorter lead times than space missions, and fill various observational niches as they are discovered. Larger colder apertures and interferometers in space are planned for the more distant future.

The detection of larger samples of submillimetre galaxies is an important goal, but it is unlikely to provide the answers to the key scientific questions

[16] http://www.mpifr-bonn.mpg.de/div/mm/apex.htm

of their power source, lifetime, spatial extent, masses, metallicities etc. To address these questions high-angular resolution imaging, to resolve individual galaxies, and multiwaveband spectroscopy, will be required. Hopefully, in 10 years, this discussion can be repeated after *Herschel* and ALMA are in service, and the true history of cool galaxies can be set out.

Acknowledgments

Parts of this contribution are based on a review on submillimetre galaxies for Physics Reports [57], including many results of the SCUBA Lens Survey with Ian Smail, Rob Ivison, Jean-Paul Kneib and David Frayer. Thanks to them for the success of the survey [240,243] and their contributions to the review. Information on the SEDs, temperatures and luminosities of dusty galaxies has been provided by Vicki Barnard and Scott Chapman. I would like to thank the organisers, especially Daniel Pfenniger and Irène Scheffre, for their very efficient production of an extremely enjoyable Easter school in Grimentz, and to Kate Quirk and members of the editorial team for proofreading the manuscript. This research has made use of the NASA/IPAC Extragalactic Database (NED) which is operated by the Jet Propulsion Laboratory, California Institute of Technology, under contract with the National Aeronautics and Space Administration. Funding for aspects of this research is provided by the US National Science Foundation, the Research Corporation and the Caltech Tolman Emolument.

References

1. T. Abel, G.L. Bryan and M.L. Norman: Science **295**, 93 (2002)
2. T. Abel, M.L. Norman, P. Madau: ApJ **23**, 66 (1999)
3. K.L. Adelberger, C.C. Steidel: ApJ **544**, 218 (2000)
4. D. Alloin, S. Guilloteau, R. Barvainis, R. Antonucci, L. Tacconi: A&A **321**, 24 (1997)
5. O. Almaini, A. Lawrence, B.J. Boyle: MNRAS **305**, L59 (1999)
6. O. Almaini et al.: MNRAS **338**, 303 (2003)
7. B. Altieri et al.: A&A **343**, L5 (1999)
8. P.B. Alton, E.M. Xilouris, S. Bianchi, J. Davies, N. Kylafis: Astron. & Astrophy. **356**, 795 (2000)
9. P.B. Alton, J. Lequeux, S. Bianchi, D. Churches, J. Davies and F. Combes: A&A **366**, 451 (2001)
10. P. Andreani and A. Franceschini: MNRAS **283**, 85 (1996)
11. E.N. Archibald, J.S. Dunlop, D.H. Hughes, S. Rawlings, S.A. Eales and R.J. Ivison: MNRAS **323**, 417 (2001)
12. E.N. Archibald, J.S. Dunlop, R. Jimenez, A.C.S. Friaca, R.J. McLure and D.H. Hughes: ApJ **336**, 353 (2002)
13. I. Aretxaga, D.H. Hughes, E.L. Chapin, E. Gaztanaga and J.S. Dunlop: MNRAS submitted (2003) astro-ph/0205313
14. C. Armand, B. Milliard and J.M. Deharveng: A&A **284**, 12 (1994)

15. S. Arnouts, S. D'Odorico, S. Christiani, S. Zaggia, A. Fontana and E. Giallongo: A&A **341**, 641 (1999)
16. D. Bacon, A. Refregier and R. S. Ellis: MNRAS **318**, 625 (2000)
17. A. J. Baker, D. Lutz, R. Genzel, L. J. Tacconi and M. D. Lehnert: A&A **372**, L37 (2001)
18. J. C. Baker et al.: MNRAS **308**, 1173 (1999)
19. X. Barcons et al.: A&A **382**, 522 (2002)
20. J. M. Bardeen, J. R. Bond, N. Kaiser and A. S. Szalay: ApJ **304**, 15 (1986)
21. A. J. Barger, L. L. Cowie, D. B. Sanders, E. Fulton, Y. Taniguchi, Y. Sato, K. Kawara and H. Okuda: Nature **394**, 248 (1998)
22. A. J. Barger, L. L. Cowie and D. B. Sanders: ApJ **518**, L5 (1999)
23. A. J., Barger, L. L. Cowie, I. Smail, R. J. Ivison, A. W. Blain and J.-P. Kneib: AJ **117**, 2656 (1999)
24. A. J. Barger, L. L. Cowie and E. A. Richards: AJ **119**, 2092 (2000)
25. V. E. Barnard and A. W. Blain: MNRAS submitted (2003)
26. V. E. Barnard et al.: MNRAS **338**, 1 (2003)
27. M. Bartelmann and P. Schneider: Physics Reports **340**, 291 (2001)
28. M. W. Bautz et al.: ApJ **543**, L119 (2000)
29. R. H. Becker, R. L. White and D. J. Helfand: ApJ **450**, 559 (1995)
30. R. H. Becker et al.: AJ **122**, 2850 (2001)
31. K. Bekki, Y. Shioya and I. Tanaka: ApJ **520**, L99 (1999)
32. D. J. Benford, P. Cox, A. Omont, T. G. Phillips and R. G. McMahon: ApJ **518**, L65 (1999)
33. D. J. Benford, B. Maffei, S. H. Moseley, F. Pajot, T. G. Phillips, C. Rioux and R. A. Shafer: AAS **198**, 0510 (2001)
34. C. L. Bennett et al.: ApJ, **464**, L1 (1996)
35. A. J. Benson, C. S. Frenk, C. M. Baugh, S. Cole and C. G. Lacey: MNRAS **327**, 1041 (2001)
36. E. Berger, S. R. Kulkarni and D. A. Frail: ApJ **560**, 652 (2001)
37. E. Berger, L. L. Cowie, S. R. Kulkarni, D. A. Frail, H. Aussel and A. J. Barger: ApJ submitted (2003) astro-ph/0210645
38. R. A. Bernstein, W. L. Freedman and B. F. Madore: ApJ **571**, 56 (2002)
39. M. Bersanelli et al.: *COBRAS/SAMBA* SCI(96)3, (ESA, Paris 1996)
40. E. Bertin, M. Dennefeld and M. Moshir: A&A **323**, 685 (1997)
41. P. N. Best: MNRAS in press (2002) astro-ph/0203318
42. A. W. Blain: MNRAS **290**, 553 (1997)
43. A. W. Blain: MNRAS **297**, 511 (1998)
44. A. W. Blain: MNRAS **304**, 669 (1999)
45. A. W. Blain: MNRAS **309**, 955 (1999)
46. A. W. Blain: MNRAS **330**, 219 (2002)
47. A. W. Blain: ApJ **570**, L51 (2002)
48. A. W. Blain and M. S. Longair: MNRAS **264**, 509 (1993)
49. A. W. Blain and M. S. Longair: MNRAS **265**, L21 (1993)
50. A. W. Blain and M. S. Longair: MNRAS **279**, 847 (1996)
51. A. W. Blain and T. G. Phillips: MNRAS **333**, 222 (2002)
52. A. W. Blain, R. J. Ivison and I. Smail: MNRAS **296**, (1998) L29
53. A. W. Blain, V. E. Barnard and S. C. Chapman: MNRAS **338**, 733 (2003)
54. A. W. Blain, A. Jameson, I. Smail, M. S. Longair, J.-P. Kneib and R. J. Ivison: MNRAS **309**, 715 (1999)

55. A.W. Blain, J.-P. Kneib, R.J. Ivison and I. Smail: ApJ, **512**, L87 (1999)
56. A.W. Blain, I. Smail, R.-J. Ivison and J.-P. Kneib: MNRAS, **302**, 632 (1999)
57. A.W. Blain, I. Smail, R.J. Ivison, J.-P. Kneib and D.T. Frayer: Physics Reports **369**, 111 (2002) astro-ph/0202228
58. M.R. Blanton et al.: AJ **121**, 2358 (2001)
59. J.R. Bond, S. Cole, G. Efstathiou and N. Kaiser, ApJ **379**, 440 (1991)
60. U. Borgeest, J. von Linde and S. Refsdal, A&A **251**, L35 (1991)
61. R.G. Bower, MNRAS **248**, 332 (1991)
62. B.J. Boyle, T. Shanks, S.M. Croom, R.J. Smith, L. Miller, N. Loaring and C. Heymans: MNRAS **317**, 1014 (2000)
63. M.C. Bradford: private communication (2002)
64. S.L. Bridle, I. Zehavi, A. Dekel, O. Lahav, M.P. Hobsin and A.N. Lasenby: MNRAS **321**, 333 (2001)
65. V. Bromm, P.S. Coppi and R.B. Larson: ApJ **564**, 23 (2002)
66. C.L. Carilli and A.W. Blain: ApJ **569**, 605 (2002)
67. C.L. Carilli and M.S. Yun: ApJ **513**, L13 (1999)
68. C.L. Carilli et al.: in *'Deep Millimeter surveys: implications for galaxy formation and evolution'*: ed. by J.D. Lowenthal and D.H. Hughes, (Singapore, World Scientific 2002) pp. 207
69. J.E. Carlstrom et al.: BAAS **197**, 5501 (2000)
70. J.E. Carlstrom, G.P. Holder and E.D Reese: ARA&A **40**, in press (2002) astro-ph/0208192
71. S.M. Carroll, W.H. Press and E.L. Turner: ARA&A **30**, 499 (1992)
72. S.C. Chapman, E.A. Richards, G.F. Lewis, G. Wilson and A.J. Barger: ApJ **548**, L147 (2001)
73. S.C. Chapman et al.: MNRAS **319**, 318 (2000)
74. S.C. Chapman et al.: ApJ **548**, L17 (2001)
75. S.C. Chapman, A. Shapley, C. Steidel and R. Windhorst: ApJ **572**, L1 (2002)
76. S.C. Chapman, I. Smail, R.J. Ivison, G. Helou, D.A. Dale and G. Lagache: ApJ **573**, 66 (2002)
77. S.C. Chapman, A.W. Blain, I. Smail and R.J. Ivison: Nature in press (2003)
78. S.C. Chapman et al.: ApJ submitted (2003)
79. S.C. Chapman et al.: ApJ **570**, 557 (2002)
80. R.R. Chary and D. Elbaz: ApJ **556**, 562 (2001)
81. H.-W. Chen: ApJ **570**, 54 (2002)
82. P. Coles and F. Lucchin: *'The origin of cosmic structure'* (Wiley, Chichester 1995)
83. M. Colless et al.: MNRAS **328**, 1039 (2001)
84. F. Combes: this volume (2002)
85. J.J. Condon: ARA&A **30**, 575 (1992)
86. A.J. Connolly, A.S. Szalay, M. Dickinson, M.U. SubbaRao and R.J. Brunner: ApJ **486**, L11 (1997)
87. L.L. Cowie, A. Songaila and A.J. Barger: AJ **118**, 603 (1998)
88. L.L. Cowie, A.J. Barger and J.-P. Kneib: AJ **123**, 2197 (2002)
89. B.P. Crill et al.: ApJ submitted (2003) astro-ph/0206254
90. E. Daddi et al.: A&A **361**, 535 (2000)
91. D.A. Dale, G. Helou, A. Contursi, N.A. Silbermann and S. Kolhatkar: ApJ **549**, 215 (2001)
92. H. Dannerbauer, M.D. Lehnert, D. Lutz, L. Tacconi, F. Bertoldi, C. Carilli, R. Genzel and K. Menten: ApJ **573**, 473 (2002)

93. J.R. Deane and N. Trentham, MNRAS **326**, 1467 (2001)
94. de Bernardis P. et al.: Nature, **404**, 955 (2000)
95. A. Dey, J.R. Graham, R.J. Ivison, I. Smail, G.S. Wright and M.C. Liu, ApJ **519**, 610 (1999)
96. J.E.G. Devriendt, B. Guiderdoni and R. Sadat: A&A **350**, 381 (1999)
97. S.G. Djorgovski and M. Davis: ApJ **313**, 59 (1987)
98. S.G. Djorgovski et al.: in *'Proceedings of IX Marcel Grossmann Meeting'*: ed. by V. Gurzadyan, R. Jantzen and R. Ruffini, in press (Singapore, World Scientific 2002) astro-ph/0106574
99. S.G. Djorgovski, S.M. Castro, D. Stern and A. Mahabal: ApJ **560**, L5 (2001)
100. D. Downes and P.M. Solomon: ApJ **507**, 615 (1998)
101. D. Downes et al.: A&A **347**, 809 (1999)
102. D. Downes, R. Neri, T. Wiklind, D.J. Wilner and P.A. Shaver: ApJ **513**, L1 (1999)
103. B.T. Draine: this volume (2002)
104. J.S. Dunlop and J.A. Peacock: MNRAS **247**, 19 (1990)
105. J.S. Dunlop: in *'Deep Millimeter surveys: implications for galaxy formation and evolution'*: ed. by J.D. Lowenthal and D.H. Hughes, (Singapore, World Scientific 2002) pp. 11
106. J.S. Dunlop et al.: MNRAS submitted (2003) astro-ph/0205480
107. L. Dunne, S. Eales, M. Edmunds, R. Ivison, P. Alexander and D.L. Clements: MNRAS **315**, 115 (2000)
108. L. Dunne and S.A. Eales: MNRAS **327**, 697 (2002)
109. S. Eales, S. Lilly, W. Gear, L. Dunne, J.R. Bond, F. Hammer, O. Le Févre and D. Crampton: ApJ **515**, 518 (1999)
110. T. Ebisuzaki et al.: ApJ **562**, L19 (2002)
111. A. Eckart, T. Ott and R. Genzel: A&A **352**, L22 (1999)
112. G.P. Efstathiou: in *'Physics of the early Universe'*: ed. by J.A. Peacock, A.F. Heavens and A.T. Davis, Proc. 36th Scottish Universities Summer School in Physics (Edinburgh University Press, Edinburgh 1990), pp. 361.
113. P.R. Eisenhardt, L. Armus, D.W. Hogg, B.T. Soifer, G. Neugebauer and M.W. Werner: ApJ **461**, 72 (1996)
114. D. Elbaz et al.: A&A **351**, L37 (1999)
115. E. Ellingson, H.K.C. Lee, J. Bechtold and R. Elston: ApJ **466**, L35 (1996)
116. S.L. Ellison, L. Yan, I.M. Hook, M. Pettini, J.V. Wall and P. Shaver: A&A **379**, 393 (2001)
117. S.L. Ellison, G.F. Lewis, M. Pettini, F.H. Chaffee and M.J. Irwin: ApJ **520**, 456 (1999)
118. S.M. Fall and Y.C. Pei: ApJ **402**, 479 (1993)
119. X. Fan et al.: BAAS **197**, 2701 (2001)
120. L. Ferrarese and D. Merritt: ApJ **539**, L9 (2000)
121. D.J. Fixsen, E.S. Cheng, J.M. Gales, J.C. Mather, R.A. Shafer and E.L. Wright: ApJ **473**, 576 (1996)
122. D.J. Fixsen, E. Dwek, J.C. Mather, C.L. Bennett and R.A. Shafer: ApJ **508**, 123 (1998)
123. D.A. Frail et al.: ApJ, 565, 829 (2002)
124. D.T. Frayer, N.Z. Scoville, M. Yun, A.S. Evans, I. Smail, A.W. Blain and J.-P. Kneib: ApJ **506**, L7 (1998)
125. D.T. Frayer et al.: ApJ **514**, L13 (1999)

126. D. T. Frayer, I. Smail, R. J. Ivison and N. Z. Scoville: AJ **120**, 1668 (2000)
127. D. T. Frayer et al.: AJ submitted (2003a)
128. D. T. Frayer et al.: AJ submitted (2003b)
129. A. Fruchter, J. H. Krolik and J. E. Rhoads: ApJ **563**, 597 (2001)
130. M. Fukugita, C. J. Hogan and P. J. E. Peebles: ApJ **503**, 518 (1999)
131. S. Furlanetto and A. Loeb: ApJ **556**, 619 (2001)
132. K. Ganga, B. Ratra, S. E. Church, N. Sugiyama, P. A. R. Ade, W. L. Holtzapfel, P. D. Mauskopf and A. E. Lange: ApJ **484**, 517 (1997)
133. M. Garrett: A&A **384**, L19 (2002)
134. W. K. Gear, S. J. Lilly, J. A. Stevens, D. L. Clements, T. M. Webb, S. A. Eales and L. Dunne: MNRAS **316**, L51 (2000)
135. K. Gebhardt et al.: ApJ **539**, L13 (2000)
136. R. Genzel and C. J. Cesarsky: ARA&A **38**, 761 (2000)
137. A. M. Ghez, M. Morris, E. E. Becklin, A. Tanner and T. Kremenek: Nature **407**, 349 (2000)
138. S. Ghinga, B. Moore, F. Governato, G. Lake, T. Quinn and J. Stadel: ApJ **544**, 616 (2000)
139. J. D. Goldader, G. Meurer, T. M. Heckman, M. Seibert, D. B. Sanders, D. Calzetti and C. C. Steidel: ApJ **568**, 651 (2002)
140. J. R. Graham and A. Dey: ApJ **471**, 720 (1996)
141. G. L. Granato, L. Danese and A. Franceschini: ApJ **460**, L11 (1996)
142. G. L. Granato, C. G. Lacey, L. Silva, A. Bressan, C. M. Baugh, S. Cole and C. S. Frenk: ApJ **542**, 710 (2000)
143. B. Guiderdoni, E. Hivon, F. R. Bouchet and B. Maffei: MNRAS **295**, 877 (1998)
144. J. E. Gunn and B. A. Peterson: ApJ **142**, 1633 (1965)
145. M. Haehnelt and M. J. Rees: MNRAS **263**, 168 (1993)
146. M. G. Hauser and E. Dwek: ARA&A **39**, 249 (2001)
147. A. Heger, N. Langer and S. Woosley: ApJ **528**, 368 (2000)
148. P. T. P. Ho et al.: in *'Imaging at Radio Through Submillimeter Wavelengths'*: ed. by J. Mangum. ASP Conf. Ser. vol. 217 (Astron. Soc. Pac, San Francisco 2000) pp. 227
149. G. P. Holder and J. Carlstrom: ApJ **558**, 515 (2001)
150. D. Hughes et al.: Nature **394**, 241 (1998)
151. D. Hughes et al.: MNRAS **335**, 871 (2002)
152. M. J. Irwin, R. A. Ibata, G. F Lewis and E. J. Totten: ApJ **505**, 529 (1998)
153. K. G. Isaak et al.: MNRAS **329**, 149 (2002)
154. R. J. Ivison, I. Smail, J.-F. Le Borgne, A. W. Blain, J.-P. Kneib, J. Bezecourt, T. H. Kerr and J. K. Davies: MNRAS **298**, 583 (1998)
155. R. J. Ivison, I. Smail, A. J. Barger, J.-P. Kneib, A. W. Blain, F. N. Owen, T. H. Kerr and L. L. Cowie: MNRAS **315**, 209 (2000)
156. R. J. Ivison, J. S. Dunlop, I. Smail, A. Dey, M. C. Liu and J. R. Graham: ApJ **542**, 27 (2000)
157. R. J. Ivison, I. Smail, D. T. Frayer, J.-P. Kneib and A. W. Blain: ApJ **561**, L45 (2001)
158. R. J. Ivison: private communication (2002)
159. A. Jenkins et al.: MNRAS **321**, 372 (2001)
160. R. Jimenez, P. Padoan, J. S. Dunlop, D. V. Bowen, M. Juvela and F. Matteuci: ApJ **532**, 152 (2000)

161. M. E. Jones et al.: Nature **365**, 320 (1993)
162. M. Joy et al.: ApJ **551**, L1 (2001)
163. G. Kauffmann and S. D. M. White: MNRAS **261**, 921 (1993)
164. K. Kawara et al.: A&A **336**, L9 (1998)
165. C. Kiss, P. Abraham, U. Klaas, M. Juvela and D. Lemke: A&A **380**, 388 (2001)
166. J.-P. Kneib, R. S. Ellis, I. Smail, W. J. Couch and R. M. Sharples: ApJ **471**, 643 (1996)
167. R. Kneissl, M. E. Jones, R. Saunders, V. Eke, A. N. Lasenby, K. Grainge and G. Cotter: MNRAS **328**, 783 (2001)
168. E. Komatsu et al.: ApJ submitted (2003) astro-ph/0302223
169. J. Kovac, E. M. Leitch, C. Pryke, J. E. Carlstrom, N. W. Halverson and W. L. Holtzapfel: Nature **420**, 772 (2002)
170. K. Kraiberg-Knudsen et al.: ApJ, submitted (2003)
171. P. Kroupa: Science **295**, 82 (2002)
172. C. Lawrence: in: *'Evolution of large scale structure: from recombination to Garching'*: ed. by A. J. Banday, R. K. Sheth and L. N. da Costa (ESO, Garching 1999) pp. 13
173. A. T. Lee et al.: ApJ **561**, L1 (2001)
174. E. M. Leitch, S. T. Myers, A. C. S. Readhead and T. J. Pearson: BAAS **189**, 116.01 (1996)
175. G. F. Lewis, S. C. Chapman, R. A. Ibata, M. J. Irwin and E. J. Totten: ApJ **505**, L1 (1998).
176. G. F. Lewis, C. Carilli, P. Papadopoulos and R. J. Ivison: MNRAS **330**, L15 (2002)
177. S. J. Lilly, O. Le Fèvre, F. Hammer and D. Crampton: ApJ **460**, L1 (1996)
178. S. J. Lilly, S. A. Eales, W. K. P. Gear, F. Hammer, O. Le Fèvre, D. Crampton, J. R. Bond and L. Dunne: ApJ **518**, 641 (1999)
179. U. Lisenfeld, K. G. Isaak and R. E. Hills: MNRAS **312**, 433 (2000)
180. A. Loeb: PRD **65**, 047301 (2002)
181. A. Loeb: in *'Lighthouses of the Universe'* (Springer, New York, 2002) astro-ph/0108432
182. A. Loeb and R. Barkana: ARA&A **39**, 19 (2001)
183. M. S. Longair: in *'The Deep Universe'* (Springer, New York 1993)
184. M. S. Longair: *'Galaxy Formation'* (Springer, Heidelberg 1998)
185. D. Lutz et al.: A&A **378**, 70 (2001)
186. J. D. MacMillan and R. N. Hendriksen: ApJ **569**, 83 (2002) astro-ph/0201153
187. P. Madau, H. C. Ferguson, M. E. Dickinson, M. Giavalisco, C. C. Steidel and A. Fruchter: MNRAS **283**, 1388 (1996)
188. J. Magorrian et al.: AJ **115**, 2285 (1998)
189. R. Maoli et al.: A&A **368**, 766 (2001)
190. J. C. Mather et al.: ApJ **420**, 439 (1994)
191. J. C. Mather et al.: (1998) astro-ph/9812454
192. P. Mazzei and G. de Zotti: ApJ **426**, 97 (1994)
193. P. Meszaros: A&A **37**, 225, (1974)
194. G. R. Meurer, T. M. Heckman and D. Calzetti: ApJ **521**, 183 (1999)
195. C. Mihos: preprint (2000) astro-ph/9903115
196. I. F. Mirabel et al.: A&A **333**, L1 (1998)
197. N. R. Mohan, A. Cimatti, H. J. A Röttgering, P. Andreani, P. Severgnini, R. P. J. Tilanus, C. L. Carilli and S. A. Stanford: A&A **383**, 440 (2002)

198. T. W. Murphy, B. T. Soifer, K. Matthews and L. Armus: ApJ **559**, 201 (2001)
199. C. B. Netterfield, N. Jarosik, L. Page, D. Wilkinson and E. Wollack: ApJ **445**, L69 (1995)
200. S. J. Oliver, M. Rowan-Robinson and W. Saunders: MNRAS **256**, P15 (1992)
201. A. Omont, P. Cox, F. Bertoldi, R. G. McMahon, C. Carilli and K. G. Isaak: A&A **374**, 371 (2001)
202. S. Padin et al.: PASP **114**, 83 (2002)
203. L. Page et al.: ApJ submitted (2003) astro-ph/0302220
204. J. A. Peacock and S. Dodds: MNRAS **280**, L19 (1996)
205. J. A. Peacock: *'Cosmological Physics'* (Cambridge University Press, Cambridge 1998)
206. J. A. Peacock et al.: MNRAS **318**, 535 (2000)
207. F.R. Pearce, A. Jenkins, C.S. Frenk, S.D.M. White, P.A. Thomas, H.M.P. Couchman, J.A. Peacock and G. Efstathiou: MNRAS **326**, 649 (2001)
208. T. J. Pearson et al.: ApJ submitted (2002) astro-ph/0205388
209. P. J. E. Peebles: *'Physical Cosmology'* (Princeton University Press, Princeton 1996)
210. W. J. Percival et al.: MNRAS **327**, 1297 (2001)
211. W. Percival and L. Miller: MNRAS **309**, 823 (1999)
212. S. Perlmutter et al.: ApJ **517**, 565 (1999)
213. M. Pettini, S. L. Ellison, C. C. Steidel and D. V. Bowen: ApJ **510**, 576 (1999)
214. M. Pettini, A.E. Shapley, C.C. Steidel, J.-G. Cuby, M. Dickinson, A. F. M. Moorwood, K. L. Adelberger and M. Giavalisco: ApJ **554**, 981 (2001)
215. T. G. Phillips ed.: *'Andvanced Technology MMW, Radio, and Terahertz Telescopes'* SPIE vol. 3357 (1998)
216. W.H. Press and P. Schechter: ApJ **187**, 425 (1974)
217. J.-L. Puget, A. Abergel, J.-P. Bernard, F. Boulanger, W.B. Burton, F.-X. Désert and D. Hartmann: A&A **308**, L5 (1996)
218. J.-L. Puget et al.: A&A **345**, 29 (1999)
219. W. T. Reach et al.: ApJ **451**, 188 (1995)
220. D.E. Reichart: ApJ **521**, L111 (1999)
221. Y. Rephaeli: ARA&A **33**, 541 (1995)
222. E. A. Richards: ApJ **533**, 611 (2000)
223. M. Rowan-Robinson et al.: Nature **351**, 719 (1991)
224. M. Rowan-Robinson: ApJ, **549**, 745 (2001)
225. K. Sakamoto, N. Z. Scoville, M. S. Yun, M. Crosas, R. Genzel and L. Tacconi: ApJ **514**, 68 (1999)
226. D. B. Sanders and I. F. Mirabel: ARA&A, **34**, 749 (1996)
227. R. Sari and T. Piran: ApJ **517**, L109 (1999)
228. W. Saunders, M. Rowan-Robinson, A. Lawrence, G. Efstathiou, N. Kaiser, R. S. Ellis and C. S. Frenk: MNRAS **242**, 318 (1990)
229. P. Schechter: ApJ **203**, 297 (1976)
230. P. Schneider, J. Ehlers and E. E. Falco: *'Gravitational Lenses'* (Springer Verlag, Berlin 1992)
231. J. Scott, J. Bechtold, A. Dobrzycki and V. P. Kulkarni: ApJS **130**, 67 (2000)
232. S. E. Scott et al.: MNRAS **331**, 817 (2002)
233. U. Seljak: ApJ **460**, 549 (1996)
234. S. Serjeant, S. Rawlings, M. Lacy, R. G. McMahon, A. Lawrence, M. Rowan-Robinson and M. Mountain: MNRAS **298**, 321 (1998)

235. A. E. Shapley, C. C. Steidel, K. L. Adelberger, M. Dickinson, M. Giavalisco and M. Pettini: ApJ **562**, 95 (2001)
236. P. Shaver, I. M. Hook, C. A. Jackson, J. V. Wall and K. I. Kellermann: in *'Highly Redshifted Radio Lines'*: ed. by C. L. Carilli, S. J. E. Radford, K. M. Menten and G. I. Langston (PASP, San Francisco 1999) pp. 163
237. R. Sheth and G. Tormen: MNRAS **323**, 1 (2001)
238. L. J. Sievers et al.: BAAS **199**, 3402 (2001)
239. J. Silk and M. J. Rees: A&A **331**, L1 (1998)
240. I. Smail, R. J. Ivison and A. W. Blain: ApJ **490**, L5 (1997)
241. I. Smail et al.: MNRAS **308**, 1061 (1999)
242. I. Smail, R. J. Ivison, F. N. Owen, A. W. Blain and J.-P. Kneib: ApJ **528**, 612 (2000)
243. I. Smail, R. J. Ivison, A. W. Blain and J.-P. Kneib: MNRAS **331**, 495 (2002)
244. I. Smail, S. C. Chapman, R. J. Ivison, A. W. Blain, T. Tanaka, T. M. Heckman, J. S. Dunlop and K. Sekiguchi: MNRAS in press (2003) astro-ph/0303128
245. G. P. Smith, T. Treu, R. Ellis, I. Smail, J.-P. Kneib and B. L. Frye: ApJ **562**, 635 (2001)
246. I. A. Smith, R. P. J. Tilanus, R. A. M. J. Wijers, N. Tanvir, P. Vreeswijk, E. Rol and C. Kouveliotou: A&A **380**, 81 (2002)
247. B. Soifer and G. Neugebauer: AJ **101**, 354 (1991)
248. R. S. Somerville, J. R. Primack and S. M. Faber: MNRAS **320**, 504 (2001)
249. A. Songaila, L. L. Cowie and S. J. Lilly: ApJ **348**, 371 (1990)
250. V. Springel, S. D. M. White, G. Tormen and G. Kauffman: MNRAS **328**, 726 (2001)
251. T. Stanev and A. Franceschini: ApJ **494**, 159 (1998)
252. S. A. Stanford, D. Stern, W. van Breughel and C. De Breuk: ApJS **131**, 185 (2000)
253. C. C. Steidel, M. Giavalisco, M. Pettini, M. Dickinson and K. L. Adelberger: ApJ **624**, L17 (1996)
254. C. C. Steidel, K. L. Adelberger, M. Giavalisco, M. Dickinson and M. Pettini: ApJ **519**, 1 (1999)
255. R. A. Sunyaev and Y. B. Zeldovich: ARA&A **18**, 537 (1990)
256. A. S. Szalay et al.: BAAS **198**, 9701 (2001)
257. A. C. Taylor et al.: MNRAS in press (2002) astro-ph/0205381
258. M. Tegmark and G. P. Efstathiou: **281**, 1297 (1996)
259. T. Totani, Y. Yoshii, F. Iwamuro, T. Maihara and K. Motohara: ApJ **558**, L87 (2001)
260. N. Trentham and A. W. Blain: MNRAS **323**, 547 (2001)
261. N. Trentham, A. W. Blain and J. Goldader: MNRAS **305**, 61 (1999)
262. S. Veilleux: in *'AGN Surveys'*: ed by R. F. Green, E. Ye. Khachikian and D. B. Sanders: in press (ASP vol. 284, San Francisco, 2003) astro-ph/0201118
263. J. Vernet and A. Cimatti: A&A **380**, 409 (2001)
264. M. Volonteri, F. Haardt and P. Madau: ApJ **582**, 559 (2002)
265. T. M. Webb et al.: ApJ **582**, 6 (2002)
266. G. J. White et al.: A&A **342**, 233 (1999)
267. S. D. M. White and C. S. Frenk: ApJ **379**, 52 (1991)
268. T. Wiklind and F. Combes: Nature **379**, 139 (1996)
269. R. E. Williams et al.: AJ **112**, 1335 (1996)
270. R. E. Williams et al.: BAAS **191**, 8508 (1997)

271. C. D. Wilson, N. Scoville, S. C. Madden and V. Charmandaris: ApJ **542**, 120 (2000)
272. A. Wootten: ed. *'Science with the Atacama Large Millimeter Array (ALMA)'* (Astron. Soc. Pac, San Francisco 2001)
273. L. Yan, P. J. McCarthy, W. Freudling, H. I. Teplitz, E. M. Malumuth, R. J. Weymann and M. A. Malkan: ApJ **519**, L47 (1999)
274. M. S. Yun and C. L. Carilli: ApJ **568**, 88 (2002)
275. M. S. Yun, N. A. Reddy and J. Condon: ApJ **554**, 803 (2001)
276. M. S. Yun, N. Z. Scoville, J. J. Carrasco and R. D. Blandford: ApJ **479**, L9 (1997)

Molecules in Galaxies at All Redshifts

Françoise Combes

Observatoire de Paris, DEMIRM, Place de L'Observatoire 61, 75014 Paris, France

1 How to Observe the H_2 Component?

1.1 The H_2 Molecule

General Properties: H_2 is a symmetrical molecule, and has no electric dipole. Consequently, the rotational levels will not be coupled except by the quadrupole. The radiative transitions are then quadrupolar, very weak, and with selection rule $\Delta J = \pm 2$, which means that the odd and even-J levels will be independent. Since the parity of J is related to the orientation of the nuclear spins, this results in the existence of two "different" molecules, the para-H_2, in which the nuclear spins of the two protons are anti-aligned, and the resulting spin $I = 0$, and the ortho-H_2, for which the spins of the two protons are parallel and $I = 1$, see Fig. 1.

Indeed, because of the Pauli exclusion principle, the wave function representing the molecule must be antisymmetric. For the para-H_2, the spin wave function is anti-symmetric, and the nuclear wave function must be symmetric. Since the two electrons are paired in the molecule, their total spin and angular momentum is zero. The angular momentum of the molecule is then only due to the rotation. The fundamental state is thus $J = 0$, for which there is a symmetric wave-function. For the ortho-H_2, on the contrary, the spin wave function being symmetrical, the nuclear radial wave function must be antisymmetric, and the fundamental state of ortho-H_2 is $J = 1$.

In addition, H_2 is the lightest molecule, with a very small inertial moment, and therefore the energy levels and their separation are very high. The energy difference between the ground states of the para and ortho-hydrogen is already $170\,\mathrm{K}/k$, and the first coupled level is $512\,\mathrm{K}/k$ above ground: this is the first line, at 28μm, which is possible to observe. This means that the H_2 molecule, when very cold at the temperature of usual molecular clouds, around $10 - 15\,\mathrm{K}$, will not radiate at all.

H_2 is the most stable form of hydrogen at low temperature. It is dominant in planetary atmospheres. Its main formation process today, in normal molecular clouds in galactic disks, is on dust grains. However its formation is still possible in primordial gas devoid of heavy elements and dust, through ion-atom reactions in the gaseous phase, involving $H+H^-$ (e.g. Palla et al. 1983).

Its destruction is through photodissociation by UV photons (Lyman band). It is however shielded by the atom HI, since the photodissociation continuum

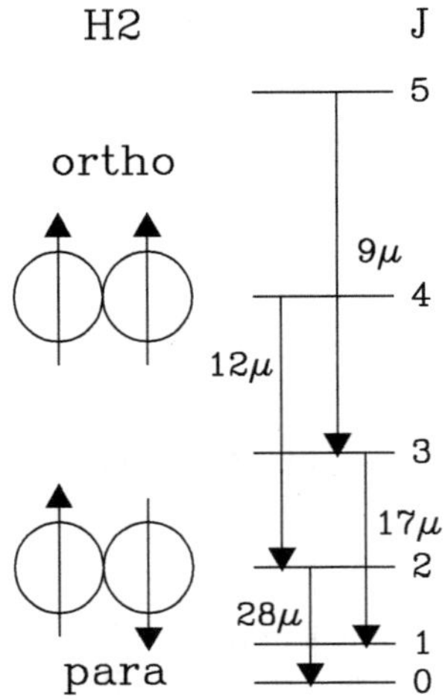

Fig. 1. Schematic of the first levels of the H_2 molecule. The even-J levels are the para-hydrogen, while the odd-J levels are from the ortho-hydrogen. The coupled levels obey $\Delta J = \pm 2$, for quadrupolar transitions, and the two species are not radiatively coupled. The four first rotational lines of hydrogen S(J) (where J is the lower state) are represented – wavelengths in μm, $S(3) = 9$, $S(2) = 12$, $S(1) = 17$ and $S(0) = 28$.

of H_2 starts at 14.7 eV, and its photo-ionization at 15.6 eV (while the HI ionization starts at 13.6 eV).

Self-shielding begins at low column densities, of about $10^{20}\,\mathrm{cm}^{-2}$ in standard UV field. H_2 will be therefore be still present, while other molecules such as CO would be already photo-dissociated.

Ortho-para Transitions: Non-reactive collisions cannot induce the ortho-para transitions (because of no dipole between the two states), but reactive collisions with protons (with an exchange of protons) may induce transitions for which ΔJ is odd. The rate coefficient for the $1 \to 0$ transition has been estimated to be $10^{-10}\,\mathrm{cm}^3/\mathrm{s}$ by Black & Dalgarno (1976). Gerlich (1990) has studied in detail the reaction $H^+ + H_2$ and confirmed a rate of $2 \cdot 10^{-10}\,\mathrm{cm}^3/\mathrm{s}$, nearly constant with temperature. At high temperature, the abundance ratio should be in favor of the ortho-H_2, with an o/p ratio of 3, but at the very low temperature of the interstellar medium, this ratio should tend to zero as $\exp(-170/T)$. However, the equilibrium might not be reached, since the formation of H_2 can involve exothermic reactions, and the new H_2 is formed in an excited state, or through catalytic recombination of H atoms on the surface of grains. There is a large energy of formation, 2.25 eV/atom, and the formation in the para state is not favored. The transformation o-H_2 $\to$ p-H_2 being very slow, it is not likely that the thermal equilibrium is obtained, the more so that the ortho-para coupling via proton-exchange favors the o-H_2 (Gerlich, 1990). Observationally, abnormal ratios are indeed observed with the ISO satellite (e.g. Rodríguez-Fernández et al. 2000).

1.2 Infrared Lines of H_2

From the ground, it is possible to detect easily only the vibrationally excited lines around 2.2 μm. These lines involve levels about 2500 K above the ground state, and are excited by fluorescence, through high-UV radiation near starbursts, or by shock waves, due to violent turbulence. This is the case in the ULIRGs galaxies NGC 6240 and Arp 220 for instance, where the H_2 $v = 1 \rightarrow 0$ $S(1)$ line is significantly brighter than the Brγ line, contrary to "normal" starbursts (van der Werf 2001). These powerful and numerous shocks indicate large gas infall and fueling of these ultra-luminous galaxies.

As for the pure rotational lines, they require space observations. The ISO satellite has detected the first lines $S(0)$ to $S(3)$ in many galactic sources, and in some external galaxies. In the edge-on galaxy NGC 891, for instance, Valentijn & van der Werf (1999) have shown that the two first lines are detected quite far from the center of the galaxy, and the radial profile is declining much slower than the CO luminosity profile (see Fig. 2). Although these lines are emitted by warm H_2, they are a tracer of the more abundant cold molecular hydrogen, independent of metallicity, which might reveal precious in the outer parts of galaxies. The derived H_2 column densities are much larger than that of HI.

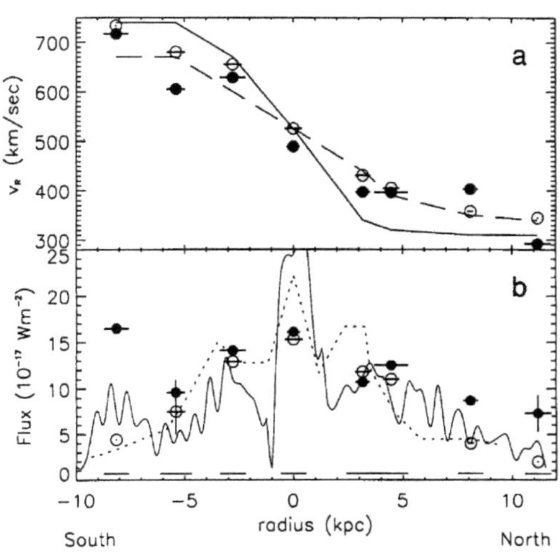

Fig. 2. Major axis profiles in the edge-on galaxy NGC 891 of the two fundamental rotation lines of H_2: $S(0)$ (*filled circles*) and $S(1)$ (*open circles*), compared to the CO profile (*full line*); a) velocity profiles b) integrated line strengths (from Valentijn & van der Werf 1999).

1.3 UV Lines of H_2

Electronic lines of H_2 in the UV have very high transition probabilities, and therefore are easy to observe. The interstellar gas is detected in absorption in front of nearby stars. However, it is possible to probe only low-extinction line of sights ($A_V < 1.5$), and the lines are very quickly saturated. The Copernicus satellite was the first in the '70s to calibrate the H_2 column densities with reddening (Savage et al. 1977), and also the H_2/CO ratio, by observing both molecules in UV absorption.

A wealth of new data is now brought by the FUSE satellite. Column densities as low as $N(H_2) = 10^{14}\,\mathrm{cm}^{-2}$ are currently detected. Molecular hydrogen is ubiquitous in our Galaxy (Shull et al. 2000; Rachford et al. 2001) in translucent and diffuse clouds. The molecular fraction in these clouds is of the order of a few percents, and the $f(H_2)$ fraction is then increasing steeply towards dense molecular clouds, entirely molecular. The absorption towards the Magellanic Clouds reveals reduced H_2 abundances, due to high UV field (Tumlinson et al. 2002, see Fig. 3). H_2 in absorption has also been detected in High Velocity Clouds (HVC), while the latter are not detected in CO (Richter et al. 2001).

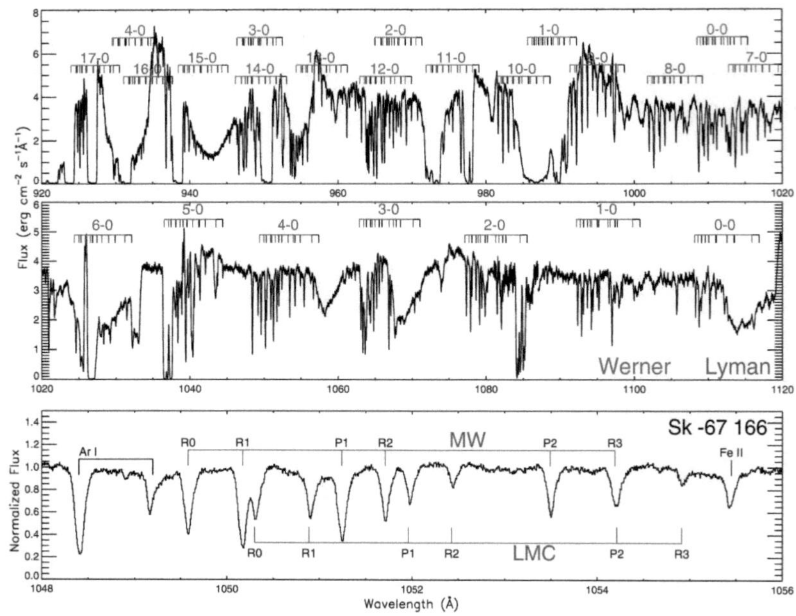

Fig. 3. FUSE spectrum in front of an LMC star, with a column density of $N(H_2) = 5.510^{15}\,\mathrm{cm}^{-2}$ in $J = 0 - 3$. On the same line of sight, absorptions are coming from the Milky Way (MW) and Large Magellanic Cloud (LMC) and are identified. With this technique, only diffuse or translucent clouds can be probed, where the molecular fraction is a few percents (from Tumlinson et al. 2002).

Table 1. Critical density $n_{\rm crit}$ for which $C_{ul}/A_{ul} = 1$.

Molecule	CO	NH_3	CS	HCN
μ [Debye]	0.1	1.5	2.0	3.0
$n_{\rm crit}$ [$\rm cm^{-3}$]	$4 \cdot 10^4$	$1.1 \cdot 10^5$	$1.1 \cdot 10^6$	$1.6 \cdot 10^7$

1.4 The CO Tracer

Since the cold molecular hydrogen does not radiate, the best tracer in galaxies is the most abundant molecule after H_2, i.e. the carbon monoxide CO. It has a small dipole ($\mu = 0.1$ Debye), and therefore requires only mild density to be excited in its rotational lines, that fall in the millimeter range. The molecule is relatively heavy so that the first level $J = 1$ is at 5.2 K above the ground state, and is therefore perfectly fitted to probe the cold molecular clouds. The CO abundance with respect to H_2 is of the order of 10^{-5}. The excitation is done by collisions with H_2 essentially. The spontaneous de-excitation rate A_{ul} is proportional to μ^2, and therefore a weak dipole indicates that the critical density $n_{\rm crit}$ for excitation, where the collisional rate C_{ul} becomes equal to the radiative rate A_{ul}, is small. The competition between collisional excitation and radiative transitions, is quantified by the ratio C_{ul}/A_{ul} which is proportional to $n(H_2)\,T^{1/2}/(\nu^3\mu^2)$. Some values for critical densities relative to current molecules are plotted in Table 1. HCN and CS, for instance, are called high density tracers, since their emission is detected towards the densest parts of the clouds. Various tracers have thus to be used, CO for the wide scale more diffuse and extended medium, HCN, CS, for the dense cores, etc...

The widely used CO lines ($J = 1-0$ at 2.6 mm, $J = 2-1$ at 1.3 mm) are generally optically thick, for "standard" clouds of solar metallicity. This complicates the interpretation in terms of H_2 column densities and mass of the clouds. More precisely, each individual molecular cloud is optically thick. However, a giant molecular cloud is composed of many sub-structures, and towards a given external galaxy, there are many clouds along a line of sight. The "macroscopic" optical depth is in general not very large, due to velocity gradients: clouds do not hide each other at a given velocity, and therefore it is possible to "count" the clouds.

Assuming that there exists a proportionality relation between the observed CO intensity and the H_2 column density, several calibration procedures have been tried:

- Observing the isotopic molecule ^{13}CO, the UV lines in absorption.
- Statistics and counting the "standard" clouds.
- Using the Virial hypothesis.

Isotopic Species: Since the CO lines are optically thick (in average $\tau_{CO} \sim 10$), it might be interesting to observe the much less abundant isotope ^{13}CO. At the solar radius, the ^{12}CO/^{13}CO abundance ratio is $\sim$ 90, and in individual molecular clouds, it is frequent that the ^{13}CO lines are optically thin. The observed average ratio between integrated ^{12}CO and ^{13}CO intensities is of the order of 10.

For diffuse line of sights, it has been possible to calibrate the CO/H_2 ratio, through successive calibrations, knowing ^{13}CO and H_2 in the solar neighbourhood by direct observations of these lines in UV absorption. However, the ratio does not apply to dense molecular clouds, representing most of the molecular mass in the Galaxy. Also, when maps of the two isotopes are compared at millimetric wavelengths for well studied clouds, it appears that the ^{13}CO does not probe the entire extent of the clouds, but only the densest regions. This could be due to problems of excitation, self-shielding, or selective photo-dissociation.

Statistical Count of the Clouds: A more empirical method is to assume that the molecular content of external galaxies is composed statistically of the same "standard" clouds, of typical masses of $10^3\,M_\odot$. For usual spatial resolutions, there are numerous clouds in the beam, of the order of 10^4 or 10^5 clouds. They are not likely to overlap, since they are well separated in velocity. The filling factor is the product of the surface factor f_s and a velocity factor f_v and it is assumed that $f_s f_v \ll 1$. This is the case actually, since the antenna temperature observed in nearby galaxies is of the order of $T_A^* \sim 0.1\,K$ for nearby galaxies, while it is $\sim 10\,K$ for an individual typical cloud. This method of "counting" the number of clouds in the beam justifies a constant factor between the integrated CO intensity I(CO) and N(H_2).

Virial Method: On more physical grounds is the "virial" method. Most molecular clouds are gravitationally bound, as has been checked with detailed studies of individual nearby clouds in the Milky Way. There is a good correlation between the virial mass and the CO luminosity, as shown in Fig. 4, but the slope is not 1, which means that it is not possible to calibrate the conversion ratio for all clouds. If the clouds are not statistically similar from one galaxy to the next, the conversion factor could be significantly wrong.

Assuming statistically that each cloud contributes to the same antenna temperature T_A^* in average, reflecting the excitation temperature of the gas, the width of the spectrum ΔV gives the cloud mass M through the virial relation,

$$\Delta V^2\, r \sim GM \;, \tag{1}$$

where r is a typical size for the clouds.

The conversion ratio N(H_2)/I(CO) can then be computed as a function of average brightness $T_R = T_A^*/\eta$ (with η the beam efficiency), and average

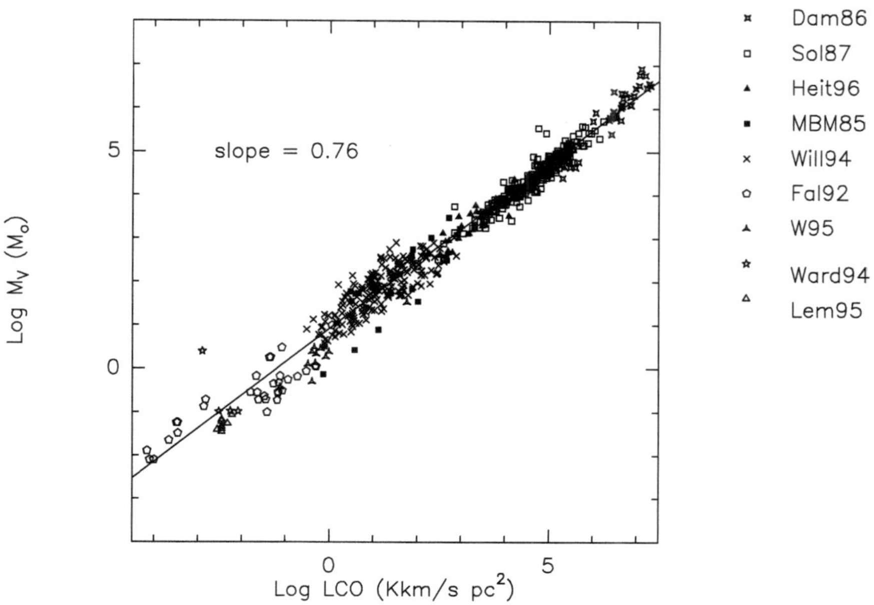

Fig. 4. Virial mass versus CO luminosity relation taken from various sources: Dam 86: Dame et al. (1986); Sol 87: Solomon et al. (1987); Heit 98: Heithausen et al. (1998); MBM 85: Magnani et al. (1985); Will 94: Williams et al. (1994); Fal 92: Falgarone et al. (1992); W 95: Wang et al. (1995); Ward 94: Ward-Thompson et al. (1994); Lem 95: Lemme et al. (1995). An indicative line of slope 0.76 is drawn.

density of clouds n. Let us consider N clouds in the beam, of diameter d, with a projected area $a = \pi d^2/4$, and a velocity dispersion ΔV. If the area of the beam θ is written $A = \pi(\theta D)^2/4$, then the integrated CO emission is:

$$I(\mathrm{CO}) = A^{-1} N \frac{\pi d^2}{4} T_R \Delta V \ . \tag{2}$$

The mean H_2 surface density is $N(H_2) = A^{-1}N(\pi d^3/6)n$, and the conversion ratio $X = N(H_2)/I(\mathrm{CO}) = \frac{2}{3} nd/(T_R \Delta V)$.

From the Virial hypothesis $\Delta V \sim n^{1/2} d$ and the conversion ratio varies as $n^{1/2}/T_R$. This factor is about $2.8 \cdot 10^{20}\,\mathrm{cm}^{-2}/(\mathrm{km/s})$ for $T_R \sim 10\,\mathrm{K}$ and $n \sim 200\,\mathrm{cm}^{-3}$.

This simple model predicts a low dependence on metallicity for the conversion ratio, since the clouds have high optical thickness and are considered to have top-hat profiles (no changes of sizes with metallicity).

However, for deficient galaxies such as the Magellanic Clouds, where clouds can be resolved, and the virial method individually applies, the conversion factor appears highly dependent on metallicity. This can be explained considering that the size of CO-emitting clouds (where the optical depth $\tau \sim 1$) varies strongly with metallicity.

Let us assume for instance an isothermal profile for the cloud density (not too far from the reality), with $\rho \propto r^{-2}$, then the column density varies as $N(\mathrm{H_2}) \propto r^{-1}$, and the diameter of clouds, where optical depth becomes unity, is varying linearly with the abundance of CO, or in the metallicity Z =(O/H). The surface filling factor of clouds varies as Z^2, and the conversion ratio varies with metallicity more rapidly than linearly. This variation could be even more rapid if the C abundance is proportional to oxygen C/O $\sim$ O/H in galaxies, and CO/$\mathrm{H_2}$ $\sim$ $(\mathrm{O/H})^2$.

In external galaxies, the ratio between HI and derived $\mathrm{H_2}$ masses $M(\mathrm{H_2})/M(\mathrm{HI})$ appears to vary indeed as $(\mathrm{O/H})^2$ (Arnault et al. 1988; Taylor et al. 1998). On the contrary, in the very center of starbursts galaxies, an over-abundance of CO could lead to overestimate the molecular content. The issue is not clear nor definite, since there are many varying parameters, with contradictory resulting effects: T_R is larger in starbursts, but the density $n(\mathrm{H_2})$ too (as shown by the high-density tracers such as HCN), and $N(\mathrm{H_2})/I(\mathrm{CO})$ varies as the ratio $n^{1/2}/T_R$.

In addition, chemical peculiarities have been observed in starbursts. ^{12}C is a primary element, while ^{13}C is secondary. The isotopic ratio ^{12}CO/^{13}CO is enhanced, mimicking more optically thin gas. But a more "normal" C^{18}O allows to disentangle the effects (Casoli et al. 1992; Henkel et al. 1993).

1.5 The Dust Tracer

Another tracer helps to determine the molecular content in galaxy disks, the continuum emission of dust in the millimeter domain. At $\lambda = 1$mm, the dust emission is in the Rayleigh-Jeans domain

$$B(\nu, T) \sim \frac{2kT}{\lambda^2} \qquad (3)$$

The flux is quasi-linear in temperature (between 20 and 40K), and in most cases, the dust emission is optically thin. The emission is however expected to be linearly varying with metallicity Z, and Z decreases exponentially with radius. The problem of tracing the molecular component in the outer parts of galaxies remain.

When the molecular component dominates in galaxies, the CO emission profile follows the dust profile (for example NGC 891, Fig. 5). When the HI dominates, on the contrary, the dust does not fall as rapidly as CO with radius, but follows a little more the HI, while still falling with radius (for example NGC 4565, Fig. 5). This reveals that CO might be a poor tracer of $\mathrm{H_2}$, even more dependent on metallicity than the dust emission.

The excitation effects also combine to metallicity, and might explain why CO emission drops more rapidly than dust with radius.

The rotational ladder of the CO molecule, and in particular the CO(2–1) line tells us about the excitation. For optically thick lines, at sufficient density,

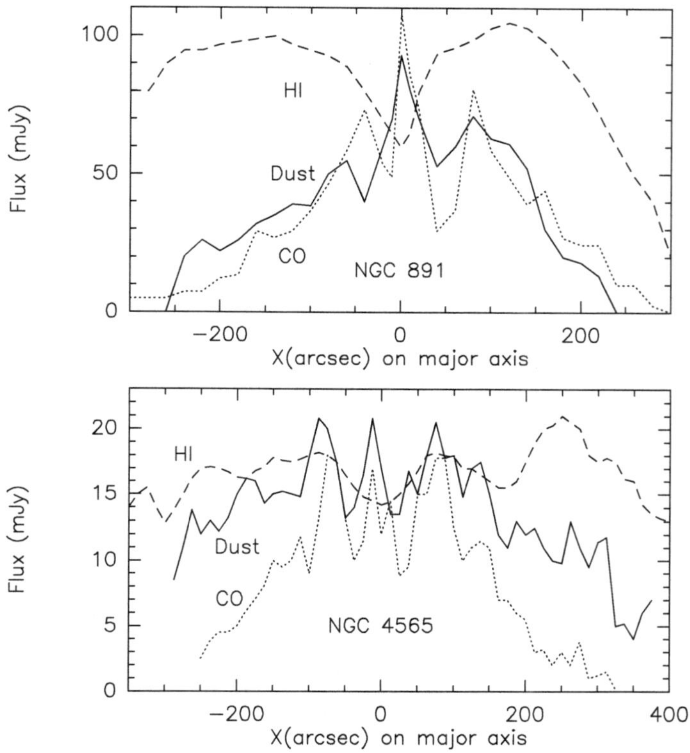

Fig. 5. Dust emission radial profiles in NGC 891 and NGC 4565, compared to the CO and HI radial profiles, from Guélin et al. (1993), and Neininger et al. (1996).

the CO(2–1)/CO(1–0) ratio is of the order of 1. For optically thin gas, when the two lines have the same excitation temperature T_{ex}, the ratio could reach the asymptotic value of 4.

In most of the disk of galaxies, except towards the very center, the CO lines appear subthermally excited. The intensity ratio between the CO(2–1) and CO(1–0) line is in average below 1, of the order of 0.6 (like in the Milky Way, Sakamoto et al. 1995). This means that the medium is not dense enough to excite the CO molecule to the second level (the more so as the critical density is larger for larger J). Even if the kinetic temperature is sufficient to excite the upper level of the $J = 2 - 1$ transition, the density is not, and $T_{ex}(2 - 1) < T_{ex}(1 - 0)$. Molecular clouds being denser in the nuclei, due to larger tidal forces that select only dense clouds, the CO(2–1)/CO(1–0) ratio is there of the order of 1 (Braine & Combes 1992, Fig. 6).

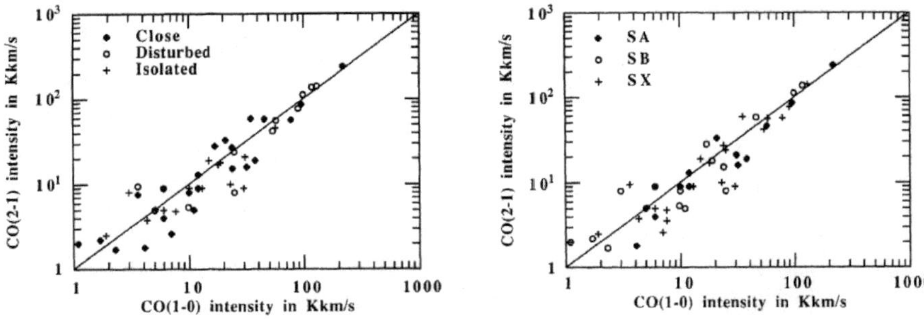

Fig. 6. (*Left*) The CO(2−1) intensity versus the CO(1−0) intensity in a sample of external galaxies, according their environment, or interacting state. The CO(2−1)/CO(1−0) ratio is very nearly one, whatever the state. (*Right*) The CO(2−1)/CO(1−0) ratio does not depend on the presence of bars, not on morphological type (from Braine & Combes 1992).

1.6 Conclusion

The cold H_2 molecule is invisible, and unfortunately the molecular clouds component, which represents the bulk of the mass of the interstellar medium in most spiral galaxies, is difficult to estimate.

The CO molecule, still our best tracer of H_2, is not quite reliable, because of metallicity effects (non-linear dependence), but also photo-dissociation in UV-bright environments, sub-thermal excitation in diffuse media, etc...

It is very important to have other tracers, like mm dust emission, or dense core tracers, HCN, HCO^+, and isotopes. The H_2 pure rotational lines, emitted by the "warm" H_2, could be a precious tracer of the colder H_2, independent of metallicity.

2 Molecular Component in the Milky Way

2.1 CO Surveys of the Milky Way

The CO lines being our best tracer of the cold molecular hydrogen, many CO surveys of the Milky Way have been reported in the literature; they consist of either a high-resolution (small beam, 1') statistical sampling of the galactic plane (with small surface mapped), to a more complete, low-resolution (large beam 8'), high-coverage surveys, like the Columbia-CfA survey. Most of them have been done in the ^{12}CO(1-0) line, but other were specialized in the isotopic line ^{13}CO(1-0) or the excited line CO(2-1). Here are some of the main surveys:

- CfA-Harvard survey, telescope of 1.2m diameter, beam 9', sampling 0.12°, Northern and Southern hemisphere (Dame et al. 1987, 2001), sky coverage of about 50% until $|b| < 32°$, cf. Fig. 7.

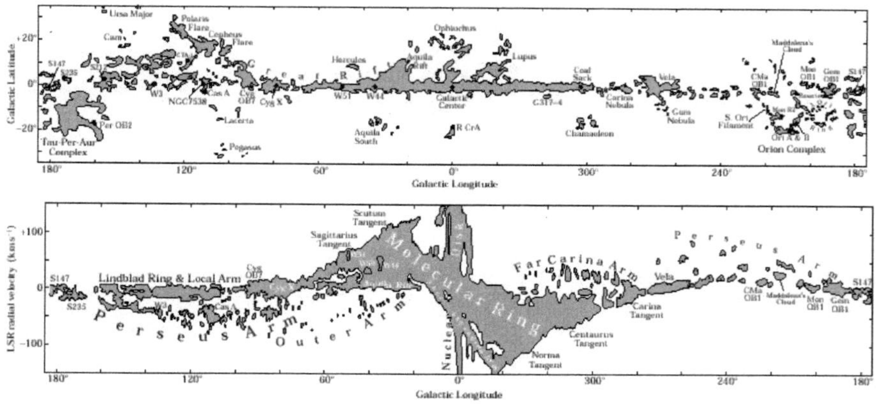

Fig. 7. (*Top*) Velocity integrated CO map of the Milky Way (with mostly 9' resolution over most of the map). The main molecular clouds are indicated. (*Bottom*) Longitude-velocity map of CO emission integrated over 4 degrees in latitude centered on the Galactic Plane, from Dame et al. (2001).

- Bell Labs survey, telescope of 7 m, beam 1.7', CO, ^{13}CO, CS in galactic center (Bally et al. 1987, 1988).
- NRAO-Kitt Peak survey, telescope of 12 m, beam 60", fraction of sky covered 10^{-3} (Burton & Gordon 1978; Solomon et al. 1979).
- Massachusetts-Stony-Brook survey, FCRAO telescope of 14m, beam 45", sky coverage 10^{-2} (Sanders et al. 1986).
- CO(2-1) survey with Nobeyama 60 cm telescope, beam 9', Sakamoto et al. (1995); the average CO(2-1)/CO(1-0) ratio found for the Galaxy is 0.66.
- ^{13}CO surveys also made in Bordeaux (beam 4.4', Jacq et al. 1988), Columbia (9' Bronfman et al. 1988), etc.

2.2 CO Distribution and Spiral Structure of the Milky Way

Since we are located in the plane of the Galaxy itself, the derivation of distances along a given line of sight is complex. The main method to obtain distances is kinematical, from the observed radial velocity of the emission, with the help of a model for the rotation of the Milky Way. It is assumed in general that the rotation is circular, and streaming motions are ignored, which could alter significantly the derived 3D-view of the Galaxy (e.g. Combes 1991).

The rotation curve is first obtained from the observation of terminal velocities, at the points where the circular velocity is parallel to the line of sight (see Fig. 8). The locus of these tangent points is a circle of diameter the segment Sun-Galactic Center.

For each line of sight of longitudes l lower than 90°, there are two points with the same projected radial velocity, but one at a near distance and the

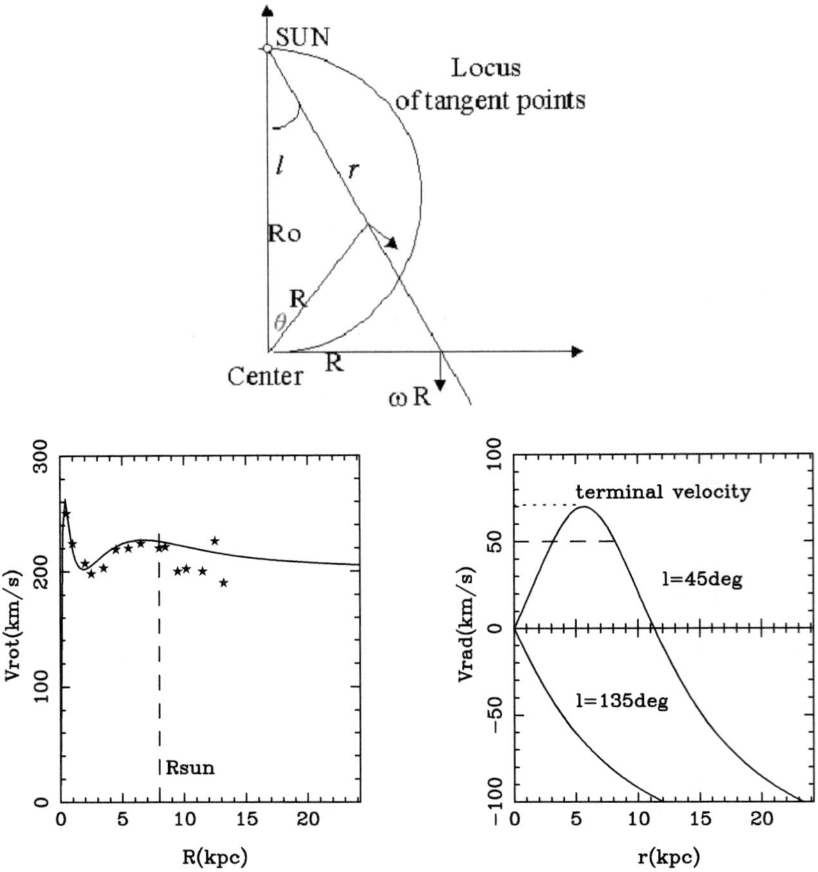

Fig. 8. (*Top*) Schematic diagram showing that two points on the line of sight may have the same projected radial velocity with respect to the observer. There is ambiguity of kinematic distances between a near point (inside the circle) and a far point (outside). The circle is the locus of points where the circular velocity is exactly parallel to the line of sight, and the observed radial velocity maximum (terminal velocity of tangent points). (*Bottom left*) Rotation curve of the Milky Way: the stars are data points (cf. Fich & Tremaine 1991), and the curve a model fitted to the data. (*Bottom right*) The observed radial velocity as a function to the distance to the observer, for a longitude of 45° where there is distance ambiguity, and for $l = 135°$ where there is not.

other outside the locus of tangent points. The observed radial velocity as a function of the longitude l and the distance to the point R is (Fig. 8):

$$V_{\rm rad}(r,l) = R_\odot \sin l \, (\Omega(R) - \Omega_\odot) \ , \tag{4}$$

where Ω is the angular rotational velocity. To remove the degeneracy, the latitude or height above the plane can play a role statistically, the nearest points appearing at a higher latitude. Also the distance of close-by stars, determined by their spectrum or by absorption (in front or behind the clouds) may be used to raise the distance ambiguity.

Radial Distribution: Once the distance of the emitting material is known, it is easy to invert the observed longitude-latitude map and find the galacto-centric ppmap, and the radial distribution in the Galaxy. One has to remind that all these steps are model dependent, and in particular the radial distribution and the spiral structure are considerably perturbed by the ignored streaming motions.

The most striking features of the radial distribution are (see Fig. 9):

- the large concentration in the center (inside 1 kpc in radius, corresponding to the decoupled circumnuclear disk),
- the relative depletion around 2 kpc,
- the Galactic molecular ring between 4 and 8 kpc, which concentrates most of the disk emission. This ring might be decomposed in spiral arms, but the problem is to disentangle the actual distances, in the spiral arm region where streaming motions are important,
- the quasi exponential radial decrease afterward.

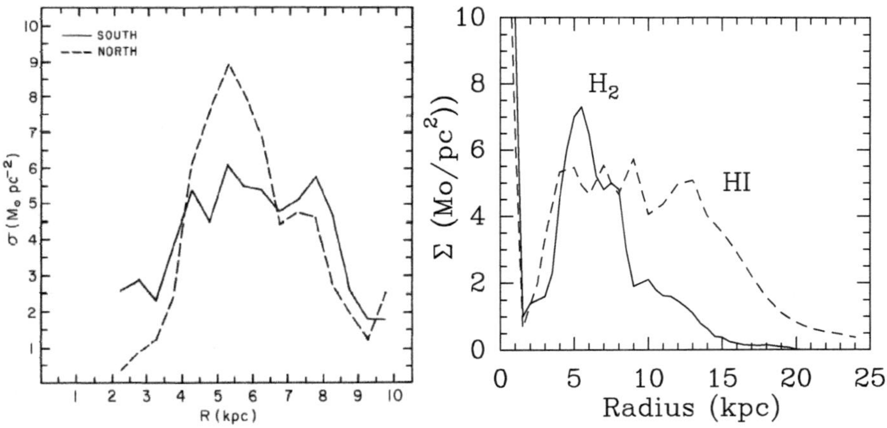

Fig. 9. (*Left*) Radial distribution of H_2 surface density derived from CO emission, in the southern and northern hemisphere, from Bronfman et al. (1988). (*Right*) Comparison between the HI and H_2 derived surface densities, as a function of galactic radius. The data inside $R_\odot = 8.5$ kpc are from a compilation of the various surveys presented here (see Burton 1992), and outside the solar circle, from Wouterloot et al. (1990).

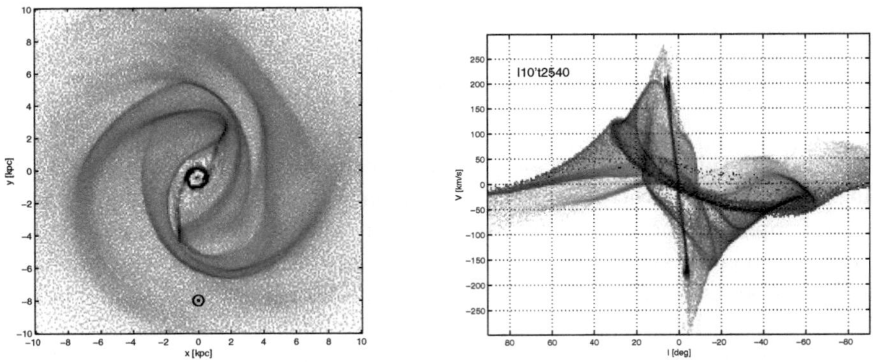

Fig. 10. An N-body+SPH simulation from Fux (1999) reproducing most of the observed features in the region $|l| \leq 35°$. (*Left*) Face-on projection of the gas distribution, with the location of the observer indicated by the $\odot$ symbol. The bar inclination angle is 25° and the corotation radius 0.55 relative to the Sun's galactocentric distance. (*Right*) $l - V$ diagram based on the gaseous particles within $|b| < 2°$.

Spiral Structure: To determine the overall spiral structure of our Galaxy is a difficult task, although there is evidence of strong and contrasted spiral arms in the $l - V$ diagram, for instance: there are strongly depleted region and a large contrast, and this cannot occur in a smooth axisymmetric disk (see Fig. 7).

There is also evidence of a barred structure inside the solar circle, through the highly non-circular motions. If the motions are assumed only circular, these elliptical streamlines are interpreted as radial expansion due to explosions (for a long time, peculiar features were dubbed "expanding", such as the expanding molecular ring in the center, or the expanding 3 kpc arm).

Since it is too model dependent to deproject the observed $l - b - V$ data into a face-on view of the Milky Way, the best tool to better understand its structure is to project realistic simulations in the $l - V$ diagram. Satisfying models have been published by Mulder & Liem (1986) or Fux (1999), as shown in Fig. 10.

2.3 The Inner Galaxy

The center of the Galaxy has always been a big puzzle: large forbidden velocities are observed; they do not fall to zero towards zero longitude, as expected for a disk in pure circular motion. Curiously, although the hypothesis of a bar potential had been proposed (Peters 1975) and was realistic, since more than two thirds of spiral galaxies are barred, the explosion and expansion model was more discussed (Oort 1977; Sanders & Bania 1976).

Now the bar has been directly seen in the near-infrared image of COBE-DIRBE, through a perspective effect: the near-side of the bar being signifi-

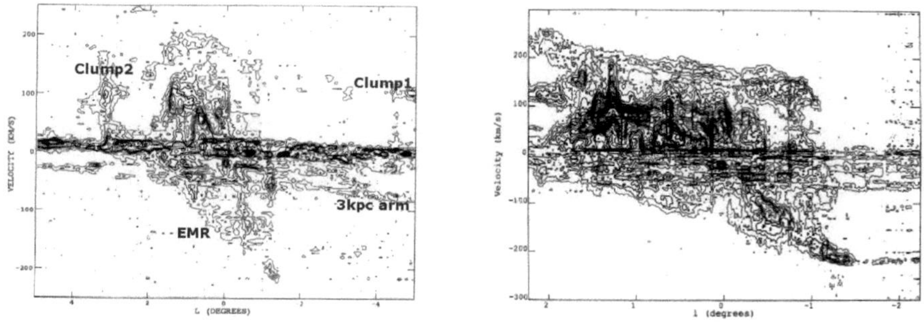

Fig. 11. Molecular gas in the inner Galactic disk, showing the circumnuclear disk: (*Left*) $l - V$ diagram in ^{13}CO, integrated over latitudes $|b| < 0.6°$, from Bally et al. (1988). Note the "expanding molecular ring or EMR", Clumps 1 and 2, and 3 kpc arm features. The main lane at $V = 0$ is coming from the molecular ring in the Galactic disk, farther from the Galactic center. (*Right*) $l - V$ diagram in ^{12}CO, integrated over latitudes $|b| < 0.1°$, from Bally et al. (1987). The central nuclear disk forms a characteristic "parallelogram" feature.

cantly less distant from us than the more remote side, its height above the plane is larger (Dwek et al. 1995).

Moreover, the characteristic features in the $l-V$ diagram, the non-circular motions giving rise to a peculiar parallelogram geometry (Fig. 11) can be explained by the shape of the periodic orbits in a bar potential: the x_1 orbits, parallel to the bar, are highly elongated, and may also reveal a cuspy shape at their extremities. Inside the parallelogram, the diagonal feature could be due to the x_2 orbits, that are perpendicular to the main bar, and of nearly circular shape (Binney et al. 1991). Taking into account the perspective effect can also explain the apparent offset for the center of the parallelogram (Fux 1999, Fig. 12). The simulations show also that the center of the bar may wander around the $(l, b) = (0, 0)$ position (Fux 1999), and part of the offset could be intrinsic, and non-stationary. The 3 kpc "expanding" arm is one of the conspicuous arm around the bar.

If the parallelogram is interpreted as the leading dust-lanes characteristic of barred galaxies, the high velocity Clumps 1 and 2 or velocity-elongated features at low longitudes in the $l - V$ diagram (Fig. 11; Bania et al. 1986, Stark & Bania 1986) are gas lumps crossing the dust-lane shocks. The bar is inclined with respect to our line of sight to the Galactic center by 25°, the corotation radius is at 4.5 kpc, which is also the approximate radius of the bar (Weinberg 1992; Fux 1999). Other features remain to be interpreted such as the inclination of the nuclear disk with respect to the main plane, and the lopsidedness.

The very large observed velocities in the circumnuclear disk, together with the relative depletion just outside (at about 2 kpc radius) strongly suggest that the nuclear disk is decoupled from the main disk, as is frequently the

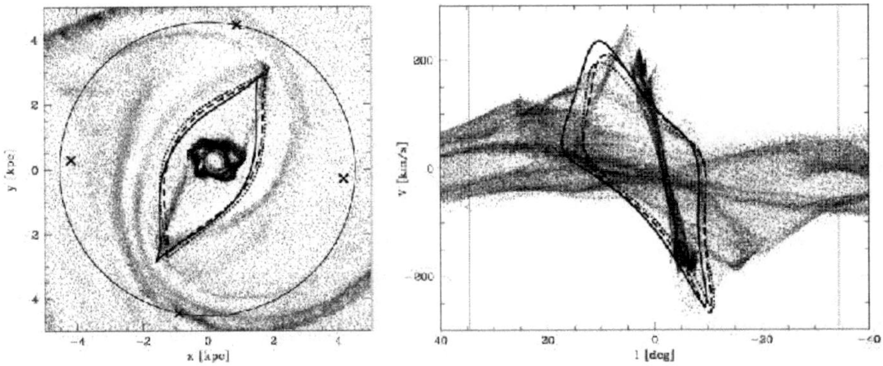

Fig. 12. The interpretation of the parallelogram feature in the central Galaxy, from the N-body+SPH simulation by Fux (1999). (*Left*) Face-on view of the gas distribution, with the identification of some x_1 and x_2 orbits. (*Right*) $l-V$ diagram based on the gaseous particles within $|b| < 2°$, and reproducing the same orbits.

case in external galaxies (Rubin et al. 1997; Emsellem et al. 2001). It is even likely that a secondary bar is rotating at a different pattern speed from the primary bar, and is dynamically decoupled. Such a nuclear bar has been seen in the near-infrared image from 2MASS by Alard (2001).

Nature of Molecular Clouds in the Inner Galaxy: The giant molecular clouds of the circumnuclear disk near the center have distinct physical parameters from those of the galactic disk. They are denser, by two orders of magnitude ($\gg 10^4\,\mathrm{cm}^{-3}$) as revealed by high density tracers such as HCN or CS. This can be explained by the fact that tidal forces from the galaxy potential are larger towards the center. Lower density clouds would have been destroyed by these forces, and are not present.

To have an order of magnitude of the variation with galactic radius R of tidal forces, let us differentiate the gravitational radial acceleration V^2/R; if we can make the approximation of a flat rotation curve $V \sim$ const, then the tidal force on a cloud of diameter d is:

$$\frac{V^2}{R^2}\left(\frac{d}{2}\right) . \tag{5}$$

The limiting density for a cloud of mass M_c is reached when this tidal force is equal to the internal gravity force $\sim 4GM_c/d^2$. The minimum density of clouds is thus:

$$\rho_c = \frac{3}{4\pi G}\frac{V^2}{R^2} = 10^3\,\mathrm{cm}^{-3}\left(\frac{200\,\mathrm{pc}}{R}\right)^2 . \tag{6}$$

Below this critical density, clouds are sheared off to the diffuse medium. In addition to a larger density, clouds have also a higher velocity dispersion in the center, due to the Toomre criterion for overall gravitational stability.

2.4 Vertical Distribution

The vertical distribution of molecular clouds together with their vertical velocity dispersion, is a good diagnostics of the dynamics of the clouds. Since we are inside the plane, these quantities are somewhat uncertain. It is clear from the (l, b) map that the molecular plane is much thinner than the HI plane. However there are projection effects for the local gas, and the warp in the outer plane for the HI. The vertical velocity dispersion σ_g is obtained in the Milky Way, at the tangent points (Malhotra et al. 1994, 1995), where the crowding of matter allows a better statistics. In external galaxies, it is more easily obtained in face-on objects, while the actual thickness h_g is observed in edge-on systems.

Vertical Equilibrium: In the hypothesis of a thin isothermal infinite disk, completely self-gravitating, the expected thickness h_g is related to the velocity dispersion σ_g and the surface density μ_g by:

$$h_g(r) = \frac{\sigma_g^2(r)}{2\pi\mu_g(r)} \tag{7}$$

and the density profile is then a sech^2 law.

If the gas clouds are considered as test particles in a potential of larger scale height $K_z z$, the density has then a Gaussian profile,

$$\rho_g = \rho_0 \exp\left(-\frac{K_z}{2\sigma_g^2} z^2\right) , \tag{8}$$

with a characteristic height,

$$h_g(r) = \frac{\sigma_g(r)}{\sqrt{K_z}} , \tag{9}$$

with $K_z = 4\pi G\rho_t$ (ρ_t being the total density).

Observed in the HI gas, the velocity dispersion is constant with radius, and equal to 10 km/s (12 km/s in the center) (Dickey et al. 1990). This is best seen in face-on galaxies. In the molecular component, also $\sigma_g = \mathrm{const}$, in the face-on galaxies NGC 628 and NGC 3938. The surprising observation is that both dispersion (atomic and molecular) are about equal (Combes & Becquaert 1997). This is not compatible with two different independent components with different thicknesses (about 60 and 220 pc for the H_2 and HI respectively). This result is obtained after subtraction of the expected

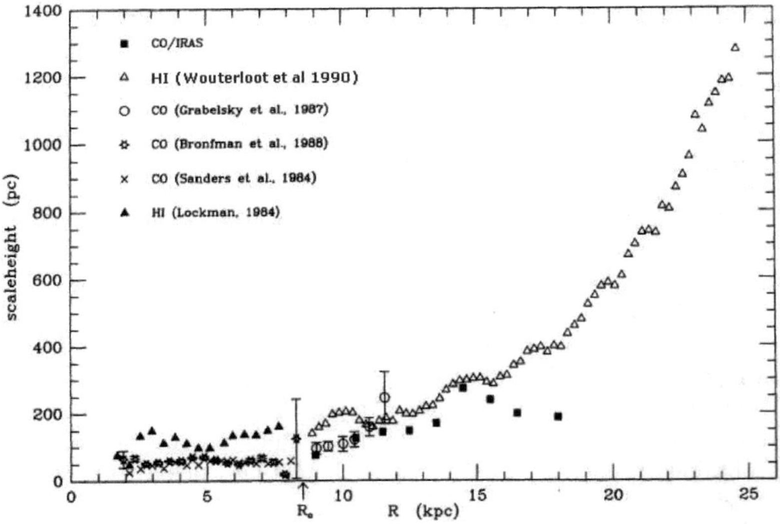

Fig. 13. The scale-height (half width at half maximum) of the atomic and molecular components as a function of galactic radius, revealing the flaring of the galactic plane, from Wouterloot et al. (1990).

linewidth due to the systematic gradient (rotation), and appears the same when the isotopic molecule ^{13}CO is considered.

In the Milky Way, the modelling of h_g and σ_g at the tangential points (Malhotra 1994) allows to deduce the azimuthal velocity dispersion. Again, no variation with radius is detected (except towards the galactic center), which should help to identify the heating processes (star formation, gravitational instabilities, ...). As for the dispersion value, there is some scatter in the literature, from 4 to 11 km/s, according to which scale is considered, and whether clumpiness of molecular clouds is taken into account or not. In average, in the Milky Way, the dispersion is of 8km/s, averaged over $200 - 400$ pc, and the scale-height of 75 pc in the molecular ring, The molecular plane is flaring, and the scale-height is slightly increasing with radius, at about the same rate as the HI component (Fig. 13).

The shape of the vertical density law is not Gaussian, as expected from a non-self gravitating medium, but tails of small clouds are observed (Magnani, Blitz, Mundy 1985). The centroid of the plane departs from $z = 0$ more than the scale height, due to corrugations and warp.

The dust layer is slightly thicker than the molecular component, but much thinner than the HI plane (Sodroski et al. 1987).

Phase Transition Between HI and H_2: If the velocity dispersions of atomic and molecular gas are comparable, this might be explained by a model in which both are two different phases of the same gas component, sharing the same dynamics. CO is observed more concentrated to the plane, since the gas is becoming molecular above a certain critical density, that is reached only below a certain height above the plane. Molecules at $z = 0$ starts with the same velocity and will reach the same maximum height above the plane, but have more chances to end their orbit in an atomic phase.

The phase transition appears rather sudden, depending on pressure, UV radiation, density (Imamura & Sofue 1997). The HI thickness of 200 pc is not explained through the turbulent velocity ($\sigma_g = 9\,\mathrm{km/s}$), but HI needs extra support to keep its height (Malhotra 1995).

From the constancy with radius of the gas velocity dispersion, and the observed linear flaring, the surface density of the plane can be deduced: and exponential law is compatible with the data, which means a constant mass-to-light ratio (since the stellar density is exponential).

Origin of the Heating: Star formation is contributing significantly in the center of the optical disk, which explains the increase of gas velocity dispersion there.

Elsewhere, in quiescent areas, self-regulated gravitational instabilities, following the Toomre criterion for disk stability, must be the main heating source.

Flaring of the Plane: The HI thickness increases linearly with radius R, $h_g = h_0 + 0.045R$, from Merrifield (1992). This is not only visible in HI, but also in the molecular plane (see Fig. 13). The flaring is due to the decrease of the total density in the plane, and the corresponding decrease of the restoring force.

The CO (and therefore H_2) follows the flare, and also the warp Grabelsky et al. (1987).

Warping of the Plane: The warping of the plane is spectacular in HI, it is mostly on one side, and asymmetrical. Corrugations are observed with amplitude larger than the scale-height h_g. The CO follows the warp, and the molecular component appears even more warped, being thinner.

This phenomenon is also observed in external galaxies, in particular M31, where CO is detected in some positions with 2 velocities, at each crossing of the warped plane. These velocities are compatible with the position-velocity models from Henderson (1979), and the characteristic figure-8 shape (see Brinks & Burton 1984).

High Velocity Clouds: High Velocity Clouds (HVC) are detected essentially in HI, no CO emission has been detected until now. The origin of HVC might be multiple, a large fraction of them belonging to the Magellanic

Stream, of low metallicity. Some H_2 has been detected in HVC through UV absorption lines (Richter et al. 2001; Tumlinson et al. 2002).

The non-detection of CO emission is likely due to the low metallicity of this gas, as confirmed by Wakker et al. (1999). Whatever its origin, the infall of gas at $Z = 0.1$ solar is required to explain the chemical evolution of the Galaxy, and in particular the G-dwarf problem (for which an infall rate of $1\,M_\odot$ is necessary).

2.5 Gamma-Ray Surveys

In the Milky Way, the emission of gamma-rays of high energy ($E > 100$ Mev) is a tracer of all interstellar matter, whatever its phase. Nucleons (HI, H_2, HII, ...) interact with cosmic rays to produce pions, that disintegrate in gamma-rays. Early surveys showed that the CO/H_2 conversion ratio must not be constant throughout the Galaxy (Wolfendale et al. 1977).

The distribution in the Galaxy of high-energy gamma-rays extends radially much more than what is expected from their sources (cosmic rays are accelerated by supernovae), and more than the gas itself (Bloemen 1989). Cosmic rays are thought to diffuse in the halo, and their distribution is obtained through gamma-rays themselves, assuming the gas distribution known (Strong et al. 1988).

Today, the lack of gamma-rays in the center is confirmed by EGRET on GRO satellite. In addition, an excess towards high latitude has been detected, as a halo of gamma-rays above the plane (Dixon et al. 1998). Several interpretations have been proposed, either in terms of an excess of nucleons in the halo (de Paolis et al. 1999) or inverse-Compton emission (Strong & Mattox 1996; Strong et al. 1999). The latter is supported by the evolution of the gamma-ray spectrum from the inner galaxy to high latitudes.

2.6 Conclusion

The interstellar medium of the Milky Way is composed of comparable amounts of H_2 and HI gas, with an estimated molecular mass of $M(H_2) \sim 2\text{–}3{\cdot}10^9\,M_\odot$. However the atomic and molecular components have very different radial repartitions. H_2 is centrally concentrated, in a circumnuclear disk, and after a relative hole, is then confined in a molecular ring between 4 and 8 kpc. HI is rather depleted in the center, and much more radially extended.

The molecular component is much more clumpy, and composed of a hierarchical cloudy structure, up to giant clouds of $10^6\,M_\odot$ in mass. The H_2 plane is thinner than the HI plane, but the velocity dispersions of both gas are comparable. The flaring and the warp are shared by the two components.

3 The Fractal Structure of the Molecular Gas

3.1 Self-similarity of Clouds

The molecular interstellar medium is very clumpy and fragmented. Its hierarchical structure can well be described by a fractal, because of its self-similarity. It has no characteristic scale. Fractals by definition are self-similar ensembles, that have a non-integer, i.e. fractional dimension (Mandelbrot 1975). The Hausdorff dimension D determines whether a system is homogeneous, and what fraction of space is filled. While a homogeneous system has a mass increasing as the 3rd power of the scale r (in three dimensions), a fractal medium may occupy a tiny fraction of space, and its mass is growing as $M \propto r^D$, with D lower than 3 ($D \approx 1.7$ for the interstellar medium).

The ensemble of molecular clouds, as many fractals considered in physics, are only approximations of mathematical fractals. They are self-similar only between two limiting scales, where boundary effects occur, while a pure mathematical fractal is infinite; and they are quite randomly distributed, their self-similarity being only statistical (see Fig. 14).

The self-similar structures in the interstellar medium extends over 6 orders of magnitude in scale, from about 10^{-4} to 100 pc, and these are not observed for the same molecular cloud only because of technical problems, lack of spatial resolution on one side, and difficulty of mapping too large areas on the other. The scaling relations all over the scales are however obtained by comparing various clouds observed with different resolutions, using the dynamics of distances in the Galaxy (from 50 pc to 20 kpc).

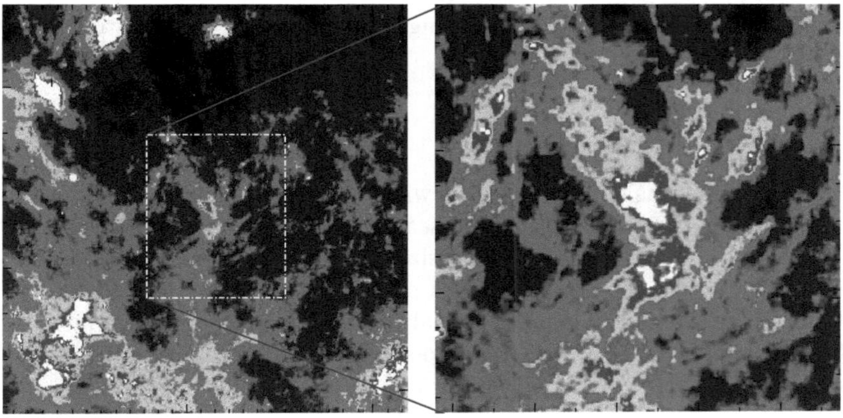

Fig. 14. (*Left*) IRAS 100 μm map of molecular clouds towards the Taurus complex, located at about 100 pc from the Sun. The far-infrared emission is from heated dust. The square is $\sim$ 4000 pc^2. (*Right*) Zoom of the central region (the square is now $\sim$ 400 pc^2).

3.2 Limits of the Fractal

The largest self-gravitating entities in the Galaxy are the Giant Molecular Clouds (GMC) of about 100 pc in diameter, and $10^6\,M_\odot$ in mass. Larger clouds cannot exist since they would be teared off by the galactic shear, i.e., the tidal forces due to the galactic potential itself. This is the high cut-off scale in the fractal structure.

As for the small-scale cut-off, it is difficult to observe directly in emission the smallest structure, due to lack of spatial resolution and sensitivity. But structures of about $10-20$ AU in size (i.e., $\sim 10^{-4}$ pc) have been observed for a long time through scattering of the quasar light (Fiedler et al. 1987, 1994; Fey et al. 1996) and also in front of pulsars (Frail et al. 1994): clumps in the electronic density diffract the light rays from remote quasars, and produce an "extreme scattering event" (ESE) lasting for a few months, in their rapid motion ($100-200$ km/s) just in front of the quasar. QSOs monitoring during several years has determined that the number of scattering structures is 10^3 times as numerous as stars in the Galaxy. The problem of stability and lifetime of these structures, with much higher pressure than surroundings, has long been debated. Heiles (1997) proposes that these TSAS (tiny scale atomic structures) are cold filaments or sheets, that appears dense in projection. The problem can be solved if they are self-gravitating (Walker & Wardle 1998); they are then of $10^{-3}\,M_\odot$ in mass, and have a gas density around $10^{10}\,cm^{-3}$. They correspond to the smallest fragments predicted theoretically (Pfenniger et al. 1994, Pfenniger & Combes 1994). These structures are now observed in a large number directly, through VLBI in the solar neighborhood, through HI absorption in front of quasars (e.g., Diamond et al. 1989; Davis et al. 1996; Faison et al. 1998). If this $10-20$ AU size is adopted for the low cut-off scale of the fractal structure, the latter ranges over 6 orders of magnitude in size, and about 10 in masses.

3.3 Scaling Laws

Power-law relations between size, linewidth and mass quantify the self-similar structure (cf. Larson 1981). These scaling relations are observed whatever the tracer. The original one is the size-linewidth relation (directly derived quantities), while the mass is only a secondary quantity, very uncertain to obtain, since there is no good universal tracer. The relation between the sizes R and the line-widths or velocity dispersion σ, can be expressed through the power-law:

$$\sigma \propto R^q \,, \tag{10}$$

with q between 0.3 and 0.5 (e.g., Larson 1981; Scalo 1985; Solomon et al. 1987; cf. Fig. 15). There are evidences that molecular clouds are virialised at least at large scale; it is more debated at small scales, where there is no good tracer of the molecular mass (in particular the conversion factor between CO and

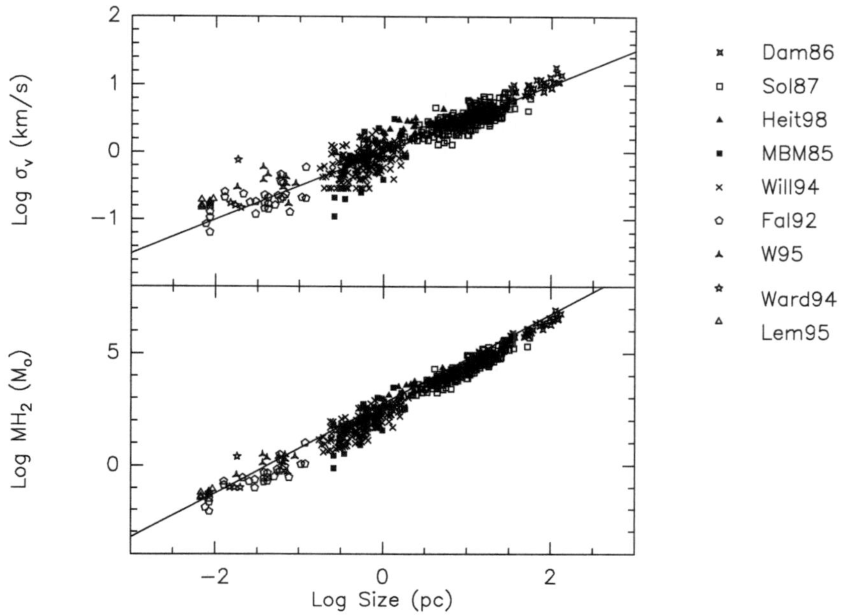

Fig. 15. (*Top*) Size-linewidth relation taken from various sources: Dam 86: Dame et al. (1986); Sol 87: Solomon et al. (1987); Heit 98: Heithausen et al. (1998); MBM 85: Magnani et al. (1985); Will 94: Williams et al. (1994); Fal 92: Falgarone et al. (1992); W 95: Wang et al. (1995); Ward 94: Ward-Thompson et al. (1994); Lem 95: Lemme et al. (1995). An indicative line of slope 0.5 is drawn. (*Bottom*) Mass-size relation deduced from the previous one, assuming that the structures are virialised. The line drawn has a slope of 2.

H_2 mass could be higher, and the mass at small-scale underestimated). If the virial is assumed at all scales, then:

$$\sigma^2 \propto \frac{M}{R} , \tag{11}$$

and the size-mass relation follows:

$$M \propto R^D , \tag{12}$$

with D the Hausdorff fractal dimension between 1.6 and 2. The mean density over a given scale R decreases thus as $1/R^\alpha$, where α is between 1 and 1.4. High Latitude Clouds (Heithausen 1996; Magnani et al. 1985) also follow these relations.

When considering the estimation of the fractal dimension, it has to be taken with caution, because of the observational biases. The main tracer used is the CO molecule, which is optically thick. Radiative transfer in a fractal is complex, and the peculiar systematic effects are unknown. Some simple simulations of fractals show already large trends (Pfenniger & Combes 1994).

The fractal structure, and its self-similarity, is broken in regions of star formation. The break is observed as a change of slope in mass spectrum power laws at about 0.05 pc in scale in the Taurus cloud, corresponding to the local Jeans length (Larson 1995). In other regions, either this break does not occur, or is occurring at larger scales (Blitz & Williams 1997; Goodman et al. 1998).

The observations of molecular clouds reveal that the structure is highly hierarchical, smaller clumps being embedded within the larger ones. This structure must be reminiscent of the formation mechanism, through recursive Jeans instability for instance. Since we have no real 3D picture, it is however difficult to ascertain a complete hierarchy, or to determine the importance of isolated clumps and/or a diffuse intercloud medium. An indicator of the 3$^{\rm rd}$ dimension is the observed radial velocity, which is turbulent without systematic pattern. It has been possible, however, to build a tree structure where each clump has a parent for instance for clouds in Taurus (Houlahan & Scalo 1992).

Filamentary Structure: The scaling relations do not characterize completely the cloud ensemble, and for the same fractal dimension, many geometries are possible. Molecular clouds have a filamentary structure, with aspect ratio close to 10 or larger. About 30% of the mass is contained in the filaments, which appear like abrupt transitions, like shock waves; their confinement could be due to gravity.

To characterize the self-similar structures in projection, several authors have computed the relation between the surface and the perimeter of these structures. This method has been used in observed 2D maps, like the IRAS continuum flux, or the extinctions maps of the sky. In all cases, this method converge towards a fractal dimension D_2, i.e., the perimeter P and area A are related by

$$P \propto A^{D_2/2} \ . \tag{13}$$

Falgarone et al. (1991) find a dimension $D_2 = 1.36$ for CO contours both at very large (degrees) and very small scales (arcmin), and the same is found for IRAS 100 μ contours in many circumstances (e.g., Bazell & Desert 1988). Comparable dimensions (D_2 between 1.3 and 1.5) are found with any tracer, for instance HI clouds (Vogelaar & Wakker 1994).

The projection of a fractal of dimension D may not be a fractal, but if it is one with dimension D_p it is impossible a priori to deduce its fractal dimension, except that (Falconer 1990):

$$D_p = D \quad \text{if} \quad D \leq 2 \ , \tag{14}$$

$$D_p = 2 \quad \text{if} \quad D \geq 2 \ . \tag{15}$$

The mass spectrum of interstellar clouds is another information, characterizing further the cloud ensemble. It has been estimated as:

$$\frac{dN}{dm} \propto m^{-\gamma} \tag{16}$$

with $\gamma \sim 1.6 - 1.9$ (Heithausen et al. 1996, 1998).

This power law is not related to the fractal dimension, but to the geometry of the fractal, and its hierarchical character. If the fractal is entirely hierarchical, the power-law expected is 2, since $m\,N(m) = \text{const}$, or

$$n(m)\, d\log m \propto m^{-1}\, d\log m \ , \tag{17}$$

and all the mass is included in the smallest fragments.

Formation by Recursive Jeans Fragmentation: The hierarchical fractal structure yielded by recursive fragmentation can be simulated schematically, in order to estimate the resulting filling factor, and the biases introduced by observing with a limited spatial resolution (e.g., Pfenniger & Combes 1994). Clouds are distributed according to the radial distribution of an isothermal sphere in r^{-2}, and fragments also have positions selected randomly according to this radial distribution, and so on, hierarchically. The number N of fragments at each level, and the fractal dimension D chosen are sufficient to determine the ensemble. The recursive relations between two successive scales L and $L-1$ are $M_L = NM_{L-1}$, $r_L^D = Nr_{L-1}^D$ and the size ratio between to imbricated spheres is $a = r_{L-1}/r_L = N^{-1/D}$. A sample of these fractal projections is shown in Fig. 16, for two dimensions 2.2 and 1.6. Typically $N = 10$ and $L = 9$.

The resulting surface filling factor is a strong function of D, and is less than 1% at $D = 1.7$. This low filling fator will lead to mass underestimate when observed with low spatial resolution. The large range of scales is also a source of bias, and of overestimates of the fractal dimension: the small scales are not resolved, and observed maps are smoothed out. This process, which makes the fractal look as a more diffuse medium of larger fractal dimension, can also lead to underestimates of the mass by factors more than 10 (e.g., simulations in Pfenniger & Combes 1994, and Fig. 16).

3.4 Turbulence

The interstellar medium is highly turbulent, its Reynolds number $R_e = vl/\nu$, where v is the velocity, l a typical dimension, and ν the kinematic viscosity, is of the order of $\sim 10^9$, because of its very low viscosity. The advection term $\mathrm{v} \cdot \nabla \mathrm{v}$ dominates the viscous term in the fluid equation.

In a first approach, the interstellar structures may be interpreted in terms of incompressible turbulence. Indeed the observed size-linewidth relation is

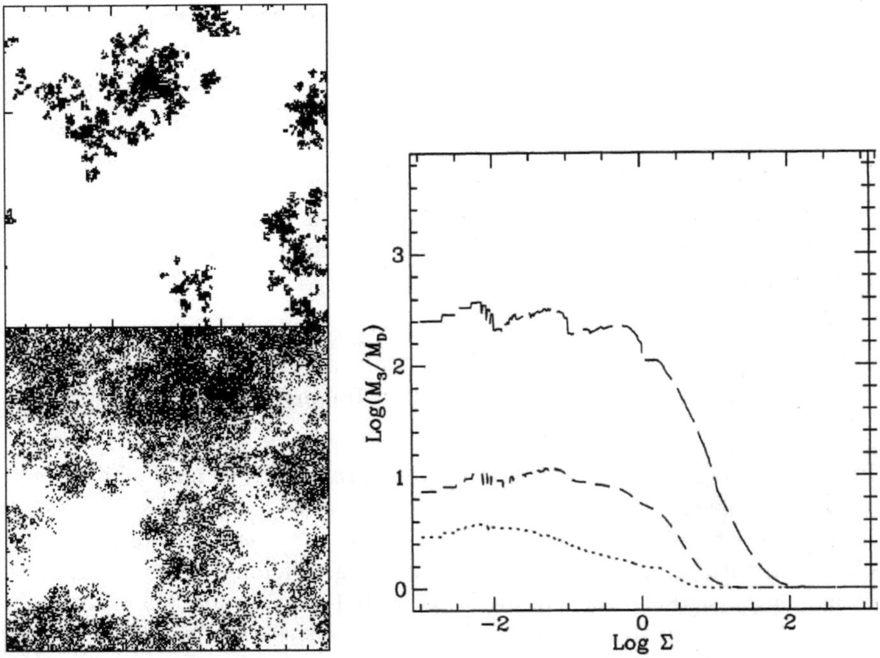

Fig. 16. (*Left top*) Simulations of a fractal of dimension $D = 1.6$ built by recursive fragmentation. (*Left bottom*) Same for a fractal of dimension $D = 2.2$, the number of points are similar in the two cases. (*Right*) Cumulated mass as a function of surface density Σ in various simulated models with different fractal dimensions. Are plotted the ratios (in log-scale) of what would be obtained for an homogeneous medium with $D = 3$ to the model with $D = 2.5$ (*dots*), $D = 2$ (*dash*) and $D = 1.5$ (*long dash*). This shows that estimated masses can be wrong by 2 orders of magnitude (from Pfenniger & Combes 1994).

very similar to what is expected in the Kolmogorov cascade (Kolmogorov 1941). In this picture, energy is dissipated into heat only at the lower scales, while it is injected only at large scale, and transferred all along the hierarchy of scales. Writing that the energy transfer rate $v^2/(r/v)$ is constant gives the relation

$$v \propto r^{1/3} \tag{18}$$

which is close to the observed scaling law, at least for the smallest cores (Myers 1983). The source of energy at large scale could be the differential galactic rotation and shear (Fleck 1981). This idealized view has been debated (e.g., Scalo 1987): it is not obvious that the energy cascades down without any dissipation in route (or injection), given the large-scale shocks, flows, winds, etc. observed in the ISM. The medium is highly supersonic (with Mach numbers larger then 10 in general), and therefore very dissipative. Energy could be provided at intermediate scale by stellar formation (bipolar flows, stellar winds, ionization fronts, supernovae, ...). Also, the interstellar medium is highly compressible, and its behaviour could be quite different from ordinary liquids in laboratory. However a modified notion of cascade could still be applied, leading to a Burgers spectrum, with

$$v \propto r^{1/2} \tag{19}$$

(see e.g. Vazquez-Semadeni et al. 1999).

Besides, many features of ordinary turbulence are present in the ISM. For instance, Falgarone et al. (1991) have pointed out that the existence of non-Gaussian wings in molecular line profiles might be the signature of the intermittency of the velocity field in turbulent flows. Also the curves obtained through 2D slicing of turbulent flows have the same fractal properties as the 2D projected images of the ISM; their fractal dimension D_2 obtained from the perimeter-area relation is also 1.36 (Sreenivasan & Méneveau 1986).

3.5 Numerical Simulations of Turbulence

The magnetic field may play an important role at small scale, through energy exchange between the various degrees of freedom, although it cannot prevent cloud collapse, and MHD simulations have been considered. The intensity of the field has been measured through Zeeman line effects to be around a few μG in dense clouds (Troland et al. 1996; Crutcher et al. 1999), and the observations are compatible with the hypothesis of equipartition between magnetic and kinetic energy.

A large number of hydro and MHD simulations have been run, however, they suffer in general of too low a dynamical range, due to huge computational requirements. To gain in spatial resolution, 2D or even less (because of symmetries) computations are performed, but often the results cannot be generalised to 3D.

When heating and cooling processes are included, and since the corresponding time-scales are faster than the dynamical time-scales, the gas can be described by a polytropic equation of state, as $P \propto \rho^\gamma$, with γ being the effective polytropic index. The isothermal value $\gamma = 1$ is one of the possibilities, between 0 and 2 found in simulations.

From the size-linewidth relation $\sigma \propto R^{1/2}$, and the second observed scaling law $\rho \propto R^{-1}$, it can be deduced that

$$\sigma \propto \rho^{-1/2} \tag{20}$$

and therefore, if the turbulent pressure P is defined as usual by,

$$\frac{dP}{d\rho} = \sigma^2 \ , \tag{21}$$

it follows that

$$P \propto \log \rho \tag{22}$$

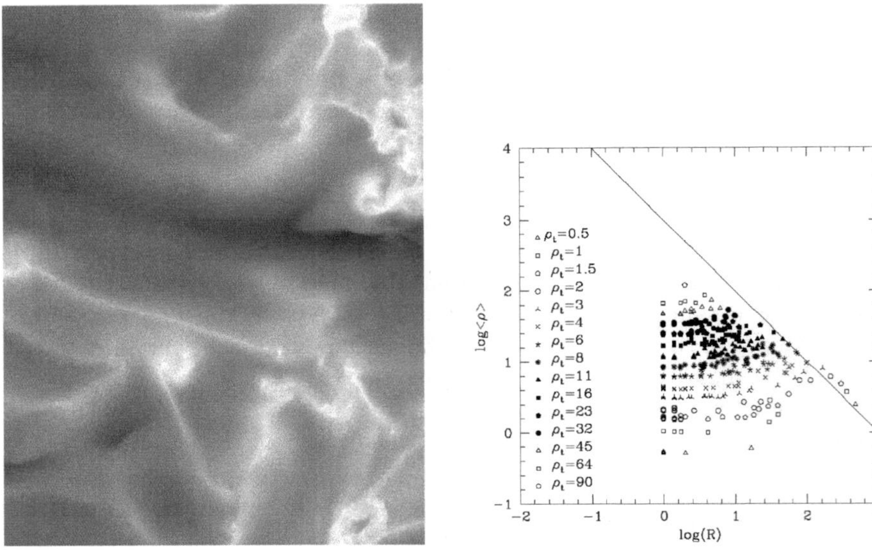

Fig. 17. (*Left*) 2D MHD self-gravitating simulations of a turbulent medium, with 800×800 resolution, from Vazquez-Semadeni et al. (1997). (*Right*) Size-density relation found in these simulations of turbulence. The straight line has power -1, as observed in the interstellar medium. In the simulations, there is a large range of scales, for a given density.

which is the logatropic equation of state, or "logatrope". This behaviour has been tested in simulations (e.g., Vazquez-Semadeni et al. 1998), but the logatrope has not been found adequate to represent dynamical processes occurring in the ISM (either hydro, or magnetic 2D simulations). The equation of state of the gas would be more similar to a polytrope of index $\gamma \approx 2$.

Vazquez-Semadeni et al. (1997) have searched for Larson relations in the results of 2D self-gravitating hydro (and MHD) simulations of turbulent ISM: they do not find clear relations, but instead a large range of sizes at a given density, and a large range of column densities (see Fig. 17); the Larson relation is more the upper envelope of the region occupied by simulated points in the $\rho - R$ diagram. They suggest that the observational results could be artifacts or selection effects (existence of a threshold in column density for UV-shielding for example).

Note also that chemical reactions network, combined with turbulence, can be the source of considerable chaos (Rousseau et al. 1998).

3.6 Self-gravity

In turbulent simulations with self-gravity, most power is on small scales due to gravitational collapse (Ossenkopf et al. 2001) in contrast to non-self-gravitating hydrodynamic turbulence where large-scale structures dominate.

Gravothermal Catastrophe: Self-gravitating gas in an isothermal regime is known to be subject to a gravothermal instability. A sphere of gas confined in a box, and in contact with a thermostat, will tend to follow an isothermal sphere distribution, if the gas is hot enough. Below a certain critical temperature, there is no equilibrium any more, and the gas heats up when being cooled down, it is the gravothermal catastrophe, caused by negative specific heat (Lynden-Bell & Wood 1968). Small sub-condensations could form, and the physics will be more complex, since in the asymptotic case of an isolated clump, it will evaporate in a large number of dynamical times. The environment is quite important, and the fact that the clumps can exchange mass and energy with surroundings as well (e.g. Padmanabhan 1990).

Gravitational instability and cloud collapse is accompanied by fragmentation in a system with very efficient cooling, and this process can provide the turbulent motions observed. The theory was first proposed by Hoyle (1953) who showed that the isothermal collapse of a cloud led to recursive fragmentation, since the Jeans length decreases faster than the cloud radius. Rees (1976) has determined the size of the smallest fragments, when they become opaque to their own radiation. They correspond roughly to the smallest scales observed in the ISM (sizes of 10 AU, and masses of $10^{-3}\,M_\odot$, see the physical parameters of the "clumpuscules" in Pfenniger & Combes 1994). The time-scale for the whole hierarchy of clouds to recursively fragment is of the same order of the free-fall time-scale of the biggest structure $\tau_{\rm ff}$. If the size-ratio between two successive scales is as before: $a = r_{L-1}/r_L = N^{-1/D}$, the time of collapse of the whole structures is:

$$\frac{\tau}{\tau_{\rm ff}} \sim 1 + \frac{1}{k} + \frac{1}{k^2} + \ldots = \frac{k}{k-1} \,, \tag{23}$$

with $k = (5\,\text{fragments})^{1/4} \sim 1.5$.

Simulations: Simulations of self-gravitating gas are very demanding, since large gradients rapidly set up, and spatial resolution must be adapted. A general rule is that the resolution is well below the Jeans length at any point, but the Jeans length shrinks along the collapse. Artificial fragmentation can sometimes happen due to artifacts (see e.g. Truelove et al. 1997). To compensate for the limited spatial range, periodic boundary conditions are used, to simulate the ISM. Klessen (1997) and Klessen et al. (1998) have considered the fragmentation of molecular clouds, in 3D, starting with an homogeneous cloud with small primordial fluctuations. These initial conditions are very similar to what is used in a cosmological background. The fluctuations are a Gaussian density field with power spectrum $P(k) \propto k^{-2}$. After one free-fall time, the gas has evolved into a system of filaments and knots, some of them contain collapsing cores. To avoid the problem of spatial resolution, the condensed cores are then replaced by sink particles, simulating therefore a low cut-off scale, cf. Fig. 18.

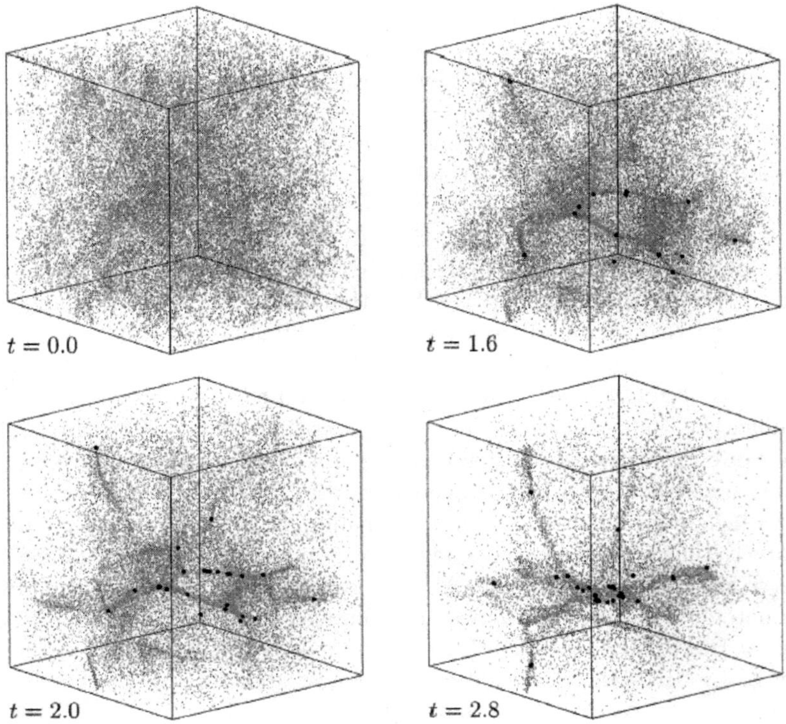

Fig. 18. SPH simulation of a molecular cloud with initially 222 Jeans masses, and initial Gaussian density fluctuations with power law $P(k) \propto k^{-2}$. The collapse results in highly condensed cores, which grow by accretion from the surrounding gas. These cores (*black dots*) contain about 10% of all the gas mass at $t = 1.6$ free-fall times, then at $t = 2.0$ and $t = 2.8$, 30% and 60% respectively (from Klessen et al. 1998).

Klessen et al. (1998) compute the mass spectrum of clumps, which looks very similar to the observed one $dN/dm \propto m^{-1.6}$ in the ISM, at least over 1.5 order of magnitude. The results however, depends still a lot on the initial conditions, and the power spectrum used (Semelin & Combes 1999). One of the problem of these simulations, in comparison with the ISM, is the absence of energy re-injection, and systematic motions. In a Galaxy, differential rotation and shear should provide both.

3.7 Molecular Fractal and IMF

Since young stars form out of molecular clouds, it is interesting to search for the mass spectrum of new stars implied by the fractal structure of the clouds, and their own mass spectrum.

Several theories have been proposed to explain the IMF, at least 4 types of models:

1. From the formation of a single star, variation of parameters to get several star masses (Larson 1973; Silk 1977; Zinnecker 1984); in particular, random variations with time and positions.
2. Stars form in clusters, and protostars interact with each other through collisions and coalescence (Bastien 1981; Murray & Lin 1996). Observations do show that most stars are born in clusters.
3. Star formation in clusters, with competitive accretion (Larson 1978; Tohline 1980; Myers 2000). A nearly uniform reservoir of gas is assumed. But dense clumps are observed (e.g. Motte et al. 1998), with a small filling factor and they don't favor accretion.
4. Observations favor stars forming in pre-formed clumps. Therefore the IMF does not give any clue in star formation, but only in cloud formation. Using random sampling of mass in a fractal hierarchy of molecular clouds gives the power-law IMF for stars, provided that a star's mass is proportional to the mass of the clump in which it forms (Elmegreen 1997). This power-law is steeper than the cloud mass spectrum because of competition for mass.

The IMF has two characteristic masses, and 3 slopes (Scalo 1998). The lower-mass characteristic scale at $0.3 - 1\,\mathrm{M}_\odot$ is much larger than the smallest cloud fragment, and must have formed by accretion and collapse of smaller clumps. The smallest mass for the fragments is that of opacity-limited clumps, of the order of $10^{-3}\,\mathrm{M}_\odot$. Accretion must operate to reach the first stellar mass, and may be form brown dwarfs in between (Rees 1976; Yoshii & Saio 1985).

The high-mass cut-off is due to a self-limiting accretion, due to stellar winds, and depends on ambient gas density and the mass of the cloud it is embedded. Stars grow until this self-limiting mass, a little larger than deuterium burning limit (Larson 1982; Shu et al. 1987). There must be a certain scatter between initial cloud masses and the final star mass, due to a series of unpredictable dynamical events, accretion, fragmentation, turbulence, self-gravity, ...

In the case of a pure hierarchy of clouds, containing equal mass per logarithmic interval, (see Sect. 3.3):

$$n(m)\, d\log m \propto m^{-1}\, d\log m \ . \tag{24}$$

The probability to have a random mass m is then:

$$m^{-1}\, d\log m \propto m^{-2}\, dm \ . \tag{25}$$

This is close to what is observed, $n(m) \propto m^{-1.85}$ (Heithausen et al. 1998).

Stars also would have this kind of law, except that they compete for gas. Once a star is formed at a certain level of the hierarchy, the higher scale has no longer any gas to form other stars. At a higher level, a star that normally would have contained much higher mass, because comprising a lot of subclumps, has now a mass lower than expected from the cloud hierarchy. This results in a steepening of the slope of the mass spectrum for the stars with respect to the clouds. Stellar clusters don't compete for mass, since they have the total mass of their components, and indeed they obey also the relation $n(m)\, d\log m \propto m^{-1}\, d\log m$.

Another parameter also steepens the IMF, the time-scale of star formation, which is larger for high masses: $\tau_{\rm ff} \propto \rho^{-1/2} \propto r^{1/2}$. Star formation is a dynamical process, and the fractal re-adjusts its structure during the process. The two factors for steepening can explain the Salpeter IMF $n(m)\, d\log m \propto m^{-1.35}\, d\log m$, instead of that of the pure hierarchy law, as shown by random sampling simulations (Elmegreen 1997, 1999, see Fig. 19).

If the simulations have no upper-mass limit, the IMF is prolonged to $M_{\rm up} = 7000\,{\rm M}_\odot$. But the maximum observed is $120 - 150\,{\rm M}_\odot$. This comes from an Eddington-limited accretion (Norberg & Maeder 2000), or oscillations and instabilities in massive stars. There could also be a problem of tim-

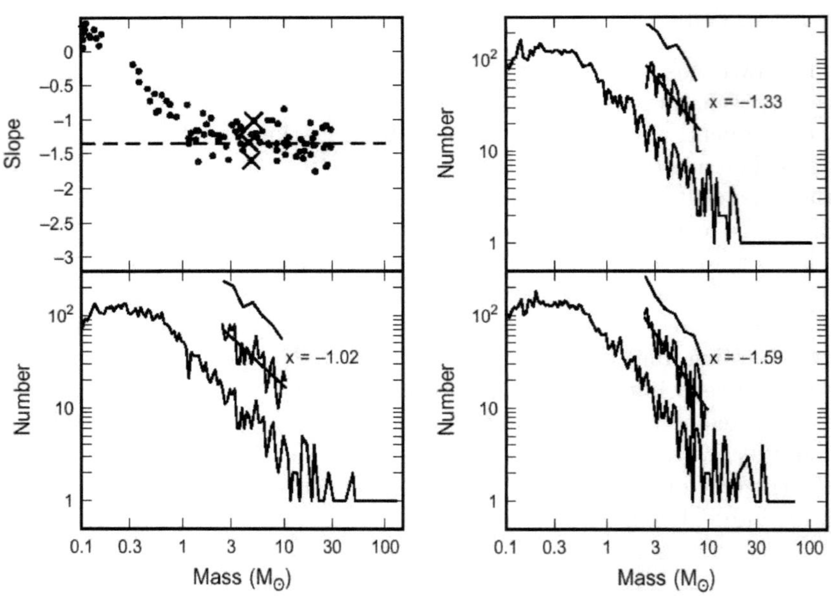

Fig. 19. Simulations of random sampling of mass in a fractal hierarchy of molecular clouds: (*Top left*) IMF slopes versus logarithm of mass, in 100 models that differ in the sequence of random numbers used to generate and sample the fractal cloud tree. Each IMF slope is fit using 200 stars. The Salpeter value of −1.35 is shown by a dashed line. The three values indicated by crosses have their complete IMFs shown in the other panels. The fitted portions of these IMFs are indicated by the offsets, with two different bin sizes shown (from Elmegreen 1999).

ing. After the smaller stars have formed, then a large fraction of the cloud is used up, and this leads to cloud destruction, well before a whole GMC can form a massive star.

Fractal Structure in Galactic Star Fields: Young stars just formed out of the fractal molecular clouds should reflect also a fractal structure, before diffusion and scattering occurs. Indeed archive HST images of 10 galaxies, Gaussian smoothed at various scales from 10 kpc to kpc along the spiral arms, show a hierarchical structure (Elmegreen & Elmegreen 2001). The densest structures are like the Pleiades, at the bottom of the hierarchy. The observations of NGC 2207 displayed in Fig. 20, show the relation on their sizes L in projection:

$$n(L)\, d\log(L) \propto L^{-D}\, d\log(L) \ , \qquad (26)$$

with $D = 1.12$. Models with a fractal structure in 3D, depending essentially on the center positions and not on the shape of the clouds, reproduce these observations. Note also that the number of HII regions in external galaxies obey $n(L) \propto L^{-2}\, dL$ (Kennicutt et al. 1989; Oey & Clarke 1998).

Stars form in the densest part of the fractal structure of the clouds. They keep the global fractal structure at least for a few dynamical times of the largest structures. The dynamical time is short at small scales (in $r^{1/2}$), where the stars have time to disperse, but at large scale the fractal survives.

3.8 Conclusions

Molecular clouds have a fractal structure, over 9 orders in mass and 6 orders in scale. The fractal dimension is $D = 1.7 - 2$. It has a filamentary geometry, and is highly hierarchical, almost a pure hierarchy according to the mass spectrum.

Turbulence and self-gravity are the key physical phenomena at the origin of the fractal structure. It is in this fractal structure of clouds that stars are formed, and the IMF may result from this structure.

4 Stability and Formation of the Fractal

4.1 Stability of the Molecular Disk

The conditions of stability of a dissipation-less component like the stellar one, is relatively well known. High enough velocity dispersions are required to suppress axisymmetric instabilities and even spirals, bars or z-instabilities. The latter provide heating to the medium which becomes un-responsive. The gas is dissipative and has a completely different behaviour; it is always unstable, and its velocity dispersion is fixed by regulation and feedback.

The local stability criterion has been established by Toomre (1964): stabilisation is obtained through pressure gradients (velocity dispersion c) at

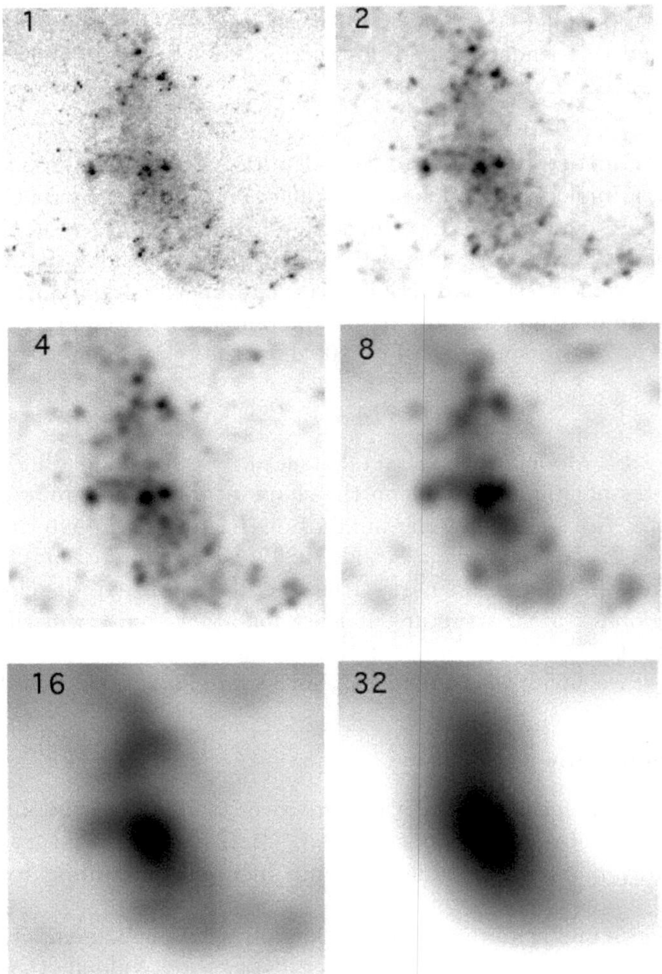

Fig. 20. Six levels of Gaussian smoothing of a star-forming patch in the galaxy NGC 2207. The number of pixels in the Gaussian smoothing function is shown in each panel (from Elmegreen & Elmegreen 2001).

small scale, smaller than the Jeans length: $\lambda_J = ct_{ff} = c/(2\pi G\rho)^{1/2}$. In presence of a self-gravitating disk, in the isothermal approximation, with surface density $\Sigma = h\rho$, the scale-height is given by $h = c^2/(2\pi G\Sigma)$. Then the Jeans length is $\lambda_J = c^2/(2\pi G\Sigma) = h$.

At large scale, the rotation stabilises through centrifugal forces, or the tidal shear prevents structure larger than L_{crit} to form. The tidal forces on a structure of size ΔR are

$$F_{tid} = \frac{d(\Omega^2 R)}{dR} \Delta R \sim \kappa^2 \Delta R \,, \tag{27}$$

(where κ is the epicyclic frequency). The internal gravity forces of the condensation ΔR are:

$$\frac{G\Sigma\pi\Delta R^2}{\Delta R^2} . \tag{28}$$

Equating this to $F_{\rm tid}$ yields the critical size, above which structures will be teared off:

$$L_{\rm crit} \sim \frac{G\Sigma}{\kappa^2} . \tag{29}$$

The Safronov-Toomre criterion is obtained in equaling $\lambda_{\rm J}$ and $L_{\rm crit}$:

$$Q = \frac{\kappa c}{\pi G\Sigma} > 1 . \tag{30}$$

For one component, a radial mode (ω, k) in a linear analysis obeys the dispersion relation (where k is the wave number): $\omega^2 = \kappa^2 + k^2c^2 - 2\pi G\Sigma k$, which means that self-gravity reduces the local frequency κ.

The tidal field of the galaxy is not always disruptive, it can also be compressive in the center, when there is a flat density core. If the mean density $\langle\rho\rangle$ of the spherical distribution is

$$\langle\rho\rangle = \frac{3M(R)}{4\pi R^3} , \tag{31}$$

then:

$$F_{\rm tid} = -\frac{d(\Omega^2 R)}{dR}\,\Delta R , \tag{32}$$

with $\Omega^2 R = GM(R)/R^2$ gives:

$$F_{\rm tid} = 4\pi G\left(\frac{2}{3}\langle\rho\rangle - \rho\right)\Delta R , \tag{33}$$

thus if the density is flat inside a certain radius (the core), the gas will be compressed, instead of being teared off. Note that the tides are always disruptive, in case of a power law density profile in $r^{-\gamma}$ with $\gamma > 1$ (Das & Jog 1999). In that case,

$$F_{\rm tid} = 4\pi G\left(\frac{2}{3-\gamma} - 1\right)\rho\,\Delta R . \tag{34}$$

Can this play a role in the formation of dense nuclear gaseous disks in starburst galaxies? In ultra-luminous galaxies (ULIRGs), the tidal field may become compressive inside 200 pc (see the case of Arp 193 in Fig. 21).

In the case of compressive tidal field, molecular clouds will not be teared off inside the core. However, the density to have a gravitationally bound cloud, leading to star formation, will be high. Indeed, an additional inward/compressive field increases the velocity dispersion required to satisfy the virial equilibrium, and hence the critical density required for stability. This minimum density is plotted in Fig. 21 for the example of NGC 4472, inside the core. Outside, since the tidal field is disruptive, the density plotted is the density below which a cloud will be disrupted by the tides.

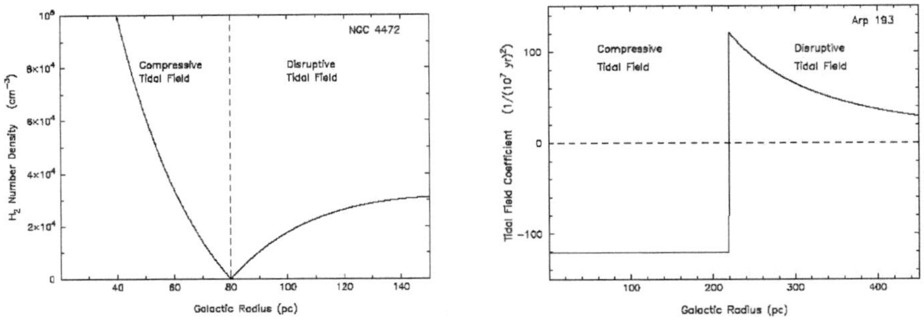

Fig. 21. (*Left*) Computed volume density for H_2 gas to exist in bound, virialized clouds versus radius for NGC 4472. The dashed line separates the outer region of disruptive tidal field from the inner region of compressive tidal field. (*Right*) Tidal field per unit mass versus radius for the ultraluminous galaxy Arp 193. The field becomes compressive inside 220 pc (from Das & Jog 1999).

Stability with Several Components: If a stability criterion is easy to derive for one component, the same is not true for a two-components fluid, since the coupling between the two makes the ensemble more unstable than each one alone. The one-component criterion can be applied separately to the stellar and gas components, where the corresponding values of Σ and c are used, leading to Q_s and Q_g. Because $c_g \ll c_s$ however, only a small percentage of mass in gas can destabilize the whole disk, even when $Q_s > 1$ and $Q_g > 1$. For two fluids with gravitational coupling, the dispersion relation yields a criterion of neutral stability $\omega^2 = 0$:

$$\frac{2\pi G k \Sigma_s}{\kappa^2 + k^2 c_s^2} + \frac{2\pi G k \Sigma_g}{\kappa^2 + k^2 c_g^2} = 1 \tag{35}$$

which gives directly an idea of the relative weight of gas and stars in the instabilities (cf. Jog 1996). At low k (long waves), essentially the stars contribute to the instability, while at high k (short waves), the gas dominates.

Stability and Vertical Structure: Stability criteria in general do not consider the thickness of galactic planes, which can be important factors of stabilisation. These can be represented by reduction factors for stars and gas $f_s = (1 + kh_s)^{-1}$ and $f_g = (1 + kh_g)^{-1}$ of scale-heights h_s and h_g respectively (cf Romeo 1992). Note that the stabilising effects depend on wavelengths $2\pi/k$. The vertical and planar heating mechanisms, and related self-regulation processes, are almost decoupled, as well as stability properties (see Fig. 22). This stabilisation is particularly important in the outer parts of galaxy disks, which are flaring.

Disks Are Marginally Stable: One striking observation involving gaseous galaxy disks is that they have structures at all scales, either small-scale

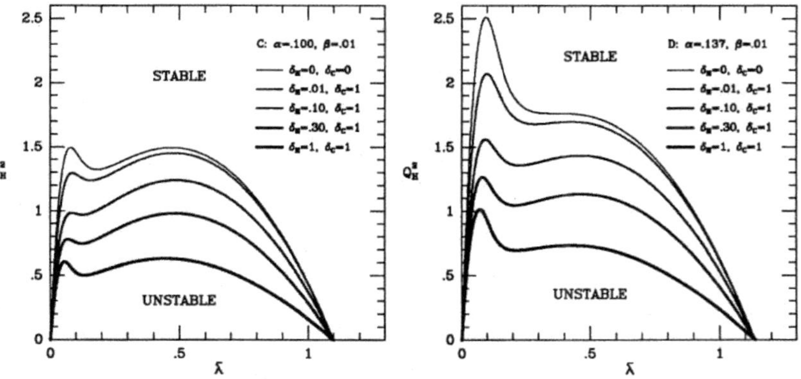

Fig. 22. Two-fluid stability curves, plotting the Q parameter as a function of reduced wavelength λ; (*Left*) Case of $\alpha = 0.1$, α being the surface density ratio between the cold (C) and hot (H) component. The various curves correspond to increasing δ_H, which is the z-velocity dispersion of the stars. $\delta_H = 0$ corresponds to the infinitely thin disk. The various curves precise how much stabilisation brings the thickening of the stellar disk. (*Right*) Same for $\alpha = 0.137$, corresponding to a larger gas fraction in a galactic disk, for instance. The low-λ instabilities are due to the gas (from Romeo 1992).

clumpiness, or spiral structure, etc. They appear therefore unstable. These instabilities could even be at the origin of the heating, since the large-scale velocity dispersion of the gas is remarkably constant with radius until the outer parts, only slightly higher in the center when there is active star formation there. It is therefore attractive to assume that disks are in some sense self-regulated by gravitational instabilities, and are at any time at the limit of stability, the gas dissipation being there to cool the disk down while instabilities are heating it.

This hypothesis is supported by the observation of the Q parameter over galactic disks, that does not show large variations, but on the contrary is almost constant over the disks. The stellar dispersion has been studied in our own galaxy and in a few external galaxies by Lewis & Freeman (1989) and Bottema (1993). The velocity dispersion decreases exponentially with radius, in parallel to the stellar surface density, and Q_s is nearly constant as a function of radius, at least for large galaxies, and vary between 2 and 3 from galaxy to galaxy.

How is established the stellar equilibrium? In a first approximation, the stellar disk can be considered as a self-gravitating infinite slab, locally isothermal (σ_z independent of z), and its density obeys the $sech^2(z/z_0)$ law in the z-direction, perpendicular to the plane. The characteristic scale-height z_0 is a function of the vertical velocity dispersion σ_z, and the stellar surface density $\Sigma_s(r) = z_0\rho_0$:

$$z_0 = \frac{\sigma_z^2}{2\pi G \Sigma_s(r)} \ . \tag{36}$$

A well-known observational result is that the stellar scale-height is a constant with radius in many edge-on external galaxies (van der Kruit & Searle 1981, 1982) or only slightly increasing (de Grijs et al. 1997). Since the radial surface density of stars is in general exponential in $\exp^{-r/h_r}$, with a radial length of h_r this means from the previous relation that the z-velocity dispersion is also exponential with radius, but with a radial length of $2h_r$. This is indeed compatible with the data, but with large uncertainties (Bottema 1993). Indeed, the z-velocity dispersion is not known for the edge-on galaxies, but the face-on ones, and the comparison requires to know the relation between z and r projection of the dispersion. In the solar neighbourhood, $\sigma_z = 0.6\sigma_r$, and this ratio is arbitrarily assumed to be valid in external galaxies too. Other assumptions is a constant mass to light ratio for the whole stellar disk, and no or little dark matter within the optical disk. These are first approximations, and it could be that in some galaxies the stellar equilibrium is reached with a scale-height slightly varying with radius (e.g. Narayan & Jog 2002).

The scale-height of the gas component is varying much more with radius, the gaseous plane is flaring first linearly, and even exponentially at large radii (Wouterloot et al. 1990). However, the vertical velocity dispersion is constant with radius in the outer parts of galaxies, where the rotation curve is flat (Dickey et al. 1990; Combes & Becquaert 1997): $c_g \approx 6\,\mathrm{km/s}$. The behaviour of $\kappa \propto 1/r$ (for a flat rotation curve) is somewhat similar to the gas surface density $\Sigma_g \propto 1/r$ (e.g. Bosma 1981), and therefore $Q_g \propto c_g\kappa/\Sigma_g$ is constant with radius in the outer parts, again supporting a regulation mechanism, in terms of gravitational instabilities (Lin & Pringle 1987).

4.2 Star Formation and Critical Surface Density

The Toomre criterion applied to gas has often been used as a criterion for star formation (Kennicutt 1989). The criterion $Q_g < 1$ for gas to be unstable can be inverted in the condition $\Sigma_g > \Sigma_{crit}$, i.e., a critical surface density for star formation. This criterion is global, not local.

So far, only global quantities have been correlated with success, when the gas surface density is averaged out over the whole galaxy, and the same for the star formation rate. The star formation tracer can vary, from the Hα flux for normal galaxies, to the Far Infrared luminosity L(FIR) for starbursts, which are highly obscured (Kennicutt 1998).

While the starbursts explore a wider range and dynamics of parameters, the relation between the global gas surface density and star formation rate (SFR) is the same for extreme and normal galaxies: it is possible to derive a "global" Schmidt law, with a power $n = 1.4$ (Kennicutt 1998)

$$\Sigma_{SFR} \propto \Sigma_{gas}^{1.4} \ , \tag{37}$$

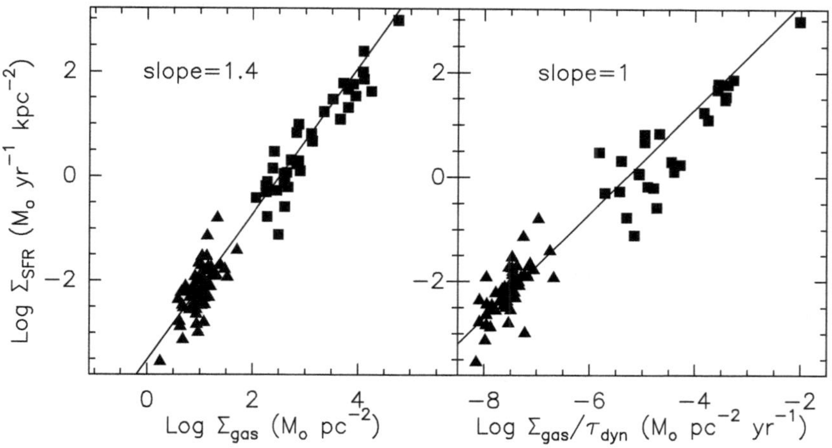

Fig. 23. Relations between Star Formation Rate (SFR) and gas surface densities, from Kennicutt (1998). (*Left*) A global "Schmidt" law, with a slope of $n = 1.4$ as the best fit. (*Right*) Linear relation between SFR and gas over dynamical time ($\tau_{\rm dyn}$) ratio. Normal galaxies are the triangles, while starburst galaxies are the squares.

(cf. Fig. 23). Another formulation works as well

$$\Sigma_{\rm SFR} \propto \Sigma_{\rm gas}\Omega \propto \frac{\Sigma_{\rm gas}}{t_{\rm dyn}} , \tag{38}$$

where Ω is the angular frequency in the galaxy, which is inversely proportional to the dynamical time-scale $t_{\rm dyn}$.

Several justifications can be found a posteriori: if the star formation is locally due to the gravitational instability of the gas, this occurs on a free-fall time-scale, and the star-formation rate is:

$$\frac{d\rho_*}{dt} = \rho_{\rm SFR} \propto \frac{\rho_{\rm gas}}{\tau_{\rm ff}} \propto \rho_{\rm gas}^{1.5} , \tag{39}$$

very close to the power $n = 1.4$; but the correlation is not observed locally. Globally, this applies also, if the star formation is due to the global gravitational instability of the gas disk, that occurs in a dynamical time-scale:

$$\frac{d\Sigma_*}{dt} = \Sigma_{\rm SFR} \propto \frac{\Sigma_{\rm gas}}{t_{\rm dyn}} , \tag{40}$$

which accounts for the second formulation. Alternatively, star formation could be triggered in marginally stable clouds, by the crossing of spiral arms, and the frequency of arm crossing is proportional to $\Omega - \Omega_p$ (Wyse & Silk 1989), or roughly to Ω far from corotation (where the clouds never cross the arms).

This second formulation might also explain the Tully-Fisher relation (Silk 1997; Tan 2000), since if $\Sigma_{\rm SFR} \propto \Sigma_{\rm gas}\Omega$, then

$$L_b \propto \Sigma_{\rm SFR} R^2 \propto \Sigma_{\rm gas} v_{\rm circ} R \tag{41}$$

with $v_{\rm circ}^2 \propto \Sigma R$ from the virial, and provided that $\Sigma_{\rm gas} \propto \Sigma_*$ is verified over the main spiral classes (Roberts & Haynes 1994), it can be deduced that $L_b \propto v_{\rm circ}^3$.

Problems with the Use of Q_g: As described before, gravitational instabilities are very efficient on a dynamical time-scale, to self-regulate the equilibrium of disks, and therefore the values of Q are almost constant over galactic disks. Only in very violent off-equilibrium cases, such as galaxy interactions, mergers, leading to starbursts, is the equilibrium transiently broken (e.g. Combes 2001).

Another difficulty is that if gravitational instabilities are necessary for star formation, they are not sufficient. There must be other essential parameters controlling the onset of star formation, since in outer parts of spiral galaxies, and in the disks of dwarf irregulars, the gas medium has formed self-gravitating structures, but without star formation.

There must be for instance a time-scale condition to have a starburst: gas must be gathered in a very short time-scale, smaller than $\sim 10^7$ yr, shorter than the onset of feedback from the first OB stars formed, through supernovae explosions and stellar winds, before the starburst can blow the gas out. Also the volumic gas density might be considered. With the same surface density, a flaring gas plane should not have the same efficiency to form stars.

4.3 Small-Scale Stability

The stability of the smallest observed molecular clumps has always been a puzzle. Their free fall time is much less than 1 Myr, and the rate of star formation should be much larger than observed. Pressure support is necessary, since magnetic fields cannot halt the collapse.

For an isothermal gas, that radiate efficiently, fragmentation cannot be stopped until the fragments are so dense that they become optically thick, and shift in the adiabatic regime. Without external perturbations, the smallest fragment when this occurs is about 10^{-3} M$_\odot$. The isothermal regime breaks down when the free-fall time $t_{\rm ff}$ becomes shorter than the Kelvin-Helmholtz time-scale t_{KH}, or time-scale to radiate the thermal energy by radiation:

$$t_{\rm ff} \sim t_{KH} = 3/2NkT/L \quad \text{with} \quad L = 4\pi f R^2 \sigma T^4 \,, \tag{42}$$

where f is a factor of the order of one, depending on departure from black-body radiation, σ is the Stefan-Boltzmann constant. This leads to a fragment mass (μ being the mean molecular weight) of:

$$M = 4\,10^{-3} T^{1/4} \mu^{-9/4} f^{-1/2} {\rm M}_\odot \,, \tag{43}$$

(e.g. Rees 1976). The average characteristics of these smallest fragments are thus $10^{-3}\,M_\odot$ in mass, $10^{10}\,cm^{-3}$ in density, 20 AU in radius, $N(H_2 \sim 10^{25}\,cm^{-2}$ in column density. For these clumps, the free-fall time is $t_{ff} \sim 1000\,yr$, but they are optically thick, in the adiabatic regime, and the pressure support ensures that their life-time is much longer. If these clumps are part of a fractal, clump-clump collisions lead to coalescence, heating, and to a statistical equilibrium with fragmentation (Pfenniger & Combes 94).

On the observational side, all situations are encountered: dense cores with isolated star formation, dense cores with clustered star-formation, dense cores without any star formation. The triggering of star formation could be due to un-balanced time-scales. An external perturbation brings a non-linear increase of velocity dispersion, due for instance to galaxy encounters. These trigger collisions, then either coalescence, or shredding and increase of velocity dispersion. If there is a time-delay between the formation of massive clouds leading to star formation, and the star formation feedback, then a starburst may be triggered.

Modelisation have been studied in the literature, trying to explore the numerous parameters: cooling of the gas, fresh supply of gas, etc.... Limit cycles appear, and often a chaotic behaviour (e.g. Hirashita & Kamaya 2000).

4.4 Gas in the Outer Parts

Observationnally, the gas in the outer parts of galaxy disks is stable with respect to star formation, although not to gravitational perturbations. Examples are to be found in HI-21cm maps, with clumpy structure and large-scale spiral structure, like in the giant spiral M101, or the dwarf irregular NGC 2915, and similar conditions are encountered in Low Surface Brightness Galaxies (LSB). These environments are evidence that star formation and gravitational instability have not exactly the same criterion. Several arguments could be advanced to explain the non-star formation: either a low volumic gas density, due to a thick disk, or a flaring in the outer parts, or the lack of external perturbations and companions for LSB.

A quite prototypical case is the dwarf galaxy NGC 2915 (Bureau et al. 1999). It is a blue compact dwarf galaxy, with a very extended HI disk, up to 22 optical scalelengths (see Fig. 24). The HI disk shows a central bar and extended spiral arms, that appear a coherent density wave. Both bar and spiral extend farther than the optical component. The velocity field reveals characteristic streaming motions. As in all dwarfs of these types, the rotation curve shows that the galaxy is dominated by dark matter all over the disk. The mass of the stars is smaller than the mass of gas, itself dominated by dark matter.

How can the bar plus spiral pattern be explained? If the dark matter is assumed as a quasi spherical halo, dominating the potential, than the disk is not self-gravitating enough to be instable to such a perturbation. The Toomre Q value varies between 8 and 20 over the disk. A solution would be

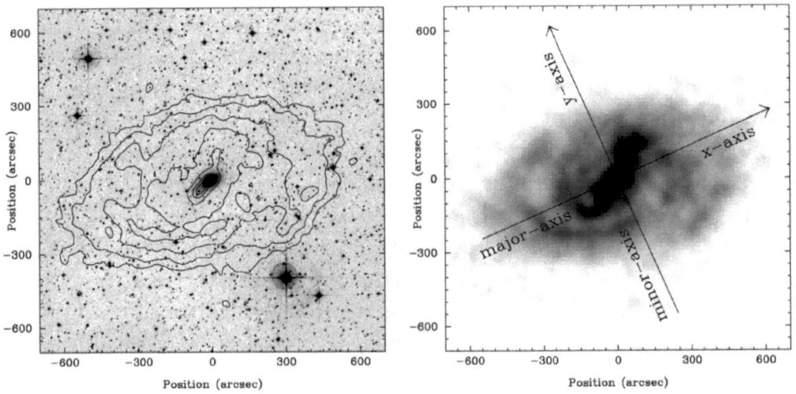

Fig. 24. (*Left*) HI-21cm map of the dwarf galaxy NGC 2915 (contours), superposed on an optical photograph from Digitized Sky Survey; (*Right*) Grey scale image of the HI map, showing better the bar and spiral structure (from Bureau et al. 1999).

to assume that a large fraction of the dark matter is baryonic, and follows the disk distribution as cold gas (Pfenniger et al. 1994, Pfenniger & Combes 1994). Multiplying the HI surface density by a factor around 10 is in general sufficient to explain the rotation curves (e.g. Hoekstra et al. 2001). This solution indeed accounts for the formation of the bar plus spiral pattern, and has also the swing parameter $X = \lambda/\lambda_{crit}$ (with $\lambda_{crit} = 4\pi^2 G\Sigma/\kappa^2$) quite optimum (Bureau et al. 1999). But the main difficulty for this solution is to explain the non-star formation.

Another possibility to trigger bar formation is to assume that the non-baryonic dark matter halo is triaxial, and slowly tumbling. CDM numerical simulations have shown that halos are sometimes triaxial, although frequently oblate and axisymmetric in the baryonic plane (Katz & Gunn 1991; Dubinski 1994). In any case, they are rotating very slowly, about an order of magnitude less than normal baryonic bars in galaxies, and the HI bar in NGC 2915 should also have a very slow pattern speed (Bureau et al. 1999). Such a model has been simulated recently by Bekki & Freeman (2002) who also require the inclination of the disk with respect to the plane including the triaxial halo's long and middle axes. Simulations have also been done of a baryonic disk dark matter, a model than can reproduce the observed pattern quite satisfactorily (Masset & Bureau, 2002).

4.5 Formation of the Structures

How to form and stabilize the hierarchical structure of the H_2 gas? Self-gravitating simulations have already shown that recursive fragmentation occurs, and forms transiently a self-similar structure, however all the mass is then collapsing into dense cores (Klessen et al. 1998).

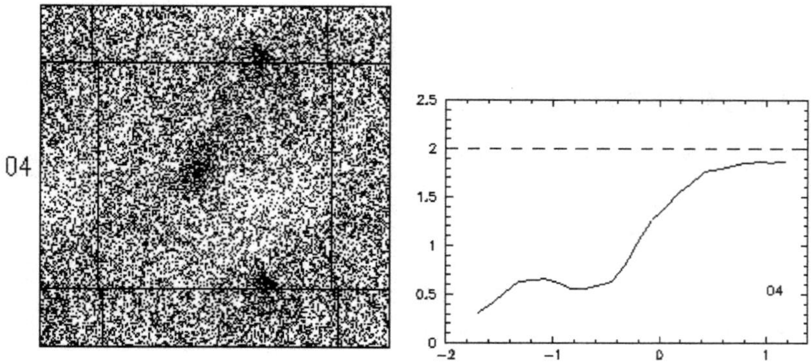

Fig. 25. (*Left*) Snapshot of a 2-D simulation of formation of structures in the galactic shear. Self-gravity and dissipation of the gas are taken into account. The galactic center is toward the bottom of the box. (*Right*) Fractal dimension computed at the same epoch, as a function of scale (from Semelin & Combes 2000).

N-body simulations were carried out (Semelin & Combes 2000), with similar initial conditions, periodic boundary conditions, and a perturbation spectrum $P(k) \propto k^{-\alpha}$, but unlike previous simulations, there is a schematical process to shift to adiabatic regime at low scale. The code used was the tree-code, with sticky particles included. The scheme to stop the dissipation and fragmentation at the smallest scale was varied, either by elastic collisions, or even super-elastic collisions, to re-inject energy at small-scales. A fractal structure is indeed observed, but is not stationary. Energy re-injection is required also at large scale, to renew the structure.

One way to re-inject energy at large scale is the differential rotation of the Galaxy. Computations of self-gravitating shearing sheets have already given interesting large-scale structures, that might be similar to fractals (Toomre 1990).

When galactic shear (and Coriolis forces) are taken into account in molecular cloud formation simulations, the structures are formed over several orders of magnitude. The characteristic spiral filaments are observed, and embedded clumps form, and can disappear and reform stochastically. Constantly the shear destroys the small clumps, that reformed again. Simulations of a shearing sheet are shown in Fig 25. The fractal dimension varies with scale and with the epoch in the simulation. It is lower at small scale (the clumps) and larger (around 1.8) just below the sheared filaments scale.

The work was generalised by Huber & Pfenniger (2001), who made the computations in 3D, taking into account the galaxy plane thickness of the gaseous component. Their cooling is simulated by a viscous force, proportional to the particle velocity. They also find the characteristic filaments due to the shear combined with self-gravity, and measure the corresponding fractal dimension of the structures. Since the filaments are much longer in the

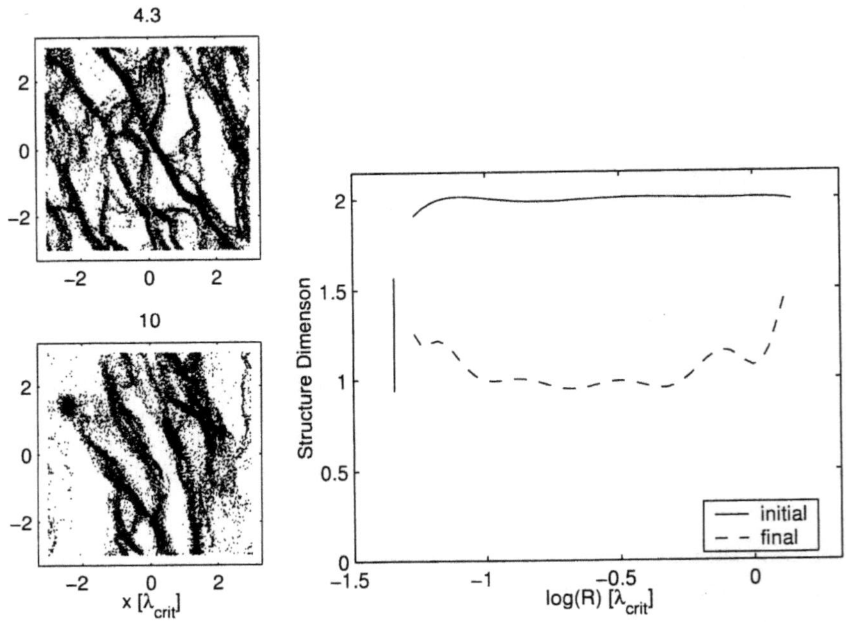

Fig. 26. (*Left*) Evolution of a shearing box, at 4.3 and 10 rotations around the galactic center, with a "medium" dissipation for the gas; (*Right*) Fractal dimension as a function of scale, for different values of the dissipation; the fractal dimension is smaller for strong dissipation (from Huber & Pfenniger 2001).

plane than thick (perpendicular to it), the fractal dimension is in fact dominated by the z-dimension plane morphology, at least at large scales. The measured fractal dimension is highly dependent on the amount of dissipation (see Fig. 26), since stronger dissipation enhances the formation of clumps in the sheared filaments. Although the dynamical range of simulations until now is too small to probe entirely a true fractal distribution, these calculations are encouraging that a fractal structure could develop at sub-kpc scales, when shearing is dominant, near the high cut-off of the ISM fractal.

Wada & Norman (1999) have also simulated a shearing galaxy disk, but with a multi-phase medium: they obtain the formation of dense clumps and filaments, surrounded by a hot diffuse medium. The quasi-stationary filamentary structure of the cold gas is another manifestation of the combination of shear and self-gravity in a cooled gas.

4.6 Conclusions

Gaseous disks, and in particular the H_2 component, are not in dynamical equilibrium or marginally, and are self-regulated by gravitational instabilities.

They are unstable at all scales and subject to spiral structure, filaments, clumpy hierarchical structure.

To explain the fractal structure of the cloud ensemble, self-gravity is required, together with injection of energy at large-scale, and may be at small scale. The galactic rotation is the main source of energy, and it takes Gyr for the gas in a galactic disk to flow slowly to the center. These radial flows can be faster in the case of external perturbations, such as galaxy interactions.

The presence of gravitational instabilities, for cloud and structure formation, is necessary but not sufficient for star formation. Other processes must be considered to trigger star formation, which would explain the large extensions of unstable gas without stars in the outer parts of galaxies.

5 H_2 in External Galaxies

5.1 H_2 Content and Morphological Types

There are several published surveys of CO emission in galaxies, in particular more than 300 galaxies have been observed in the FCRAO survey (Young & Knezek 1989), where the average molecular content is found comparable to the atomic one $M(H_2/M(HI\sim 1$ (Young & Scoville 1991). Part of the galaxies may have a bias, in particular some are selected through the 100μm IRAS flux, which is in general well correlated with CO emission. Another survey by Casoli et al. (1998) near the Coma cluster has shown on average $M(H_2)/M(HI) \sim 0.2$.

The relative molecular content varies strongly with morphological type: while the HI content is more abundant in late-type galaxies, the H_2 content has the opposite trend. The apparent H_2/HI mass ratio decreases monotonously from S0 to Sm galaxies by a factor near 10 (Young & Knezek 1989; Casoli et al. 1998), as shown in Fig. 27. This gradient towards the late types might be only due to a reduced CO emission, because of metallicity effects, and not to an intrinsic reduction of the H_2 content. When only the more massive objects are taken into account, this tendency with type disappear: there is no gradient of molecular fraction. This supports the hypothesis that the gradient is due to metallicity, which is correlated with total mass.

The relation linking mass M and metallicity Z is observed of the form $Z \propto M^{1/2}$ (e.g. Zaritsky 1993; Tamura et al. 2001), in agreement with models (e.g. Yoshii & Arimoto 1987). Such a relation is due to the fact that the star formation rate is non-linear with gas density, and deeper potential wells retain more efficiently their stellar ejecta.

The strong dependency of the CO-to-H_2 conversion factor X on metallicity has been confirmed now in many galaxies. In the Small Magellanic Cloud, X could be 10 times higher than the standard value of $2.3 \cdot 10^{20}$ mol cm^{-2} (K km/s)$^{-1}$ (Rubio et al. 1993). The effect has been seen also in Local Group galaxies, such as M31, M33, NGC 6822 and IC 10 (Wilson 1995). The physical explanation is complex, since the CO lines are optically thick,

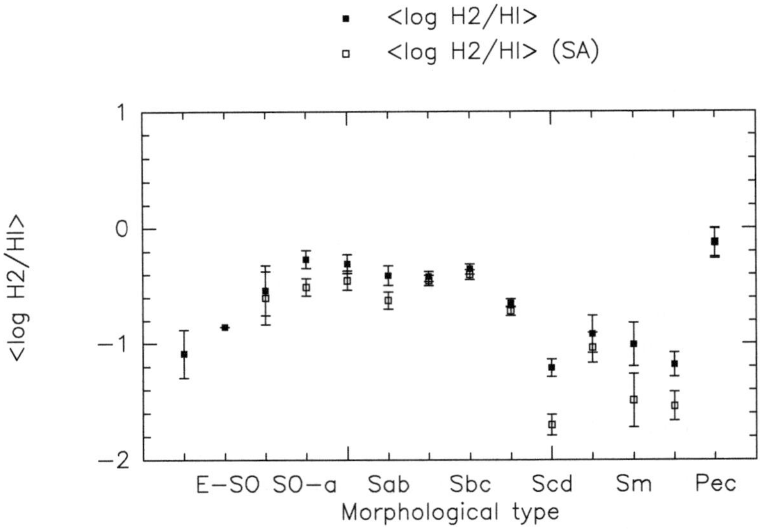

Fig. 27. H_2/HI mass ratio in galaxies as a function of morphological type; full squares: mean values with upper limits treated as detections; empty squares: mean values with upper limits taken into account using the survival analysis (SA), from Casoli et al. (1998).

but the size of the clouds (and therefore the filling factor) decreases with metallicity, both due to direct CO abundance, and UV-phtotodissociation increased by the depletion of dust. When the dust is depleted by a factor 20, there is only 10% less H_2 but 95% less CO (Maloney & Black 1988).

CO in Dwarf Galaxies: In dwarf irregular galaxies, rich in HI gas, CO emission is very low, and it is difficult to know the H_2 content. If the HI/H_2 ratio is assumed constant from galaxy to galaxy, then X varies with $Z^{-2.2}$ (Arnault et al. 1988). Dwarf Ellipticals are easier to detect than Dwarf Irregulars (Sage et al. 1992). Observations by Gondhalekar et al. (1998), Taylor et al. (1998), and Barone et al. (2000) confirm this strong dependency, and conclude that the conversion factor must be varying nonlinearly with metallicity, increasing sharply below $\sim 1/10$ of the solar metallicity [12+log(O/H) ≤ 7.9] (cf. Fig. 28).

Low Surface Brightness Galaxies: Low Surface Brightness (LSB) galaxies have in general large characteristic radii, large gas fraction (up to $f_g = 95\%$ in LSB dwarfs, Shombert et al. 2001), and are dark matter dominated. They are quite un-evolved objects. Their total gas content is similar to that of normal (HSB) galaxies (McGaugh & de Blok 1997). They have also low surface densities of HI, although with large sizes, and in their physical properties they resemble the outer parts of normal HSB galaxies.

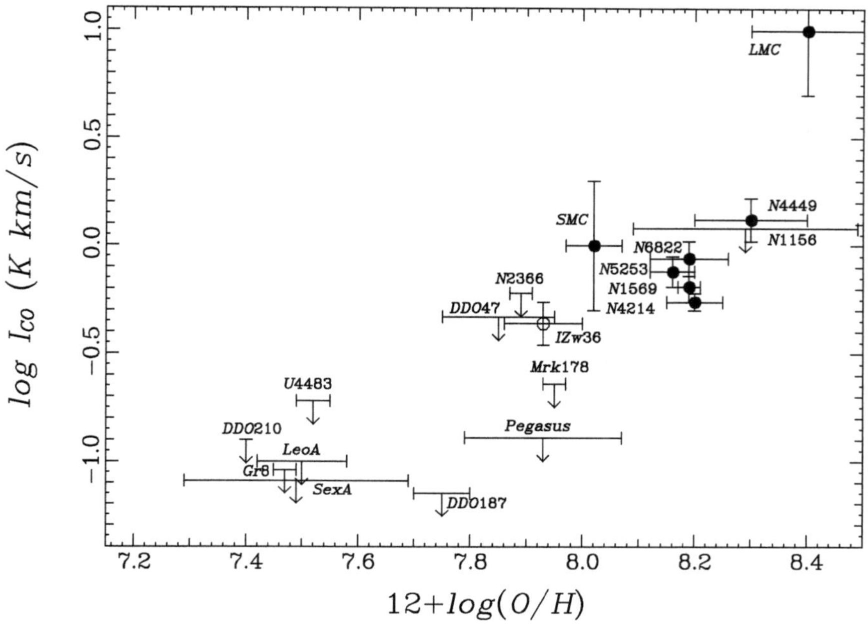

Fig. 28. Dependence of the CO integrated intensity $I_{\rm CO}$ (normally proportional to the average H_2 surface density $N(H_2)$ on the oxygen abundance, or metallicity (from Taylor et al. 1998).

Their metallicity is low, according to the correlation of metallicity with surface brightness (Vila-Costas & Edmunds 1992). It is therefore no surprise that CO emission is difficult to detect in LSB (de Blok & van der Hulst 1998). Only in edge-on LSB galaxies CO has been detected (Matthews & Gao 2001), and the derived $M(H_2)/M(\rm HI)$ is between $1 - 5 \cdot 10^{-2}$ (with standard conversion ratio). Due to their low surface density, these galaxies have a very low efficiency of star formation (van Zee et al. 1997). A gas rich galaxy is stable only at very low $\Sigma_{\rm g}$ (cf. Malin 1, Impey & Bothun 1989). The cause of their low evolution rate could be the absence of companions, since LSB live in poor environments (Zaritsky & Lorrimer 1993). It is well known that galaxy interactions, by driving in a high amount of gas, trigger star formation.

Since their discovery, the recognition that a large number of LSB exist change the luminosity function of galaxies; LSB could be a significant reservoir of baryons in the local universe (Bothun et al. 1997). Their low evolution time-scale is compatible with models (Boissier & Prantzos 2000).

It is important to note that LSB obey the same Tully-Fisher relation as normal galaxies; for the same rotational velocity V, these galaxies are twice as large in R, and according to the Virial, $M \propto V^2R$ and the mass-to-light ratio M/L increases as surface density decreases. In general, the

Tully-Fisher relation for gas-rich galaxies works much better in adding the gas mass to the total mass of the galaxies, in some sense including the gas mass in the "luminosity". It appears that the TF-relation is a link between the baryonic mass and the rotational velocity (it is the "baryonic TF", McGaugh et al. 2000).

5.2 Molecular Disks Structure

Radial Distribution in Spirals: The atomic and molecular gas (traced by CO) have very different radial distributions (cf. Fig. 29). The CO emission, like all components related to star formation, follows an exponential distribution, while the HI gas alone is extending much beyond the "optical" disk, in average $R_{\mathrm{HI}} = 2R_{\mathrm{opt}}$. The HI is often depleted in the center of galaxies, and sometimes the depletion is compensated by H_2, but it is not always the case (e.g. M31). The larger sample of radial distributions reported in the BIMA-SONG survey (Regan et al. 2001) confirm these conclusions.

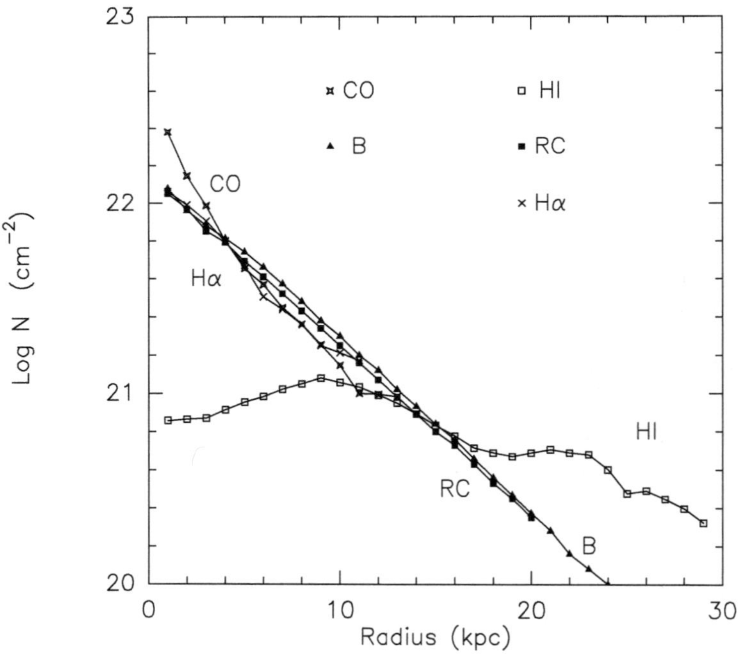

Fig. 29. Radial distributions of various surface densities in a typical spiral galaxy NGC 6946: H_2(CO) and HI column densities, Blue, Radio-continuum and Hα surface densities (adapted from Tacconi & Young, 1986).

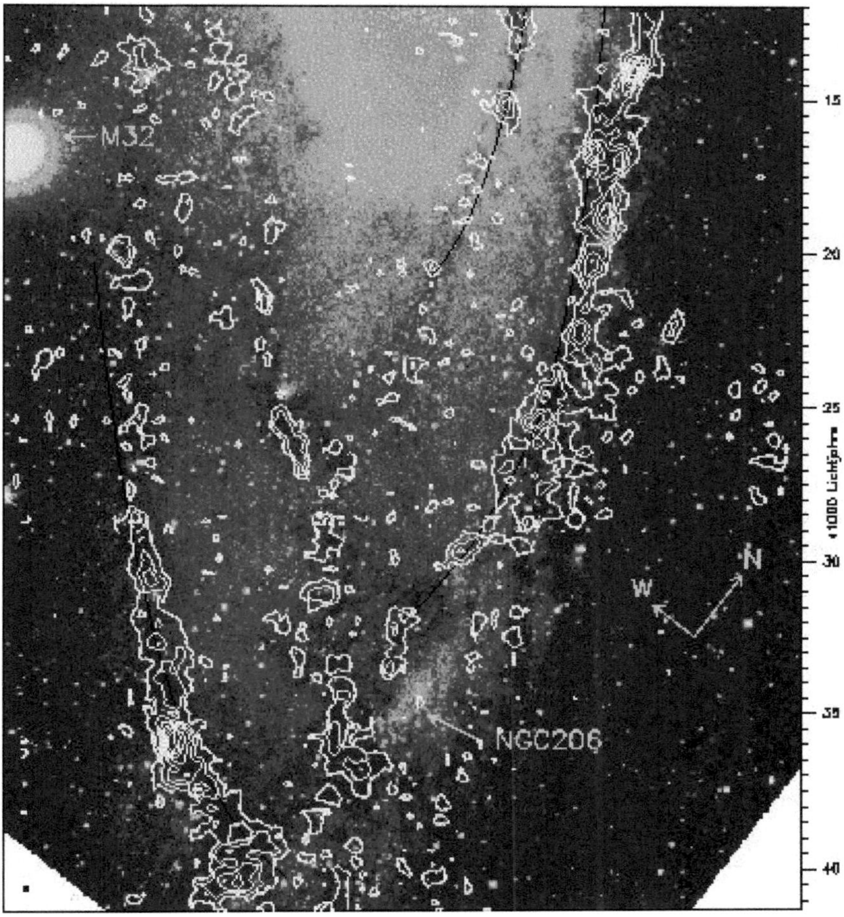

Fig. 30. Integrated CO intensity $I(\mathrm{CO})$ contours, superposed on the Digital Sky Survey (DSS) red image of M31. The black dot in the bottom left corner represents the telescope 23" beam (from Neininger et al. 1998).

Spiral Structure: The H_2 component participates even better than the HI and stellar component to the density waves. Due to its low velocity dispersion, it is more easily trapped in the spiral potential wells, and the structure is more contrasted. Large streaming motions are detected at the crossing of the spiral density wave. A good illustration is plotted in Fig. 30, with the CO(1 − 0) map of M31, with 23" (90 pc) resolution, by Neininger et al. (1998). The on-the-fly mapping procedure allows to reach a large dynamical range in spatial scales. The CO emission is particularly well correlated to the dust lanes and more generally to HI in the arms.

The particular large CO emission contrast may be explained also by other factors, such as:

- different CO excitation in arms (larger density),
- heating due to the enhanced star formation,
- enhanced GMC formation in arms,
- enhanced H_2 or CO formation (chemical time-scale of 10^5 yr),
- HI is formed out of photo-dissociation of H_2, just at the edge of the arms,
- more self-shielding of CO in the arms.

The CO emission appears to follow the dust lanes, in the concave side of spiral arms (cf. M51, Tilanus & Allen 1991), and is often gathered in clouds aligned as pearls on the spiral string.

5.3 H_2 in Barred Galaxies

CO emission is particularly useful to map the bar and rings in spiral galaxies, since HI is in general depleted in central regions, and Hα is often obscured.

Barred galaxies have more CO emission, and the derived H_2 gas is more concentrated to the circum-nuclear regions than in unbarred galaxies (Sakamoto et al. 1999). This confirms dynamical theories of transport of the gas by bars (e.g. Combes 2001). More than half of the gas in the central kpc comes from the large-scale galactic disk and is too high in surface density to have been consumed through star formation, the present rate being $0.5-1 M_\odot yr^{-1}$. There is no correlation, however, with any AGN activity (Seyfert, Liner, etc...), although bars have been invoked as a mechanism to fuel the central black holes (e.g. Mulchaey & Regan 1997).

CO maps of barred galaxies reveal all kinds of morphologies, rings, gas bars, spiral structure, characteristic "twin peaks", where the gas accumulates at the crossing of parallel and perpendicular orbits, at the start of the leading offset dust-lanes. The CO gas traces particularly well the resonant rings (e.g. Buta & Combes 1996). That the CO traces the offset dust-lanes is conspicuous in the well studied barred galaxy NGC 1530 (Reynaud & Downes 1997).

In a small sample of barred galaxies, Sheth et al. (2002) have observed a spatial offset of 300 pc in average between Hα arms and CO arms; this is interpreted as a time-delay between gas accumulation and star formation, and depends on the gas flow. The gas flows can inhibit star formation, when they are too fast (Reynaud & Downes, 1999). The star formation is favored in ring accumulations, or nuclear bars.

Two embedded bars, with different pattern speeds are sometimes required by observations of morphology and dynamics, as in the example of M 100 (Garcia-Burillo et al. 1998a, Fig. 31). Sometimes, the gas in the center is also observed at high altitude above the plane (cf. NGC 891, 5907, or the peanut-shaped galaxy NGC 4013, Garcia-Burillo et al. 1999). These might be accounted for by processes associated to star-formation, more than purely dynamical mechanisms. The only exception could be gas in retrograde orbits.

The CO gas has been detected also in counter-rotating galaxies, as in the early-type spiral NGC 3626 (Garcia-Burillo et al. 1998b), where the H_2 is

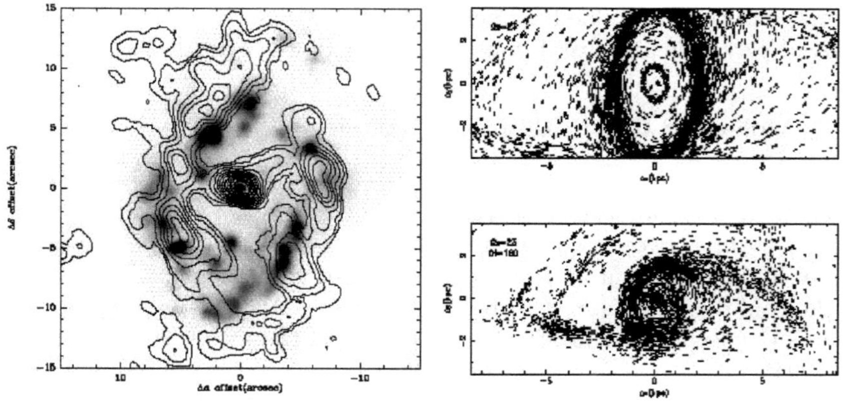

Fig. 31. (*Left*) Integrated CO emission contours obtained towards M100 (NGC 4321) with the IRAM interferometer, overlaid on an Hα grey scale image. (*Right*) Particle orbits of molecular clouds in a simulation with only one pattern speed (*top*, $\Omega = 23$ km/s/kpc), and with two different pattern speeds for the two embedded bars (*bottom*, $\Omega = 23$ and 160 km/s/kpc) from Garcia-Burillo et al. (1998a). Only in the second case, a nuclear spiral structure is obtained, similar to the one observed in CO (*left*).

concentrated in a compact nuclear disk of ~ 1 kpc radius, rotating in a sense opposite to that of the stars, and in the same sense as the HII and HI gas. The counter-rotation gives rise to 2-streams instability, and the formation of $m = 1$ leading spiral structure (example NGC 3593, Garcia-Burillo et al. 2000).

Molecular gas can be expelled out of the galactic plane in rare circumstances, in violently interacting systems: e.g. NGC 4438 in the Virgo cluster (Combes et al. 1988), NGC 4631/56 (Neininger 2000).

5.4 Molecular Gas in Polar Ring Galaxies

Polar Ring Galaxies (PRG) are due either to accreted gas from a companion or are formed in a merger of two galaxies with orthogonal disk orientations (e.g. Bekki 1998). PRG are used to determine the flattening of the dark matter, since the kinematics of the polar material probe the 3D potential.

The polar ring is due to gas resettling in the polar plane, but stars are dispersed in a wide-scale halo. The ages of the stars in the polar ring thus date the accretion/merger event. Frequently large amount of HI gas are detected in the polar planes, rotating around the central early-type galaxies. This gas is likely to come from outer parts of the interacting galaxies, and therefore its metallicity should be low, unless it has been enriched by the subsequent star formation.

CO emission has been detected in polar rings (Taniguchi et al. 1990, Combes et al. 1992, Watson et al. 1994). This could bring some insight in the formation event: metallicity coming from the recent star formation, or the merged perpendicular disk forming the polar ring has not been destroyed.

5.5 Molecular Gas in Elliptical Galaxies

Most elliptical galaxies possess accreted gas, already detected in HI (Knapp et al. 1979, van Gorkom et al. 1997). This gas is either the remnant of the merger event between two spirals that have given birth to the elliptical, or comes from subsequent accretion of small gas-rich companions. The emission of the cold dust has been detected at 100 μm by IRAS (Knapp et al. 1989), and CO molecules are also frequently detected (Lees et al. 1991, Wiklind et al. 1995).

The relative H_2 content is low (see Fig. 27), the $M(H_2)/M(HI)$ ratio is $2-5$ times lower than in spiral galaxies. The CO gas is more present in field ellipticals than in clusters. The excitation temperature, as measured by the ratio between CO$(2-1)$ and CO$(1-0)$ emissions, is low, suggesting low molecular densities. There is also a small gas-to-dust ratio, but correlated to low dust temperature, which might reveal only an excitation effect. Indeed, the comparison between H_2 mass obtained from CO emission and IRAS dust emission reveals that, at low dust temperature, the CO emission is not tracing properly the molecular content (Wiklind et al. 1995).

Maps of the CO emission reveals that the gas is rotating in thin disks, that will form the typical stellar disks seen in many ellipticals. This gas does not come from mass loss of the parent stars, since it has a much higher specific angular momentum, but must have been accreted (Young 2002).

Shells Around Ellipticals: The merging events giving birth to ellipticals are also forming shells. Shells are very thin stellar features, discovered by Malin & Carter (1983) with the unsharp masking technique. Since shells always accompany mergers and interactions, they are likely to be the vestiges of the stars dispersed in the galaxy-collision event. Simulations confirm the scenario (Quinn 1984; Dupraz & Combes 1987). The stars of the small companion, disrupted in the interaction, oscillate back and forth in the potential well of the elliptical galaxy, and wrap in phase space such as to form several density waves at their apocenter, the "shells".

Recently, HI gas has been detected in shells (e.g. Schiminovich et al. 1994), while only stars where expected to follow the shells. Normally, the diffuse gas condenses to the center in the phase-wrap process (Weil & Hernquist 1993).

But CO is now also detected in shells (Charmandaris et al. 2000, see Fig. 32). It is therefore possible that the HI gas is just coming from photo-dissociated molecular clumps, which themselves are so fragmented and compact, that they can follow nearly radial orbits throughout the center, without experiencing much dissipation. The phase wrapping process is thus still the preferred formation mechanism for these gaseous shells.

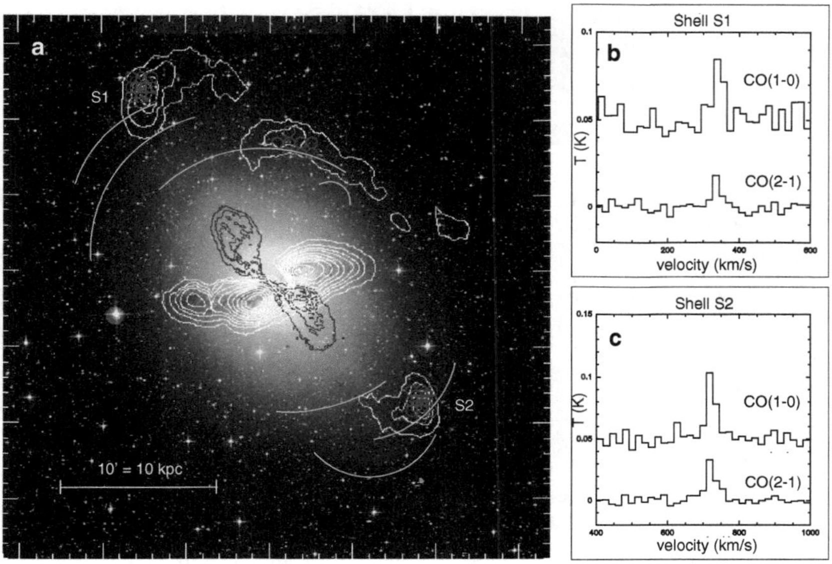

Fig. 32. (*Left*) Digitized Sky Survey optical image of Cen A with the contours of HI gas (from Schiminovich et al. 1994) superimposed in white. North is up and east is to the left, while the image scale is shown by the horizontal bar. The positions observed in CO are marked with the red circles whose size corresponds to the SEST $44''$ beam of CO$(1-0)$. The locations of the outer stellar shells are underlined by the yellow solid lines. The inner 6cm radio continuum lobes (from Clarke et al. 1992) are depicted by the blue contours. Note the jet alignment with the location of the CO detections. The outer radio lobes are far more extended. (*Right top*) CO$(1-0)$ and CO$(2-1)$ spectra towards the northern shell S1 with the temperature scale in main beam T_{mb}, (*Right bottom*) Same for the southern shell S2 (from Charmandaris et al. 2000).

5.6 CO Emission in Tidal Dwarf Galaxies

During major mergers between two spiral galaxies, long tidal tails develop, and it is frequent that the gas of the tails experience gravitational instabilities, and condenses in large complexes, giving rise to new stars: the tidal dwarf galaxies (e.g. Duc & Mirabel 1994). CO emission has been detected in these small dwarfs (see Fig. 33), rather easily with respect to their small mass and therefore expected small metallicity (Braine et al. 2000, 2001). This supports the idea that these dwarfs have been formed out of recycled gas, coming from the giant galaxies, and therefore have a high metallicity for their mass.

Is the molecular gas dragged with the tidal tail gas and recondenses in giant complexes in the tidal dwarf, or does the molecular gas itself re-form locally in the collapse? This might not be an important question, since in normal interstellar gas, the chemical time-scale to form molecules is much less than the dynamical time, and it is likely that the gas remains in part

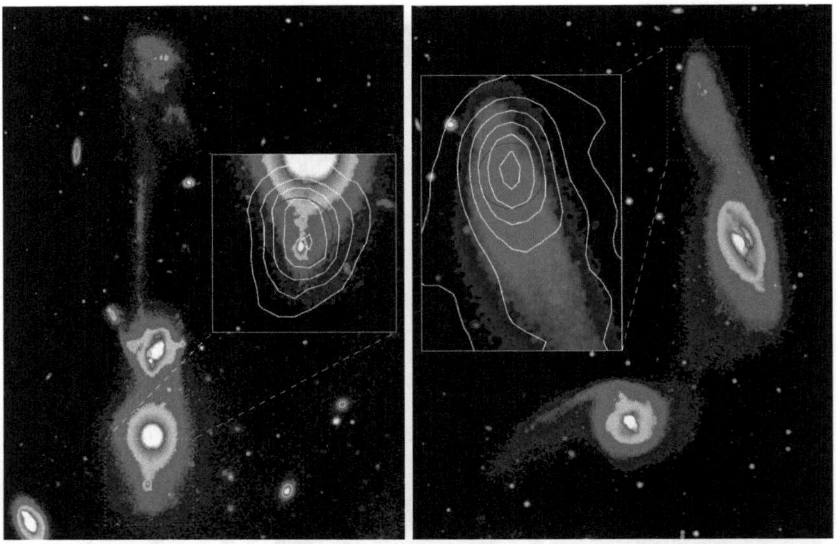

Fig. 33. (*Left*) The tidal dwarf galaxy in the interacting system Arp 105 (NGC 3561 or "the Guitar"). The zoom on the dwarf galaxy in the South shows the HI contours superposed to the V-band image. The dark circle (at the position of the HI maximum) is the central position where the CO emission has been detected, and its size is 22", the beam size of the IRAM-30 m telescope (from Braine et al. 2001). (*Right*) The tidal dwarf galaxy in the interacting system Arp 245 (NGC 2992/3). The HI contours are shown in the magnified image of the Northen dwarf. CO(1 − 0) emission has been detected at and around the dark circle position, around the HI maximum (from Braine et al. 2001).

molecular all along the tidal tails, even if it is more difficult to detect it, because of low density and low excitation. One puzzle is that the tidal dwarfs are most often located at the tips of the tidal tails detected in HI, while the simulations predict them all along the tails (Barnes & Hernquist 1992).

What is the fate of these tidal dwarfs, and could they enrich the interstellar medium of peculiar metallic dwarf galaxies? In general, they are re-accreted and merge with their parent galaxies; the tidal tails themselves have been observed to infall back to the merger remnant (Hibbard & Mihos 1995).

5.7 Conclusions

The CO emission in galaxies depends on morphological type, and has a behaviour opposite to that of HI, but this could be only a metallicity effect.

Galaxies can have large gas mass fractions, when they have low surface brightness, and are therefore stable. LSB galaxies are un-evolved, extended, un-concentrated systems, they may contain large quantities of H_2 but have low CO emission, because of their low metallicity. Their un-evolved state might be due to their poor environment, or their large intrinsic spin parameter.

CO emission is a good tracer of density waves, spirals, bars, rings. It is often associated to the thin dust lanes, tracing large-scale galactic shocks.

Its radial distribution is overall exponential, following the optical luminosity. But there are large departures from the ideal exponential, contrary to stars, due to inhomogeneities and contrasted features.

Elliptical galaxies contain cold H_2 gas, with a lower $M(H_2)/M(HI)$ ratio than in spirals. This is due either to different excitation, or a different CO/H_2 conversion ratio. This gas is coming from accretion, there is no correlation with the stellar component of the parent galaxy.

CO emission is a very useful tool to trace density, star formation, perturbations like warps, polar rings, gas dragged out of the spiral planes.

6 Molecules in ULIRGs, and High Density Tracers

6.1 Ultra-luminous Galaxies

The role of large-scale dynamics and interactions of galaxies is now well recognized for the starburst activity in galaxies (Kennicutt et al. 1987; Sanders et al. 1988). IRAS ultra-luminous galaxies are all mergers (Sanders & Mirabel 1996), and the fraction of interacting galaxies is increasing with L_{IR}/L_B.

In addition, interacting galaxies appear to have more H_2 content or at least much more CO emission, than unperturbed and isolated galaxies. The H_2 gas is also more concentrated towards the galaxy nucleus. In average, the H_2 content is multiplied by $4-5$ (Braine & Combes 1993) with respect to unperturbed systems, as shown in Fig. 34. This can be explained by the

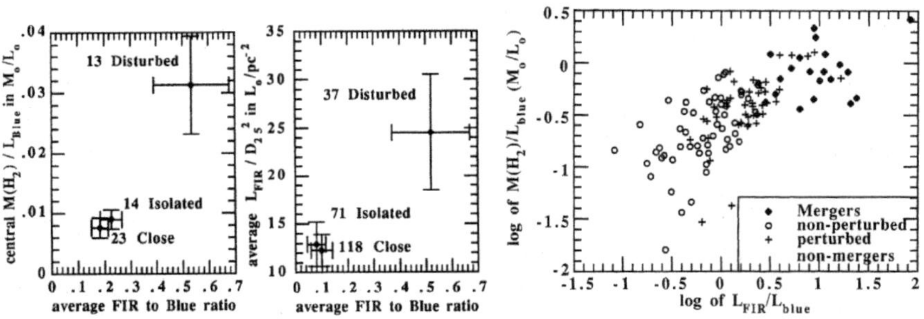

Fig. 34. (*Left*) Central H_2 mass versus FIR luminosity, both normalised to Blue luminosity, to avoid size and distance biases. Averages for the isolated, close and disturbed galaxies (number of galaxies indicated) are plotted with error bars being the uncertainty in the mean for each group. (*Middle*) Comparison of the two normalisations, by optical surface of the galaxy or Blue luminosity. (*Right*) The size-independent $M(H_2)$–FIR relation, plotted with indication of the interaction state of the galaxies. The slope is nearly unity (from Braine & Combes 1993).

gravitational torques of the interactions driving gas very quickly to the centers. This dense gas accumulation towards the circumnuclear regions can then trigger a starburst. The condition for a starburst is that the gas accumulation occurs in a time short enough that feedback mechanisms have no time to regulate the star formation. This condition is better fulfilled towards the nuclei where the dynamical time-scale is short ($\sim 10^7$ yr), and where the gas density can reach a maximum.

There is more star formation, and more H_2 content in interacting galaxies, but the star formation efficiency, which is defined by the ratio between star-formation indicators like the FIR or the Hα luminosity and the amount of gaseous fuel, is also larger (e.g. Rownd & Young 1999). The star formation efficiency does not depend on Hubble type, but is sensitive to extreme variations in the galaxy environment. The star formation efficiency is lower for HI deficient galaxies in clusters, and it is enhanced in strong interactions and mergers. Could it be that the CO/H_2 conversion ratio is different in perturbed galaxies? the gas is warmer and more excited, but it is denser too, and the conversion ratio roughly varies as $X \propto n^{1/2}/T_R$, and it is not obvious in which sense it is perturbed. If the derived molecular gas is lower, than the star formation efficiency SFE $= L_{\rm FIR}/M(\rm H_2)$ would even be larger.

High End of the Luminosity Function: At luminosities above $10^{11}\,L_\odot$, infrared galaxies become the dominant population of extragalactic objects in the local Universe ($z < 0.3$). As shown in Fig. 35, they are more numerous than optically selected starburst, Seyfert galaxies and QSO at comparable

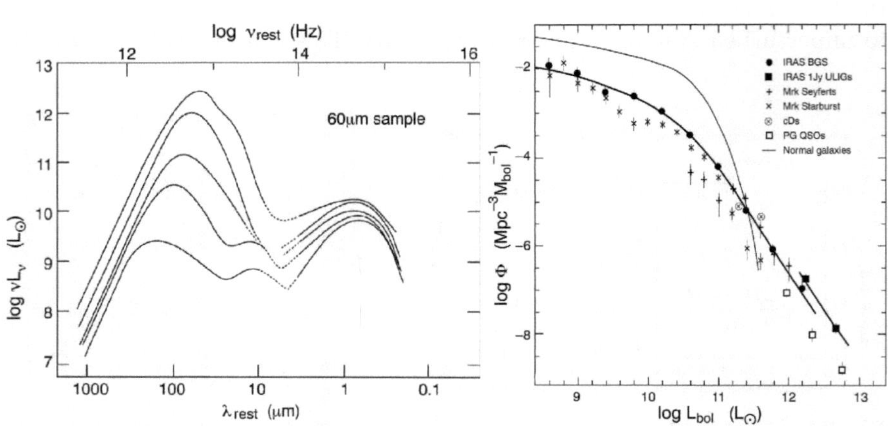

Fig. 35. (*Left*) Variation of the mean SED (Spectral Energy Distribution from submillimeter to UV wavelengths) with increasing L_{IR} for a 60 μm sample of infrared galaxies. (*Right*) The luminosity function for infrared galaxies compared with other extragalactic objects: IRAS Bright Galaxy Survey (BGS), IRAS 1-Jy Survey of ULIRGs, Markarian Seyfert, Markarian starbursts, cDs, Palomar-Green QSOs and normal galaxies, from Sanders & Mirabel (1996).

bolometric luminosity (Sanders & Mirabel 1996). Their energy is essentially coming from a starburst, although a fraction comes also from an active nucleus, whose importance increases with total luminosity (Kim et al. 1998). In the majority of objects, the infrared luminosity is due to dust heating from an intense starburst within giant molecular clouds. At the highest luminosities $L_{\rm IR} > 10^{12} L_\odot$, nearly all objects appear to be advanced mergers, fueled by enormous concentration of molecular gas that has been dynamically driven into the circum-nuclear regions.

Their SED (Spectral Energy Distribution) has the striking particularity that most of the energy is getting out in the infra-red. Since a lot of gas and dust are present, obscuration is high, and the optical light coming from star formation is absorbed, and is used to heat the dust. The energy of the starburst is then recycled, and re-radiated by the dust at lower frequency. Figure 35 shows how the infrared luminosity can vary by more than three orders of magnitude, when the optical luminosity only varies by a factor 2. The ratio $L_{\rm IR}/L_{\rm B}$ varies considerably, and is therefore an excellent indicator of starbursts. The brightest objects are the more obscured ones. The ratio between the IRAS fluxes at 60 and 100μm, $F60/F100$ also increases with infrared luminosity: the brightest objects have a hotter dust, due to more star formation in a given volume. Recent HST observations of samples of ULIRGs have confirmed how perturbed and interacting they are (Borne et al. 2000).

A remarkable feature is also the excellent correlation between the radio continuum emission at centimeter wavelength, and the far-infrared emission of galaxies. The only exceptions are the radio-loud AGN, where most of the radio emission comes from the active nuclei. The origin of the tight correlation must be that the star formation explains both emissions, the FIR coming from the dust heated by young stars, and the radio-continuum (RC) being synchrotron emission from supernovae.

When the radial distribution of both emission are considered, than the observed tight correlation is revealed in the inner regions only where OB stars and the associated Type II supernovae control the FIR and RC emission (Mayya & Regaradjan, 1997). In the outer parts heating of very small dust grains by the old disk stars provides a secondary component in the FIR emission, without associated RC emission. This explains also why galaxies with very weak star formation rate do not follow the relation so tightly. Another factor is the radial smoothing of the radio emission as a result of the propagation of energetic electrons accelerated during the supernova phase (Marsh & Helou 1995). The tight RC–FIR relation could also be used to deduce the mean magnetic field strength in galaxies, since the underlying mechanism for the correlation implies the linear correlation between the energy densities of the radiation field and the magnetic field (Lisenfeld et al. 1997).

6.2 Molecular Gas in ULIRGs

These ultra-luminous galaxies have huge quantities of H_2 gas. The gas is dense and hot, with densities of the order of 10^5 to $10^7\,\mathrm{cm}^{-3}$, and dust temperatures of $60 - 80\,\mathrm{K}$. These physical conditions are similar to what exists in star forming regions in Giant Molecular Clouds, like in the Orion complex for instance. But these regions in our Galaxy are very diluted spatially, while they are highly packed in starburst galaxies.

The large sample of ULIRGs observed in Solomon et al. (1997), reveals a tight correlation between the CO and $100\,\mu\mathrm{m}$ luminosities, suggesting that the dusty gas is close to the black body emission regime (see Fig. 37). The CO and dust emissions comes from very small sizes, since the gas giving rise to the starburst is highly confined in the circum-nuclear region, inside the central kpc. In the prototype Arp 200 for example, the main starburst is inside 300 pc. Two disks are seen in the late stage of merging, as shown by the two components in the velocity dispersion, and in the 1.3 mm continuum coming from dust emission (Fig. 36). This explains the high optical thickness, and the near black-body regime. Usually in galaxies, the dust emission at $100\,\mu\mathrm{m}$ is optically thin. However, the optical depth is varying with frequency as $\tau \propto \nu^{\beta}$, with $\beta \sim 2$, and significant optical depth may begin at $60\,\mu\mathrm{m}$.

Gas is concentrated in central nuclear disks or rings (Downes & Solomon 1998). What is the stability of these central disks? If the Toomre criterion is computed with gas only, it is Q_g slightly larger than 2, but taking into account stars also, the disk might be at the limit of stability, with a combined gas and stars criterion $Q \sim 1$. Due to the peculiarities of its physical condition, the ISM of interacting galaxies is likely to condense in large gaseous complexes, of the order of $10^8\,\mathrm{M}_\odot$ (Elmegreen et al. 1993).

Interacting galaxies have gas with enhanced velocity dispersion, due to the intense dynamical perturbations: streaming motions, asymmetries, spiral arms, strong bars, etc... This has the consequence to increase the critical Jeans scale for gravitational instabilities, and to create giant complexes. The global Jeans length is $\lambda \propto c^2/\Sigma_\mathrm{g}$, where c is the velocity dispersion and Σ_g is the gas surface density of the galactic disk. The corresponding growth time is $\tau_\mathrm{ff} \propto c/\Sigma_\mathrm{g}$, and the instabilities will occur as soon as $Q \propto c\kappa/\Sigma_\mathrm{g}$ becomes lower than 1. For the same ratio c/Σ_g, a perturbed system with elevated c and Σ_g will see the condensations of larger complexes of mass $M \propto c^4/\Sigma_\mathrm{g}$, in the same time-scale τ. These complexes with larger internal dispersions, and larger gravitational support will be less easy to disrupt through star-formation, which enhances the star-formation efficiency. The thermal Jeans length is also larger, due to the hotter gas temperature induced by the larger number of stars in the clouds, and high mass stars are favored.

This may explain the formation of giant star clusters, as the Super Star Clusters (SSC) observed in starbursting galaxies (e.g. de Grijs et al. 2001).

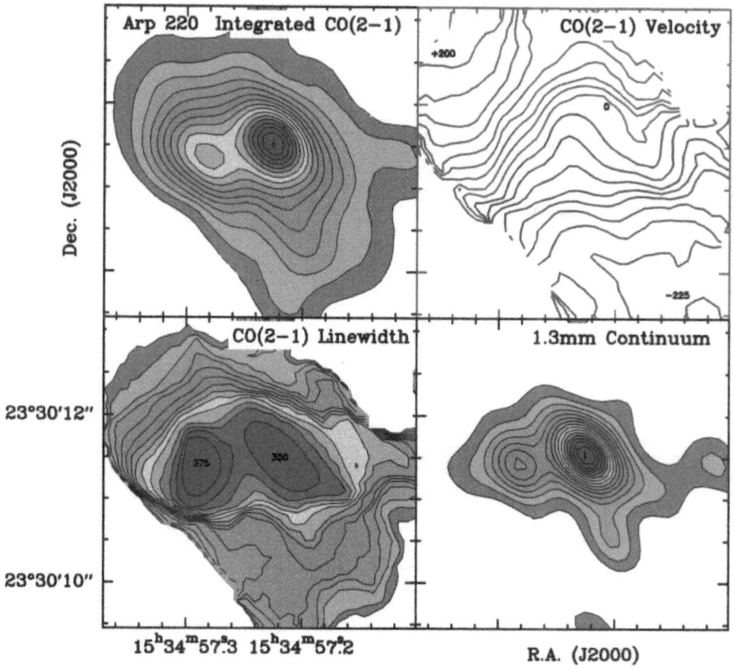

Fig. 36. CO(2 – 1) emission map towards Arp 220 obtained with the IRAM interferometer with a 0.7 x 0,5 "beam. (*Top left*) Integrated emission, (*Top right*) Velocity field, (*Bottom left*) Velocity dispersion, (*Bottom right*) 1.3mm continuum, from Downes & Solomon (1998).

Black Body Model: Due to the small sizes of nuclear gaseous disks in starbursts, the average column density reach easily $N(H_2) > 10^{24}\,\mathrm{cm}^{-2}$, and therefore the optical depth of the continuum dust emission reaches $\tau \sim 1$ at $100\,\mu$m.

In the extreme case of a black-body model, the far-infrared luminosity, for a dust temperature T_d, of a nuclear region of size R, is:

$$L_{\mathrm{FIR}} = 4\pi R^2 \sigma T_d^4 \ , \tag{44}$$

where σ is the Stefan-Boltzmann constant; there is no optical depth term, since it is assumed $\tau \gg 1$. In parallel, the CO luminosity can be expressed, in the Rayleigh-Jeans domain, at frequency $\nu \ll kT/h$:

$$L_{\mathrm{CO}} = 4\pi^2 R^2 (2k/\lambda^2) \int T_b d\nu \ , \tag{45}$$

the ratio of the two luminosities is then varying as

$$\frac{L_{\mathrm{FIR}}}{L_{\mathrm{CO}}} \propto \frac{T_d^3}{f_v \Delta V} \ , \tag{46}$$

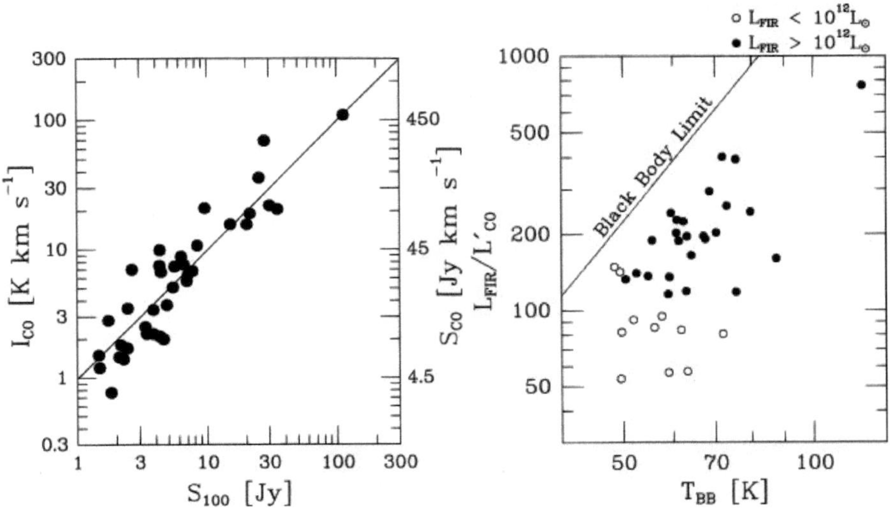

Fig. 37. (*Left*) Integrated CO(1-0) line intensity, I(CO), versus IRAS 100 μm flux density, for the ultraluminous galaxies of Solomon et al. (1997). The solid line has a slope 1. (*Right*) Far infrared to CO luminosity ratio versus the blackbody dust temperature derived from the IRAS flux ratio S60/S100. The line indicates the upper limit for the luminosity ratio for the blackbody model, from Solomon et al. (1997).

where f_v is the filling factor in velocity, and ΔV the total velocity width.

This relation corresponds to the predicted curve plotted in Fig. 37. The data points are slightly below this maximum curve, may be because the dust temperature T_d is different than the brightness temperature of the gas T_b, or because the CO and FIR are not emitted by exactly the same regions (the CO region could be larger), or the velocity filling factor is small, reducing the optical thickness of the line.

AGN or Starburst: According to mid-infrared line diagnostics or PAH abundances, about 80% of all ULIRGs are found to be predominantly powered by star formation (Lutz et al. 1998), but the fraction of AGN-powered objects increases with luminosity. Whereas only about 15% of ULIRGs at luminosities below $2 \cdot 10^{12}\,L_\odot$ are AGN powered, this fraction increases to about half at higher luminosity.

In these ULIRGs containing such large molecular masses, it is inevitable that most of the luminosity comes from the starburst, since the black hole cannot swallow all the available gas, and nuclear disks are unstable to star formation. To give orders of magnitude, if the molecular disk has a radius of 500 pc, a rotational velocity of $V_{\rm rot} = 300$ km/s, its period is 10 Myr. Assuming a FIR luminosity of $L_{FIR} = 10^{12}\,L_\odot$, and a star formation rate of $50\,M_\odot$/yr, it needs 10 rotations (or 100 Myr) for half of the gas to be turned into stars, the total H_2 mass being $5 \cdot 10^9\,M_\odot$. For this a radiation efficiency of $M_*/L_{FIR} = 5 \cdot 10^{-3}\,M_\odot/L_\odot$, or $L/M \sim 200$, has been adopted, corresponding to observations.

On the other hand, if $10^{12}\,L_\odot$ comes from accretion onto a black hole, at the efficiency of $L = 0.1\,(dm/dt)c^2$, the accretion must be only $1\,M_\odot$/yr, and therefore only 1% of the gas would be accreted on the same time-scale. The gas would remain available at 99% to form stars (Downes & Solomon 1998).

Dynamical Triggering: Numerical simulations (e.g. Barnes & Hernquist 1992) demonstrate the efficiency for gravity torques to drive the gas towards the center in a short time-scale. Mihos & Hernquist (1994) including star formation recipes like the Schmidt law in their simulations, show how easily starbursts can be obtained in galaxy interactions, and sometimes several bursts occur, according to the orbit and pericentric distances.

The self-gravity of the interacting disks play the fundamental role in galaxy interactions. Gas is driven to the nuclei, from bars and spirals triggered in its own disk by the perturbation. This explains why the mass ratio between the bulge and the disk is a more fundamental parameter than the geometry of the encounter (Mihos & Hernquist 1996).

The effects of the geometry are more visible on the direct manifestations of the tidal interaction, i.e. on the tails and debris. For the galaxy experiencing a retrograde collision, no extended tidal tail is formed. There is no resonance effects in the target galaxy disk, and less violent material perturbations; a transient leading tidal arm is instead developed. There is however a tidally-induced two-arm density wave, and the subsequent torques produce gas inflow towards the center, as for the prograde encounter.

6.3 High Density Tracers

Observing the wide-spread CO$(1-0)$ line gives an overall view of the molecular content of galaxies, but many aspects of the gas would be overlooked if other molecules, other isotopes, or other J levels were not explored as well.

Nuclei of galaxies possess denser gas than the rest of the disk; in general, giant molecular clouds must be denser to survive to the larger tidal forces. High-J levels of the CO molecule have a higher critical density to be excited, $10^5\,\mathrm{cm}^{-3}$ for the CO$(3-2)$ transition, $10^4\,\mathrm{cm}^{-3}$ for CO$(2-1)$, while only $10^5\,\mathrm{cm}^{-3}$ for CO$(1-0)$. Other high density tracers are the larger dipole molecules, such as HCN, HCO^+, CS, CH_3OH, H_2CO, OCS, etc... Some molecules trace peculiar regions, according to their formation mechanism; SiO traces shocks, supershells in starbursts for example.

The observation of isotopic molecules may highlight optical thickness, excitation problems, and also chemical evolution: primary or secondary elements can trace the age of the star formation events (as the ^{12}C and ^{13}C for instance).

A laboratory for such studies is the nearby starburst galaxy M 82: maps of all high J-levels of CO have been performed (e.g. Mao et al. 2000). Images at high-J are roughly similar in morphology although somewhat less extended than CO$(1-0)$. Two hot spots on either side of the nucleus are always conspicuous in each line, due to the edge-on nuclear ring, in this barred galaxy. Inside the ring, the ISM is depleted, this could be due to the bar torques, and also to the violence of the starburst, that expelled hot gas in a galactic bipolar flow. Radiative transfer has been computed with a LVG (Large Velocity Gradient) model, and the typical physical conditions are $N(H_2) \sim 10^{23}\,cm^{-2}$ in column density, $M(H_2)$ in the central region of a few $10^8\,M_\odot$, $n(H_2) 10^4\,cm^{-3}$ in average volume density, close to the tidal limit. The CO emission in the high-J levels comes primarily from PDR (photon-dominated regions), which is quite different from the other high density tracers. At least two components in the molecular gas are required to interpret the data: dense cores and a more diffuse medium, or intercloud. The diffuse component intervenes in the CO$(1-0)$ emission, also explains the high CI/CO ratio (Stutzki et al. 1997). This might not be representative of starburst at high redshift. Kinetic temperatures are derived between $20 - 60\,K$, relatively low for a starburst. The heating is due to star formation, cosmic rays and turbulence dissipation. This result is consistent with the weakness of CH_3OH or SiO, which are high temperature tracers.

The SiO molecule has been mapped in M 82 by Garcia-Burillo et al. (2001). SiO traces the walls of the supershells, related to the bipolar outflow, but not the star forming regions. There is in particular a vertical filament, part of an SiO chimney coincident with radio cm emission, and ionised gas ejected by the starburst. The outflow creates shocks, in which SiO is formed.

As for the excitation of the CO gas, the observations combined with radiative transfer analysis (cf. Fig. 38) are consistent with an optically thick gas. The intensity ratio between the two isotopic species $^{12}CO/^{13}CO$ is between 10 and 15 (while the abundance ratio is around 60).

Mauersberger et al. (1999) have carried out a survey of CO$(3-2)$ in 30 spiral galaxies. The intensity ratio $R(3-2/1-0)$ is between $0.2-0.7$, always lower than 1. This is what is predicted if $T_{kin} < 50\,K$ and $n(H_2) < 10^3\,cm^{-3}$. Although dense cores should have higher values, they are relatively diluted in the averaging on galactic scales. This is not the case in the nuclei of ultraluminous starburst galaxies, where the dense gas dominates.

CS and HCN are high density tracers even at low temperatures. The ratios CS/CO and HCN/CO are correlated with far-infrared luminosity L_{FIR}. While the ratios are around 1/6 in ULIRGs, they are only 1/80 in normal galaxies and in the Milky Way. The ratios depend on the spatial resolution, however, since they are larger in dense cores. The large-scale morphology is the same as in CO (spiral arms, rings, etc...). Strong starbursts have a larger fraction of dense gas, as shown on Fig. 39. Not only the HCN luminosity increases with

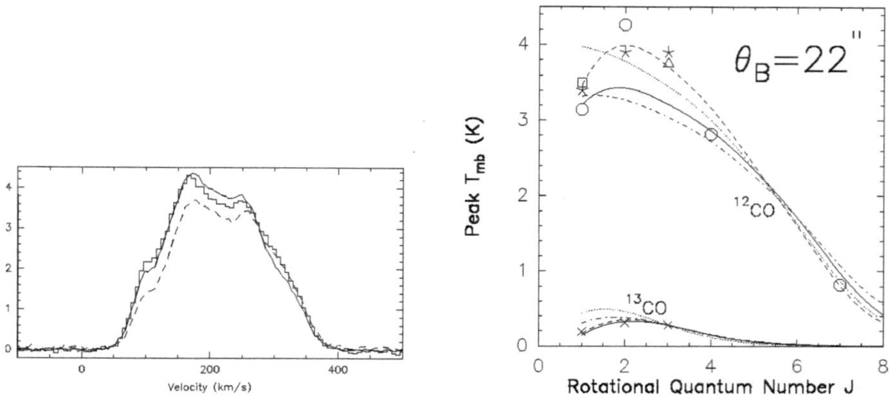

Fig. 38. (*Left*) CO(1 – 0) (*dash*) CO(2 – 1) (*solid*) and CO(3 – 2) (*histogram*) profiles toward the center of M 82. All spectra were convolved to the same spatial resolution of 21"; the temperature scale is main beam $T_{\rm mb}$, from Mauersberger et al. (1999). (*Right*) Peak line temperatures as a function of upper rotational number J for a beam size of 22", towards the center of M 82. The curves are model results from radiative transfer calculations, and are compared to the data points, from Mao et al. (2000).

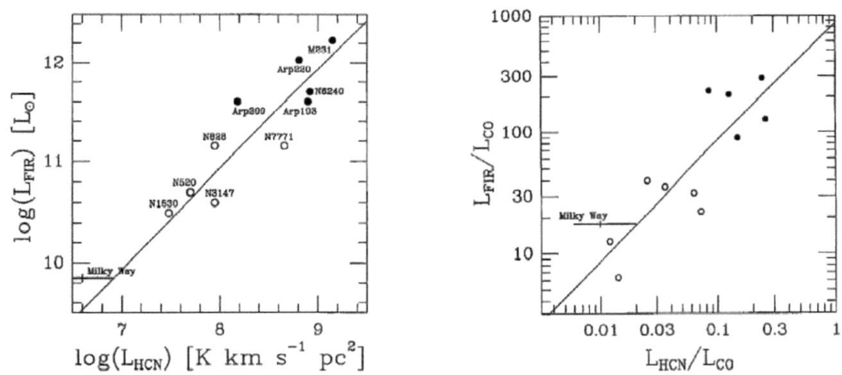

Fig. 39. (*Left*) Correlation between FIR and HCN luminosities for the Milky Way, luminous and ultra-luminous galaxies. The solid line has a slope of 1. (*Right*) FIR versus HCN luminosities, normalised to CO luminosity. The HCN/CO ratio measures the fraction of dense gas, and is larger for starburst galaxies, since the FIR/CO ratio is a measure of star formation efficiency, from Solomon et al. (1992).

$L_{\rm FIR}$, but the HCN/CO ratio, measuring the fraction of dense gas, increases with normalised $L_{\rm FIR}$.

Isotopic Molecules: The ^{12}C/^{13}C abundance ratio in the Milky Way, is estimated between 50 – 90 at the Sun radius. There is a measured radial gradient in the galaxy, and the ratio tends towards 10 – 20 in the center.

This abundance ratio is a tracer of astration, since the ^{13}C is a secondary element, formed by low-mass stars (it takes a time-scale of 1 Gyr to enrich it), while the ^{12}C is primary, formed at once in massive stars, with a time-scale less than 100 Myr. In the Galactic Center, with high astration level, there is also a deficiency of deuterium, and this has to be taken into account in observations of DCN, DCO^+, etc...

In starbursts and ULIRGS (of Arp 220 type), the $^{12}CO/^{13}CO$ intensity ratio is larger. This is not due to a low optical depth, since the $C^{18}O$ is normal with respect to ^{12}CO. The right interpretation is that ^{12}C is overproduced in the nucleosynthesis of a recent burst (Casoli et al. 1992). The $^{12}C/^{13}C$ abundance ratio has been mapped in M 82 and IC 342 by Henkel et al. (1998) from CN, HCN, HCO+ observations. The ratio is always $^{12}C/^{13}C > 40$ (much higher than in the Galactic Center). In a similar way, the oxygen and nitrogen isotopic ratios have been estimated $^{16}O/^{18}O > 100$ and $^{14}N/^{15}N > 100$. $HC^{15}N$ has been detected in LMC and NGC 4945 by Chin et al. (1999) who derived the abundance ratio $^{14}N/^{15}N = 111$, lower than in the Milky Way. They deduce that ^{15}N is synthesized by massive stars, and is a primary element. There is some controversy about ^{15}N formation, at least with respect to standard stellar evolution, and nucleosynthesis calculations since it should be destroyed in massive stars, instead of being formed in SN-II explosions. Molecular results are however consistent: ^{14}N is more secondary, and the ratio increases with time and astration.

Deuterated species have been detected in the LMC: DCN, DCO^+ (Chin et al. 1996). The ratio HCN/DCN is of the order of 20, because of strong fractionation. Although the deuterium abundance is low, $D/H = 2\cdot 10^{-5}$, deuterated molecules have lower energies in the ground state then the corresponding hydrogenated isotope. At the low temperatures of the ISM, reactions of the type:

$$HD + HCN \rightarrow H_2 + DCN \ , \tag{47}$$

are shifted towards the formation of DCN, since the deuterium in molecules is a trace with respect to the reservoir of HD.

The HNC/HCN Ratio: Useful to disentangle abundances, excitation, density or temperature, HNC (the hydrogen isocyanide) is a high density tracer as well.

HNC emission is weaker than HCN, except in ULIRGS such as Arp 220, where the intensity ratio is > 1. The reason is not very clear however, since in NGC 6240 (another prototypical starburst), it is 10 times lower. The comparison between HNC and HCN, and also HNC and HCO^+ is plotted in Fig. 40 (see Hüttemeister et al. 1995).

Other Molecules: Other molecules are useful to trace different physical conditions. OCS is detected in NGC 253 and M 82 (Mauersberger et al. 1995);

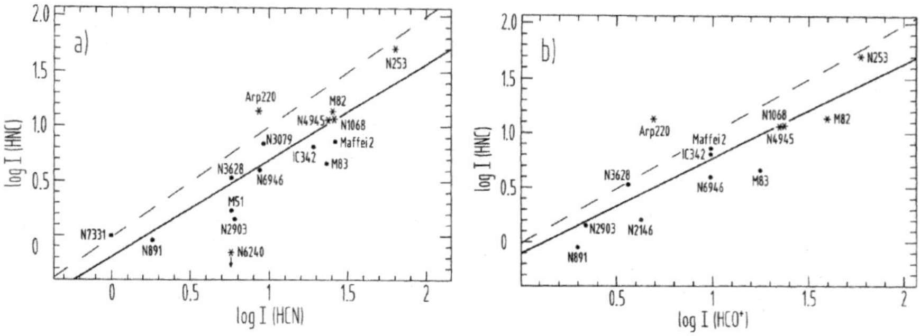

Fig. 40. (*Left*) HNC versus HCN integrated intensities, for starburst galaxies (*asterisks*), and normal galaxies (*filled circles*). The dash line indicates equal intensities, and the solid line is the better fit. (*Right*) Same, for HNC versus HCO^+, from Hüttemeister et al. (1995).

NH_3 in Maffei 2 and other galaxies (Henkel et al. 2000), with rotational temperatures of 85 K. H_2CO (formaldehyde) and CH_3OH (methanol) in NGC 253 trace high-density subthermally excited, clumpy structure. All these studies point out very different physical conditions and various chemistry, from one galaxy to the next (Hüttemeister et al. 1997).

Atomic Carbon: The CI fine structure line $^3P_1$-$^3P_0$ at 492 GHz is an important tracer of non-ionising radiation. In Arp 220, the CI line is strong, as predicted from its FIR flux, while CII (at 158 μm) emission is depleted. This could be due to higher density, optical thickness of the C^+ line and dust opacity, in the far-infrared. In a sample of about twenty galaxies detected in CI, the intensity ratio has been determined to be CI/CO = 0.2 (Gerin & Phillips 2000). The cooling efficiency is comparable for CO and CI. Normally C^+ is a better coolant than CI, except in Arp 220 and Mkn 231.

6.4 Conclusions

The molecular component has a much higher mass fraction in starbursts and ULIRGs than in normal galaxies; it is not the case for the HI component. It explains the considerable enhancement in star forming efficiency. This H_2 enhancement is not only due to CO emission enhancement because of gas excitation, density or temperature, since all gas density tracers confirm the large H_2 abundance and density.

In interacting galaxies and mergers, gravity torques produce efficient gas inflow from the external parts of galaxies towards the center. Large quantities of atomic gas can then be transformed in H_2. Ultra-luminous galaxies have dense nuclear disks, concentrating the CO and dust emission; their radiating conditions are approaching the black-body regime.

Only in strong starbursts is the H_2 gas dense enough to emit sufficiently high-J CO lines. This has important consequences for high redshift galaxies, since they can be observed only with these lines, redshifted.

Various molecules help to constrain the physical conditions of the molecular gas: density, temperature, excitation, shocks, clumpiness, chemical abundances. At least two components are required to interpret the observations: hot dense cores where stars form, embedded in an intercloud gas, a more diffuse medium, likely to be subthermally excited.

7 Molecules Traced in Absorption

7.1 Advantages of the Absorption

Absorption lines are a sensitive probe for studying the interstellar medium. Especially in distant objects, where emission lines becomes diluted with distance squared, whereas the detectability of absorption lines only depends on the observed flux of the background source. This is well known from optical spectroscopy, where the combination of sensitive detectors and large telescopes allow observation of very tenuous gas towards distant background QSO.

In principle, absorption of molecular rotational lines can be used to probe the densest and coldest part of the ISM in distant galaxies, much in the same way as optical lines probe the warm and diffuse gas. Since new stars are formed in molecular clouds, a study of this ISM component trace the star formation conditions and its history in galaxies. There are, however, several difficulties associated with the detection of such lines and it was not until 1993 that the first distant molecular absorption line system was detected. Since then a total of four such systems at redshifts between $z = 0.2 - 0.9$ have been observed.

Absorption measures are very useful, in particular in the Galaxy, when both emission and absorption can be detected along the same line of sight. Then it is possible to obtain the physical conditions, temperature T and column density N, with a coupled system of two equations linking these two quantities, while absorption and emission alone can provide only one equation.

One of the fundamental advantage of absorption studies is to benefit of spatial resolution due to the pencil beam towards the background continuum source. The QSO size in the millimeter domain is particular small (tens of μas), since only the core of the radio source is emitting, and not the radio lobes (as at cm wavelengths).

However, one must be aware that absorption is biased towards detecting cold gas, contrary to emission (e.g. Combes & Wiklind 1996). In the Rayleigh-Jeans domain, simple radiative transfer on the line of sight relates the signal received, in antenna temperature T_A^*, to the excitation temperature $T_{\rm ex}$ of the observed medium, the background radiation $T_{\rm bg}$, and the optical depth τ:

$$T_A^* = (T_{\rm ex} - T_{\rm bg})\left(1 - {\rm e}^{-\tau}\right) \ . \tag{48}$$

Emission is detected when $T_{\rm ex} > T_{\rm bg}$. For the atomic line HI at 21cm for instance, there is a large influence of the stimulated emission ("negative absorption") since the ΔT between the two hyperfine levels $F = 1,0$ of the H-atom ground state is only $\sim 0.7\,{\rm K}$. The optical depth of the $F = 1$ level is related to the total HI column density N by a relation:

$$\int \tau \, dv \propto \frac{N}{T} \ , \tag{49}$$

where the optical depth has been integrated over the whole velocity profile. In general, the atomic gas is optical thin, so that $\tau \ll 1$, and therefore the equation, together with the transfer equation, simplifies to:

$$N \propto \int T_{\rm ex} \tau \, dv \propto \int T_A \, dv \ , \tag{50}$$

for HI emission, since the background temperature $T_{\rm bg}$ has been neglected. The column density is then independent of temperature. On the contrary, for absorption, the background temperature is equal to the powerful radio source, in front of which $T_{\rm ex}$ is negligible, and therefore the absorbing signal has an amplitude equal to $T_A = -T_{\rm cont}\tau$ in the line. The optical depth and therefore the intensity of the absorption varies as $1/T$ for a given column density. This is a strong bias to see cold gas in absorption.

In the millimeter range, the molecules have a whole ladder of rotational levels, the emission then depends on temperature since the column density in the emitting (upper) level $N_{\rm u}$ is a function of the total column density and the rotational (excitation) temperature:

$$\frac{N_{\rm u}}{N_{\rm tot}} = \frac{g_{\rm u}}{Z} \, {\rm e}^{-E_{\rm u}/kT} \ , \tag{51}$$

where $g_{\rm u}$ is the statistical weight of the upper level, and Z the partition function. The upper column density is, for emission:

$$N_{\rm u} \propto \int T_{\rm ex} \tau \, dv \propto \int T_A \, dv \ , \tag{52}$$

if $\tau \ll 1$, and

$$N_{\rm tot} \propto T N_{\rm u} {\rm e}^{E_{\rm u}/kT} \ . \tag{53}$$

For absorption:

$$\int \tau \, dv \propto \frac{N}{T_{\rm ex}} \left(1 - {\rm e}^{-h\nu/kT}\right) \ , \tag{54}$$

also strongly weighted by the temperature $T_{\rm ex}$.

Since collisional excitation requires $4 \cdot 10^4\,\mathrm{cm}^{-3}$ for CO, and $1.6 \cdot 10^7\,\mathrm{cm}^{-3}$ for HCN, even in hot (in terms of kinetic temperature) but diffuse media, the excitation temperature is very low, and tends to $T_{\mathrm{ex}} = T_{\mathrm{bg}} = 2.76\,\mathrm{K}$. Therefore, molecular absorption is weighted towards detecting the diffuse media.

The utility of molecular absorption lines is also to detect all high density tracers but this time in diffuse media, and therefore explore their spatial extension. Due to their high dipole, these molecules populate more their ground levels, and are not detected in emission in diffuse media. Their velocity integrated opacity is directly proportional to the square of the permanent electric dipole moment. This means that a molecule like HCO^+, which is $\sim 10^{-4}$ less abundant than CO, but has a dipole moment $\sim 10^2$ times larger than that of CO, can be as easily observed.

Finally, one of the main advantage of absorption is that, due to the small extent of the background continuum source, the signal detected will be only proportional to the opacity, which is not spatially diluted. Once the background QSO is detected in the continuum, there is no distance dependence to detect the absorption line. Molecular absorption lines are as easy to detect at $z \approx 0$ as at $z \approx 1$, except that at small distances, emission can make absorption line measurements more difficult. The following will illustrate this, by showing HCO^+, HCN and CS absorption from Cen A at a distance of $\sim 4\,\mathrm{Mpc}$ (Wiklind & Combes 1997a) and from PKS 1830–211 at a distance of $\sim 4000\,\mathrm{Mpc}$ (Wiklind & Combes 1996a). Both set of spectra have been obtained with similar integration time.

7.2 Molecular Absorption in the Galaxy

Absorption lines in the millimeter is more difficult to observe than in the cm range, since continuum sources are weaker (the flux is going as $S \propto \nu^{-\alpha}$) and smaller. They are also more difficult to observe than at high redshift, since there is the confusion with emission. It is only with the advent of mm interferometers that the tiny absorption regions have been separated from emission. The literature is thus relatively recent, beginning with the molecular absorption in the Milky Way detected in front of BL Lac by Marscher et al. (1991).

When looking at all possible background source to detect molecular absorption in the Milky Way, one realizes the small filling factor in surface of the molecular component even of the diffuse CO medium: only 9% of the 3C sources observed have CO emission in front of them (Liszt & Wilson 1993). Among these line of sights, 60% show absorption lines.

The surprise of the first observations is that many molecules were detected, even towards very diffuse media, with extinction of only $A_V \sim 1\,\mathrm{mag}$: already a very abundant chemistry exists at very low column densities, which was not expected (Lucas & Liszt, 1994). A variety of lines, such as ^{13}CO, CN, HCO^+, HCN, HNC, C2H, $N2H^+$ are reported with line ratios quite variable

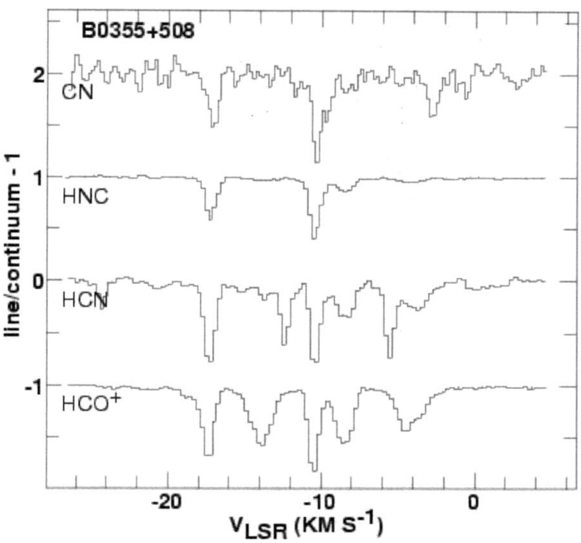

Fig. 41. CN, HNC, HCN and HCO^+ absorption spectra toward B 0355+508. CN and HCN have well separated hyperfine components. The HNC spectrum, in which the hyperfine structure only widens and does not create new components, helps to distinguish which of the HCO+ features are present in the CN spectrum (from Liszt & Lucas 2001).

from one line of sight to the other (Lucas & Liszt 1994, 1998). Figure 41 illustrates the richness in molecules (Liszt & Lucas 2001). Among a survey of 30 lines of sight, HCO^+ absorption is detected 30% as often as HI absorption, and even more frequent than CO (Lucas & Liszt 1996).

The bigger surprise is the strength of HCO^+ absorption, in these diffuse media; there is not only the higher critical density to see HCO^+ in emission, explaining why HCO^+ is so "cold", but the chemistry has to be revised to be consistent with observations. Some lines are very optically thick (^{13}CO is detected) and for others $\tau \ll 1$, and the hyperfine lines of HCN are observed in the canonical ratio 5:3:1 as expected. The lines are quite narrow, of the order of $\Delta V = 0.5 - 1\,\mathrm{km/s}$.

Absorption lines are sometimes variable over a year time-scale, suggesting the presence of moving clumpy material, of sizes $10 - 100\,\mathrm{AU}$ in front on the continuum source (Marsher et al. 1993). Also spatial fluctuations in the chemistry only have been invoked (Liszt & Lucas 2000).

To explain the large variations from region to region of the abundances of CO versus HCO^+, variable by factors up to 20, several propositions have been studied; one of them is the chaotic chemistry, where only a tiny difference in initial conditions can change completely the final result. In certain conditions, there are double steady state chemical solutions, between which the medium experiences bi-stability (Le Bourlot et al. 1993). Two phases can then be

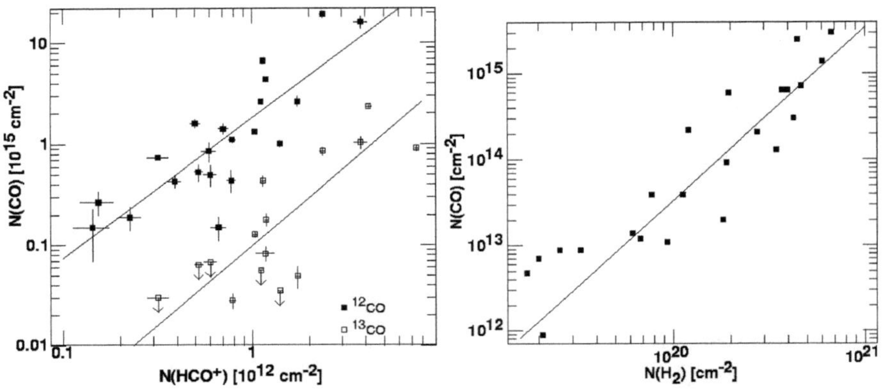

Fig. 42. (*Left*) Variation of ^{12}CO and ^{13}CO column density with HCO+. The slopes of the regression lines are 1.4 for ^{12}CO and 1.6 for ^{13}CO. (*Right*) CO and H_2 column densities determined in the UV, taken from the compilation of Federman et al. (1994, revised in 1995). The slope of the regression line is 2.02. (from Liszt & Lucas 1998).

found, either low ionisation or high ionisation for instance, and very close regions in space could end up in one or the other solution, randomly.

Another proposition is that CO can form rapidly from HCO^+ in diffuse clouds. H_2 can form at relatively low density. Whenever H_2 is there, HCO^+ is present with a minimum abundance $HCO^+/H_2 = 2 \cdot 10^{-9}$, and then CO forms by recombination of HCO^+. There is evidence of a rather sudden CO turn on, when the column density of H_2 crosses the value $10^{20} - 10^{21}$ cm^{-2}, as shown in Fig. 42. The fact that HCO^+ is much more extended than CO is supported by its good linear correlation with OH: $X(HCO^+) = 0.03 - 0.05X(OH)$, even at low column density.

The CO molecule then forms later, at higher column densities, when C^+ recombines. Eventually, diffuse clouds have similar chemical abundances as in dark clouds, except for the CO molecule. N(CO) increases abruptly when N(HCO^+) reaches $1-2 \cdot 10^{12}$ cm^{-2}. The correlation between these two column densities observed in absorption reveals a non-linear slope of 1.5. Already, by UV absorption measurements, Federman et al. (1994) had found a non-linear relation, with a power-law slope of ~ 2 between CO and H_2 column densities (cf. Fig. 42). This may explain the observed line-width of HCO^+ spectra, always larger by at least 15% than the CO line-width: the HCO^+ is like an envelope around the dense molecular clouds, where CO is absent. Surprisingly the ^{13}CO absorption line is relatively strong. Fractionation, much more efficient than selective photodissociation, is certainly the cause, through reactions like:

$$^{12}CO + {}^{13}C^+ \quad \rightarrow \quad {}^{13}CO + {}^{12}C^+ \qquad (55)$$

7.3 Extragalactic Molecular Absorptions

When the line of sight of a radio-loud QSO crosses a galaxy, and also a molecular cloud (a quite rare event), absorption lines are detected in the millimeter and centimeter rotational lines of many molecules.

These molecular absorption objects are the continuation at high column densities ($10^{21} - 10^{24}\,\mathrm{cm}^{-2}$) of the whole spectrum of absorption systems, from the Lyman-α forest ($10^{12} - 10^{19}\,\mathrm{cm}^{-2}$) to the damped Lyman-$\alpha$ and HI 21cm absorptions ($10^{19} - 10^{21}\,\mathrm{cm}^{-2}$). It is currently thought that the Lyman-α forest originates from gaseous filaments in the extra-galactic medium, that the damped and HI absorptions involve mainly the outer parts of spiral galaxies. The molecular absorptions concern the central parts of galaxies. These molecular events are much rarer, the number N versus the column density decreases as a relatively steep power law.

Comparison with Emission: The absorption technique is much more sensitive than emission, as soon as a molecular cloud is found in front of a background radio QSO. At high redshift for instance, the detection limit is $10^{10}\,\mathrm{M}_{\odot}$, while the absorption limit does not depend on redshift, and can be as low as $1\,\mathrm{M}_{\odot}$, depending on its column density. The absorption limit is in optical depth, the background source in general is highly diluted, but its brightness can reach up to $10^{12}\,\mathrm{K}$, while the flux detected is less than 1 Jy (or less than 100 mK with current telescopes). If for galactic studies, interferometric resolutions are required, since absorption is generally buried among strong emission of local molecular clouds, single dish observations are sufficient for all extragalactic studies. The nearest absorption is Centaurus A, where both emission and absorption are detected with the same amplitude.

Centaurus A: The absorption was first detected by Eckart et al. (1990) in CO and HCO^+. Many other lines are detected in absorption only (see Fig. 43; Wiklind & Combes 1997a). After about 10 years, no temporal variation is detected. This helps to put constraints on the source size, which must be larger than 500 AU. The analysis of all the lines show that the gas is in average of low density, low excitation and low kinetic temperature. Lines are optically thin, and there is a particularly wide absorption in HCO^+, which could correspond to a nuclear disk (Wiklind & Combes 1997a).

Higher Redshift Absorptions: The first high-z absorption was detected towards the BL Lac object PKS 1413+135 (Wiklind & Combes 1994), after many unfruitful searches towards DLA (Damped Lyman-α Absorptions). Since then, 4 systems are known, but they remain rare. Half of them are gravitationally lensed objects, and true intervening absorption, PKS 1830-211 and B 0218+357. The absorbing molecular clouds belong to the lensing

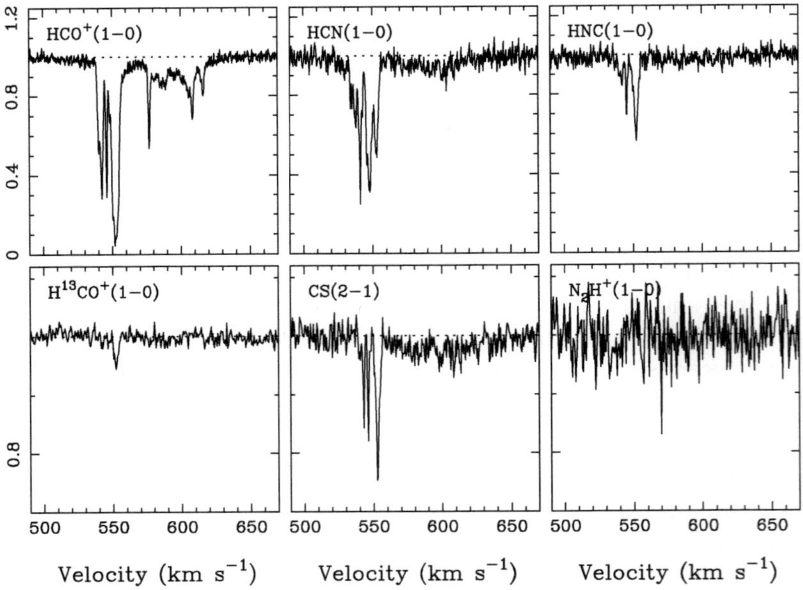

Fig. 43. Absorption lines towards Centaurus A of $HCO^+(1--0)$, $H^{13}CO^+(1-0)$, HCN(1–0), HNC(1–0), CS(2–1) and $N_2H^+(1–0)$. Only N_2H^+ remains undetected. The spectra have been normalized to a continuum level of unity (from Wiklind & Combes 1997a).

galaxy. This is a way to find molecules in normal galaxies at high redshift (not only in ultra-luminous galaxies).

Redshifts range up to $z \sim 1$, while the background QSO up to $z \sim 2$; it is difficult to find higher redshifts QSO, that are strong enough in the millimeter. Large variety of line widths, optical depths, sometimes several components, at different velocities, are detected along the same line of sight.

In addition to the four detected molecular absorption line systems, a systematic survey was carried out of about a hundred candidates, selected from flat-spectrum continuum sources. The continuum needs to be at least 0.2 Jy to allow detection of intervening molecular gas. The redshift of the absorbing candidate is known, either from previously detected HI absorption; 21 cm or damped Lyman-α, which is the case for PKS 1413+135 and B 0218+357 (Carilli et al. 1992, 1993; Wiklind & Combes 1994, 1995) or from optical lines emission of a galaxy on the line of sight to a radiosource; B 1504+377 (Wiklind & Combes 1996b). When the continuum source is strong enough, at least 1 Jy, and no redshift is known, it is possible to search for absorption lines by scanning in frequency in a manner similar to what was done for PKS 1830-210. This last method is the most promising with the new generation millimeter instruments, that will gain an order of magnitude in sensitivity. Indeed, the best candidates are the most obscured ones, where no redshift is available.

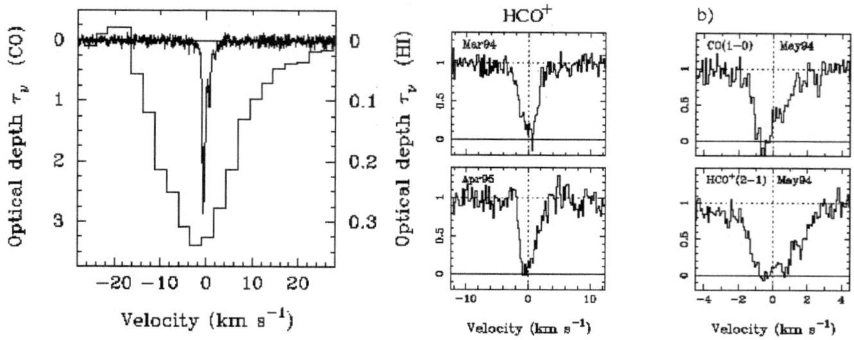

Fig. 44. (*Left*) A comparison of the CO(1 − 0) and 21 cm HI absorption towards PKS 1413+135. In both cases the optical depth is plotted. (*Middle*) $HCO^+(2-1)$ absorption lines at two different epochs. (*Right*) Normalized spectra of CO(1 − 0) and $HCO^+(2-1)$ both obtained in May 1994. Note that the relative intensities of the two velocity components have varied (from Wiklind & Combes 1997b).

PKS 1413+135 at $z = 0.247$: This is a particularly narrow absorption, with two components, each less than 1 km/s wide. There has been for a long time a debate to know whether the continuum source is a background object, or is the central AGN in the galaxy. The position of the source coincides with the center, and if it was located much behind, a gravitational deformation should occur, which is not the case. Lamer et al. (1999) conclude that the galaxy at $z = 0.247$ is indeed the host of the BL Lac. The object is highly obscured, and appears a spiral galaxy edge-on. The radio source, as well as the optical AGN are highly time-variable. The CO line is optically thick, which corresponds to what has been deduced in the optical, $N(H_2) > 10^{22}\,cm^{-2}$, $A_V > 30$ mag McHardy et al. (1994).

Some temporal variability has been detected, as shown in Fig. 44, and might suggest small-scale structure. The opacity ratio between the two components has varied by 2.3 over 2 years (Wiklind & Combes 1997b). The variability is seen in the CO, not in HCO^+. CO is more optically thick, and HCO^+ belongs more from the diffuse component.

The variations could be due to the line of sight change due to the variability of the continuum source. This source, as a BL Lac, is likely to experience superluminal motions. The radio core is unresolved at 2.3 mas or 7 pc. Assuming a size of the order of 10 μas = 0.03 pc, the proper motions of small molecular clumps in front of the background sources, at 250 km/s = 50 AU/yr are insufficient to explain the observed variability (time-scale of 100 yr would then be expected). Velocities larger than 25 000 km/s are required. The motion must come from the core itself.

The interpretation of the physical conditions of the gas is difficult. The data are compatible with either a multi-component model with similar filling factors, or with dense clumps embedded in a diffuse medium. The diffuse component accounts for most of the absorption, while the clumps comprise most of the mass.

Because of the very narrow velocity width, the cloud along the line of sight must be quite small, of the order of 1pc according to the size/line-width relation followed by molecular clouds. The density is estimated $n(H_2) \sim 10^4\,cm^{-3}$.

B 1504+377 at $z = 0.672$: Seven different velocity components are detected in absorption, in front of the continuum source, which is local to the galaxy. The large separation of 330 km/s is surprising for a line-of-sight passing by the nucleus, but this suggests the presence of a nuclear ring, with non-circular motions, and a forefront spiral arm. The abundance ratio HNC/HCN, which is sometimes used as thermometer, suggests that the kinetic temperature is equal to the excitation temperature, therefore indicating a dense medium.

HCO^+ is enhanced by $10 - 100$, as in the Galaxy; there must therefore exist both the diffuse and dense media on the line of sight (Wiklind & Combes 1996b).

B 0218+357 at $z = 0.685$: This is one of the smallest separation double source gravitational lens. One of the two images is surrounded by an Einstein ring visible in the radio continuum with VLBA measurements (Patnaik et al. 1995). The image separation is 335 mas (1.8 kpc).

This is the absorption system with the largest column density $> 10^{24}\,cm^{-2}$ (Wiklind & Combes 1995). All three CO$(2-1)$ isotopic lines are optically thick, as shown in Fig. 45. This was an excellent opportunity to search for molecular oxygen in interstellar space. This crucial molecule has not yet been detected in the Milky Way, in part because large telescopes on the ground are prevented by the atmospheric lines; yet O is more abundant than C, and chemical models predict a large abundance for O_2. Because of its large redshifts, the main O_2 lines can be observed from the ground in the B 0218 molecular clouds, without atmospheric absorption. The result of the search is an upper limit of $O_2/CO < 2\cdot10^{-3}$ at $1\,\sigma$ (Combes et al. 1997). This is much below theoretical expectations and suggests that oxygen is frozen on to dust grains, or that steady-state chemistry is never reached. In the latter case, most of the oxygen would remain atomic even in dense molecular clouds.

In a similar way, it is not possible to observe from the ground the fundamental line of water, and to get insight on its spatial extension. Only the isotopes $H_2^{18}O$ and HDO have been detected, but in dense cores, with a deduced H_2O/H_2 abundance ratio between 10^{-7} and 10^{-5}. It was thought until recently that these abundances concerned only the neighbourhood of star-forming regions, such as the Orion hot core, where water ice is evaporated from grains. However, Cernicharo et al. (1997) detected with the ISO

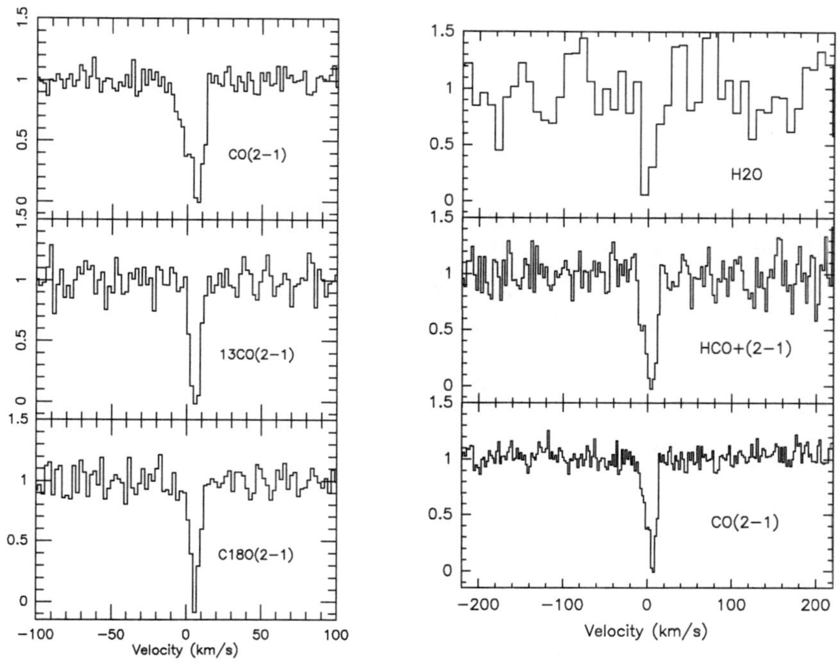

Fig. 45. (*Left*) Normalised spectra of the three isotopic molecules ^{12}CO, ^{13}CO, $C^{18}O$, in the $J = 2-1$ line towards the gravitationally lensed quasar B 0218+357. The three lines are optically thick, meaning that the column density of H_2 is of the order of $10^{24}\,cm^{-2}$ or more (from Combes & Wiklind 1995). (*Right*) Spectrum of ortho-water in its fundamental line at 557 GHz, redshifted at 331 GHz, in absorption towards B 0218+357. The line has the same width as the previously detected $HCO^+(2-1)$ and $CO(2-1)$ lines. Spectra have been normalised to the continuum level completely absorbed (from Combes & Wiklind 1997).

satellite water in absorption at 179 μ in front of Sgr B2, and this revealed that cold water was ubiquitous.

The detection with the IRAM-30 m telescope of ortho-water in its fundamental line at 557 GHz confirms this result (Combes & Wiklind 1997). The line is highly optically thick, and has about the same width as the other optically thick lines detected in absorption in this cloud (see Fig. 45). If the excitation temperature was high (as in the Orion hot core), we would have expected to detect also the excited line at 183 GHz (redshifted at 109 GHz). An upper limit on this line gives us an upper limit on T_{ex} of 10 − 15 K, and an estimation of the optical depth of the 557 GHz line of $\sim 40\,000$ (Combes & Wiklind 1997). This leads to an H_2O/H_2 abundance ratio of 10^{-5}, in the upper range of expected values.

It was also possible to try the detection of the LiH molecule at 444 GHz, again never detected until now, although fundamental for primordial cloud

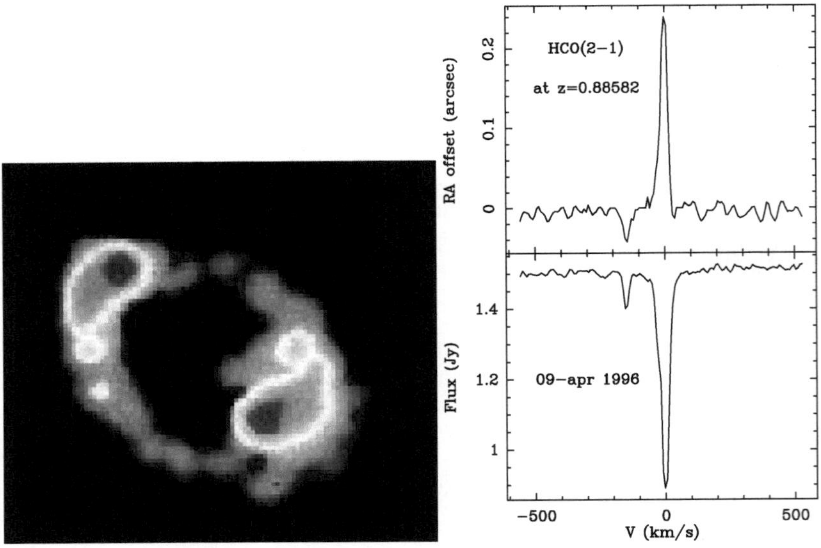

Fig. 46. (*Left*) The gravitationally lensed quasar PKS 1830-211 consists of two images, embedded in an Einstein ring, as shown in this cm radio continuum map (from Jauncey et al. 1991). (*Right*) Results of the fit of visibility data from April 1996 observations: the integrated spectrum is at bottom, and the phase-center position in right ascension at top: this shows that the $V = -147.5\,\mathrm{km\,s^{-1}}$ component is absorbed in front of the NE image (from Wiklind & Combes 1998).

and galaxy formation. A tentative detection of a very narrow line has been reported (Combes & Wiklind 1998). The LiH abundance would be $\mathrm{LiH/H_2} \sim 3 \cdot 10^{-12}$.

PKS 1830-211 at $z = 0.88582$: This gravitational lens reveals two images, embedded in an Einstein ring (see Fig. 46). Two velocity components are detected, and interferometric data show that each velocity component absorbs in front of a different gravitational image (Wiklind & Combes 1998).

The measurement of a few dozens of molecular lines in this source bring some insight in the evolution of chemical conditions with redshift. The various line ratios that have been obtained in the many absorptions systems show that there does not seem to be variations vith redshift, but there is a large scatter, even at $z = 0$ (Lucas & Liszt 1994, 1996).

Monitoring of Temporal Variations: The fact that at a given velocity, only one image is covered by a molecular cloud, allows to know the flux ratio between the two images, even with a single dish, without resolving the source. This has been used to monitor the temporal variability, and measure a time-delay between the two images, to deduce a value of the Hubble constant

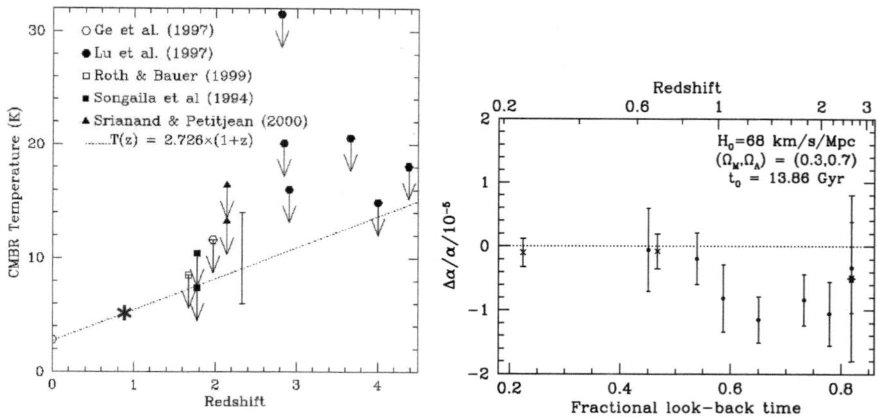

Fig. 47. (*Left*) Measurements of the Cosmic Microwave Background Radiation temperature at various redshifts (from Srianand et al. 2000). The asterisk is the mm constraint obtained from PKS 1830-211, at $z = 0.88582$. (*Right*) Variation with time of the fine structure constant α. Comparison of the mm/HI constraints (*crosses*, from PKS 1413+135 at $z = 0.247$ and B 0218+357 at $z = 0.685$) which do not show any variation, with the recent optical constraints (*star* and *dots*) (from Murphy et al. 2001).

(Wiklind Combes 2001). At the same time, the monitoring carried out over 3 years, with one observation of 1h every week, has revealed evidence for micro-lensing by either stars or molecular clouds in the lensing galaxy.

Measure of $\mathbf{T}_{mb}(\mathbf{z})$: Molecular clouds are usually very cold, with a kinetic temperature of the order of 10 – 20 K. The excitation temperature of the molecules could be even colder, close to the background temperature $T_{\rm bg}$. This occurs when the absorption involves diffuse gas, which is the case in front of PKS 1830-211 where $T_{\rm ex} \sim T_{\rm bg}$ for most of the molecules. This yields one constraint for the measurement of the cosmic background temperature as a function of z. Other measurements have been recently obtained at $z = 2.34$ from H_2, CI and C^+ lines, by Srianand et al. (2000), see Fig. 47.

Variation of Constants: Finally, the large spectral resolution (up to $R = 10^6$) reached in millimetric spectroscopy, and the narrowness of the molecular lines make it possible to improve the constraints on the variation of the fine structure constant $\alpha = e^2/hc$ with space and time.

The method of alkali doublet in optical has been refined to include many multiplets, and a positive result has been reported recently by Webb et al. (2001). Only a variation at high z is detected. Comparison of HI 21-cm absorption with corresponding molecular absorption spectra allows to constrain the variation at other redshifts (Murphy et al. 2001).

7.4 Conclusion

Absorption is a precious tool to observe cold gas at any redshift, although detection is favored towards diffuse gas, with low excitation. Much smaller masses are detected than in emission. In the Milky Way, a new chemistry has been discovered in diffuse clouds, that is nearly as rich as that of dense clouds. The evolution with z of physical and chemical conditions of the molecular component of galaxies can be investigated. Moreover, for intervening galaxies, this is the only way to detect cold gas in galaxies that are not ultra-luminous.

8 Molecules at High Redshift, Perspectives

8.1 Current State of the Art

To explore the history of star formation, and its efficiency, it is of first importance to estimate the molecular content of galaxies as a function of z. The first detection of CO lines at high redshift was done in the hyperluminous object IRAS 10214+4724 at $z = 2.28$ by Brown & Vanden Bout (1992), then confirmed by Solomon et al. (1992); it was a big surprise, since an enormous H_2 mass was first derived. It has now been revised, knowing that the flux is considerably amplified by gravitational lensing (by a factor ~ 50). It may still be surprising to find such huge amounts of CO molecules, especially since the gas is expected to have lower metallicity at high z. Theoretical calculations have shown however that in a violent and concentrated starburst, the metallicity could reach solar values very quickly (Elbaz et al. 1992).

Today, nearly 20 objects have been detected in CO lines at high z: they are often gravitationally amplified, either being multi-imaged by a strong lens,

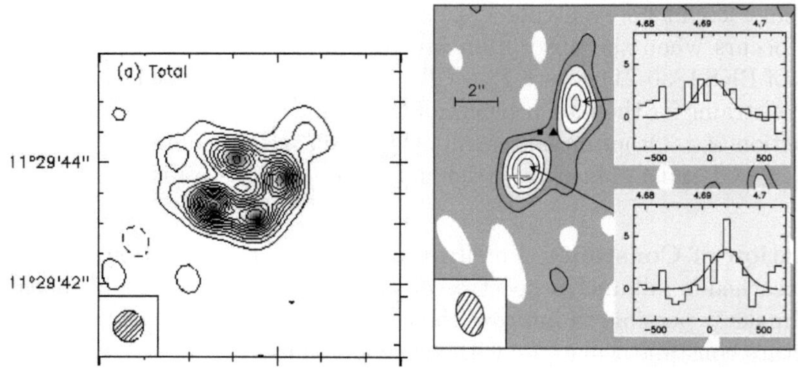

Fig. 48. (*Left*) Image of the Cloverleaf quasar H 1413+117, at $z = 2.56$, mapped in the CO(7 − 6) line with the IRAM interferometer with a 0.6" beam (from Kneib et al. 1998). (*Right*) The quasar BR 1202-07 at $z = 4.69$, mapped in the dust continuum at 1.3 mm with the IRAM interferometer. The insets show the CO(5 − 4) spectra (from Omont et al. 1996).

like the Cloverleaf quasar H 1413+117 at $z = 2.558$ shown in Fig. 48 (Barvainis et al. 1994), the lensed radiogalaxy MG 0414+0534 at $z = 2.639$ (Barvainis et al. 1998), or the possibly magnified object BR 1202-0725 at $z = 4.69$ (Ohta et al. 1996, Omont et al. 1996); or they have been intentionally searched for behind a foreground galaxy cluster, to benefit of a weak amplification, like the submillimeter-selected hyperluminous galaxies SMM 02399-0136 at $z = 2.808$ (Frayer et al. 1998), and SMM 02399-0134 at $z = 1.062$ (Kneib et al. 2002). Often several high-J CO lines are detected, revealing the high temperature of the gas (typical of a starburst at ~ 60 K). Higher temperatures are rare, as in the magnified BAL quasar APM 08279+5255, at $z = 3.911$, where the gas temperature derived from the CO lines is ~ 200 K, maybe excited by the quasar (Downes et al. 1999). There are in the list some objects that do not reveal any amplification, as the weak radio galaxy 53W 002 (Scoville et al. 1997b) at $z = 2.394$, or the radio-quiet quasar BRI 1335-0417, at $z = 4.407$ (Guilloteau et al. 1997). These objects possess the largest molecular contents known ($\sim 10^{11}$ $M_\odot$ with a standard CO/H_2 conversion ratio), as shown in Table 2. The derived molecular masses are so high that H_2 would constitute between 30 to 80% of the total dynamical mass (cf 4C 60.07, Papadopoulos et al. 2000). The standard conversion ratio might there be reduced by factors $3 - 4$ (Solomon et al. 1997).

The strategy to detect high-z CO lines is to search for millimetric continuum first, since the dust emission is much easier to detect because of the negative K-correction. All objects of the Table 2 were first detected in dust emission, either by IRAS or in the submillimeter range through the bolometer arrays SCUBA at JCMT, or MAMBO at IRAM. One advantage of the CO molecule is its relatively tight rotational ladder: at high z, the separation of the lines is much reduced, and many lines may be observed for the same object, in the millimetric range. Since most of the objects detected are amplified, it is however difficult to know the true line ratios, and therefore the excitation of the gas, since the regions emitting in the various lines may have a different spatial extension, and therefore a different lens amplification factor.

What the detections reveal up to now is that the objects are giant starbursts, with hot and dense gas. Only in these conditions, the high-J levels of CO can be excited, and therefore the CO be detected at all at high z. Indeed, when excited, the high-J lines have a much larger flux than the CO$(1-0)$ line, see Fig. 49. This is a crucial advantage of the molecular tracer, with respect to the HI-21 cm emission for instance, which is almost impossible to detect beyond $z = 0.2$.

The 16 objects detected in CO between $1 < z < 5$ are displayed in Fig. 49 together with the detection limits of present instruments. It is clearly visible that all these objects would not have been detected in CO$(1-0)$, but benefit from the higher detection performance for the high-J lines, as will be developed below. One object has been detected in the CO$(1-0)$ line with the

Table 2. CO data for high redshift objects.

Source	z	CO line	S mJy	ΔV km/s	M H_2 10^{10} $M_\odot$	Ref
SMM 02399-0134	1.062	2-1	3	500	2*	1
0957+561	1.414	2-1	3	440	0.4*	2
HR 10	1.439	2-1	4	400	7	3
F 10214+4724	2.285	3-2	18	230	2*	4
53W 002	2.394	3-2	3	540	7	5
H 1413+117	2.558	3-2	23	330	2 – 6*	6
SMM 14011+0252	2.565	3-2	13	200	5*	7
MG 0414+0534	2.639	3-2	4	580	5*	8
SMM 02399-0136	2.808	3-2	4	710	8*	9
6C 1909+722	3.532	4-3	2	530	4.5	10
4C 60.07	3.791	4-3	1.7	1000	8	10
APM 08279+5255	3.911	4-3	6	400	0.3*	11
PSS 2322+1944	4.120	5-4	12	300	25	12
BR 1335-0414	4.407	5-4	7	420	10	13
BR 0952-0115	4.434	5-4	4	230	0.3*	14
BR 1202-0725	4.690	5-4	8	320	10	15

* corrected for magnification, when estimated

Masses have been rescaled to $H_0 = 75$ km/s/Mpc. When multiple images are resolved, the flux corresponds to their sum. (1) Kneib et al. (2002); (2) Planesas et al. (1999) (3) Andreani et al. (2000); (4) Solomon et al. (1992), Downes et al. (1995); (5) Scoville et al. (1997b); (6) Barvainis et al. (1994); (7) Frayer et al. (1999); (8) Barvainis et al. (1998); (9) Frayer et al. (1998); (10) Papadopoulos et al. (2000); (11) Downes et al. (1999); (12) Cox et al. (2002); (13) Guilloteau et al. (1997); (14) Guilloteau et al. (1999); (15) Omont et al. (1996)

VLA, APM 08279+5255, at $z = 3.91$ (Papadopoulos et al. 2001). It appears then that the molecular mass may be $10 - 100$ times higher than estimated from the higher lines.

Extremely Red Objects (EROs): In the aim to search for primeval galaxies, Elston et al. (1988) had identified extremely red objects that are conspicuous only in the near-infrared ($R - K > 5$). Maybe 10% of the submm sources could be EROs (Smail et al. 1999). A proto-typical ERO at $z = 1.44$ (Dey et al. 1999) has been detected in submm continuum (Cimatti et al. 1998), and in CO lines (Andreani et al. 2000).

The SED (Spectral Energy Distribution) for a typical starburst is displayed in Fig. 50 in its rest-frame. Most of the optical flux is obscured and re-radiated in the infrared.

The submillimetric objects detected until now could be evolving in present day bright elliptical galaxies, they have the same comoving density, at $1-2\,L^*$ (Frayer et al. 1999), but they certainly are of a different nature and have a different history than the local spiral galaxies like the Milky Way.

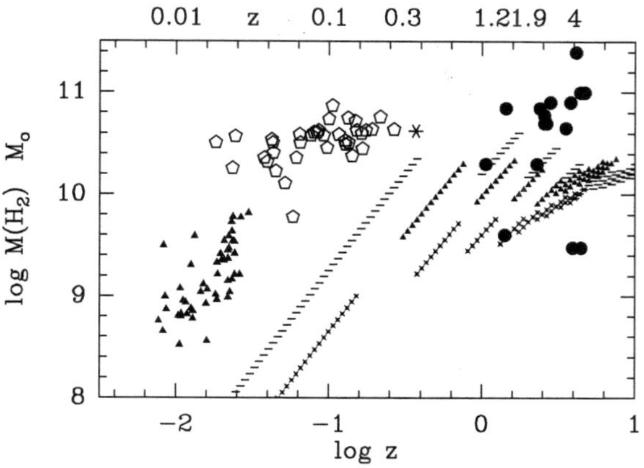

Fig. 49. H_2 masses for the CO-detected objects at high redshift (*full dots*), compared to the ultra-luminous-IR sample of Solomon et al. (1997, *open pentagons*), to the Coma supercluster sample from Casoli et al (1996, filled triangles), and to the quasar 3C 48, marked as a star (Scoville et al. 1993, Wink et al. 1997). The lines delineated by various symbols indicate the $1\,\sigma$ detection limit at the IRAM-30 m telescope of $S(\mathrm{CO}) = 1.0$ Jy km/s with the 3 mm receiver (*hyphens*), 2.0 Jy km/s with the 2 mm (*triangles*) and 1.3 mm (*crosses*) receivers, rms that we can reach in 6 h integration. Note the absence of detected objects between 0.36 and 1 in redshift. The points at high z can be detected well below the sensitivity limit, since they are gravitationally amplified.

8.2 Modelling of a Starburst

The physical conditions of the molecular gas is only known in local starbursts, and the first predictions will be done in assuming the same densities, gas concentration and star formation efficiency for the high-redshift starburst objects. The active region is generally confined to a compact nuclear disk, sub-kpc in size (Scoville et al. 1997a; Solomon et al. 1990, 1997). The gas is much denser here than in average over a normal galaxy, of the order of $10^4\,\mathrm{cm}^{-3}$, with clumps at least of $10^6\,\mathrm{cm}^{-3}$ to explain the data on high density tracers (HCN, CS...). As already shown before, the interpretation of the data requires at least two density and temperature components, and a simple model has been computed with two temperatures, at 30 and 90 K (cf. Combes et al. 1999). The total molecular mass considered is $6{\cdot}10^{10}\,\mathrm{M}_\odot$ and the average column density $N(\mathrm{H}_2)$ of $10^{24}\,\mathrm{cm}^{-2}$, typical of the Orion cloud center.

Going towards high redshift ($z > 9$), the temperature of the cosmic background T_{bg} becomes of the same order as the interstellar dust temperature, and the excitation of the gas by the background radiation competes with that of gas collisions. It might then appear easier to detect the lines (Silk & Spaans 1997), but this is not the case when every effect is taken into account. To have an idea of the increase of the dust temperature with z, the

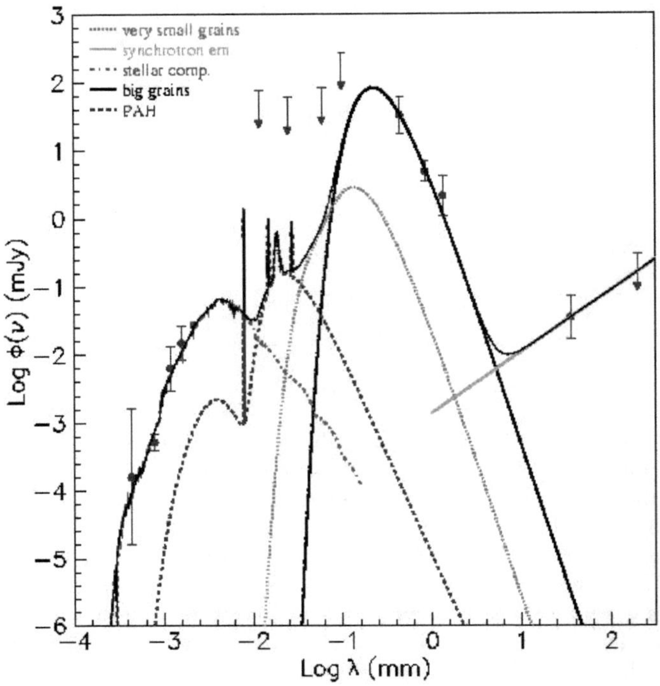

Fig. 50. Spectral modelling applied to one known local starburst (NGC 6090). The stellar population library is derived from the PEGASE package (Fioc & Rocca-Volmerange 1997) for different ages and metallicities. The stellar light is obscured according to the extinction law constrained by nearby starbursts by Calzetti et al. (2000). The absorbed flux is reemitted in the IR. Relying on Désert et al. (1990), this flux is distributed into 3 dust components: big grains, small grains and PAH (from Melchior et al. 2001).

simplest assumption is to consider the same heating power due to the starburst (this comes from the assumed similar geometry for the gas, and similar star-forming efficiency). At steady state, the dust must then radiate the same energy in the far-infrared that it receives from the stars and the background. The stellar power is thus proportional to the quantity $T^6_{\rm dust} - T^6_{\rm bg}$, if the dust is optically thin, and its opacity varies in ν^β, with $\beta = 2$. In these conditions, the energy re-radiated by the dust is proportional to $T^6_{\rm dust}$, while the the energy it received from the cosmic background is proportional to $T^6_{\rm bg}$. Since β can also be equal to 1 or 1.5, or the dust be optically thick, the quantity to keep constant with z is then $T^4_{\rm dust} - T^4_{\rm bg}$, which is in some sense the other extreme; this in fact does not change fundamentally the results.

Simple Two Component Model: The central starburst can be considered as the accumulation of very active star-forming complexes, such as Orion, typically $8.6 \cdot 10^7$ clouds of $700\,M_\odot$ each. The total mass is $6 \cdot 10^{10}\,M_\odot$ concentrated in a circum-nuclear region of 1 kpc in diameter. The two cloud components have both low filling factors. The dense and hot component, corresponding to the star forming cores, has a surface filling factor of $f_s = 1$, but a velocity filling factor $f_v = 0.03$, such as the product $f_s f_v \ll 1$. Each cloud is modeled with an individual velocity dispersion of 10 km/s, embedded in the rotational gradient of the galaxy, 300 km/s. The warm and more diffuse component has $f_s = 100$, but still $f_s f_v = 3$ only, it only begins to be macroscopically optically thick. The computed dust temperature for the warm component begins to differ from 30 K only at $z = 5$, and is 34 K, 58 K and 85 K at $z = 10$, 20 and 30 respectively.

LVG Approximation: To compute the excitation of the CO molecules, and the population of the various J levels, a very simple approximation, justified by the turbulence inside the clouds, is to assume that there exists a large velocity gradient (LVG) in the medium, to reduce the optical depth, and to limit the regions that can absorb the photons to the immediate neighborhood of their emission.

It is in addition verified that most of the clouds, which are individually optically thick in the CO lines, are not overlapping on the line of sight, at a given velocity, and this is due to the low velocity filling factor. There is only mild overlap, that is taken into account in the simple model. For the sake of comparison, a model is also considered of a spherical homogeneous medium, at $T = 50$ K, density $10^3\,cm^{-3}$, $N(H_2) = 3.5 \cdot 10^{24}\,cm^{-2}$.

Model Results: With the two-component disk model, the excitation is almost LTE (high excitation), as shown in Fig. 51. It is not the case for the homogeneous sphere model. The radiative excitation by the background temperature at high z is clear in the figure.

The predictions of the line and continuum intensities as a function of redshift and frequencies are plotted in Fig. 52. The CO flux increases with J as the frequency to the square power, since the line is optically thick, and in the Rayleigh-Jeans domain. But there is no negative K-correction, as in the continuum, and it is always more difficult to detect the CO lines at higher redshift, although the high-J levels help. In the continuum, as seen on Fig. 52, the slope of the curves are of the 4th power, instead of 2 nd, and they overlap at high z.

The model has been computed with lower than solar metallicity, and therefore a lower abundance than the standard $CO/H_2 = 10^{-4}$ (see Fig. 53). The flux is first slow to decrease because of the slight macroscopic optical depth at the beginning, but then falls rapidly.

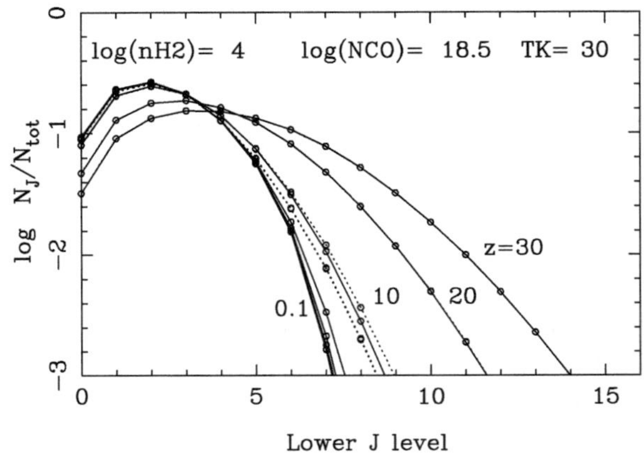

Fig. 51. Relative populations of the CO molecules, for the warm cloud component. The 8 curves correspond to redshifts $z = 0.1$, 1, 2, 3, 5, 10, 20, 30. The value of T_K indicated is for $z = 0$, and the temperature increases significantly from $z = 5$. The full curves are the actual distributions, and the dotted curves correspond to the LTE ones, at the corresponding T_K (from Combes et al. 1999).

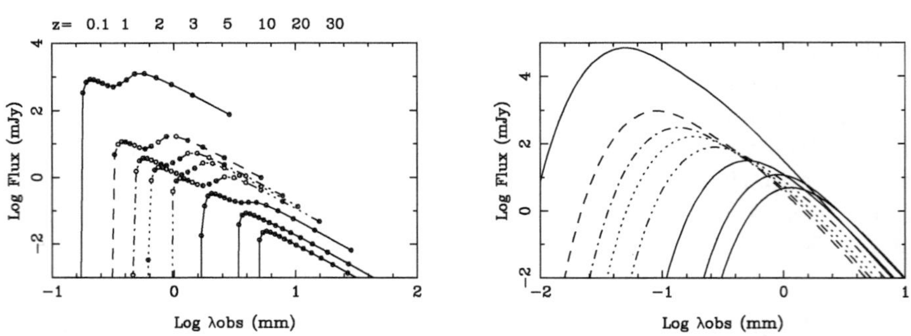

Fig. 52. Expected flux for the two-component (30 and 90 K) cloud model, for various redshifts $z = 0.1$, 1, 2, 3, 5, 10, 20, 30, and $q_0 = 0.5$. (*Left*) are the CO lines, materialised each by a circle (they are joined by a line only to guide the eye). (*Right*) is the continuum emission from dust. It has been assumed here that $T^6_{\rm dust} - T^6_{\rm bg}$ is constant with z (from Combes et al. 1999).

Comparison with Continuum: The CO emission does not benefit from a negative K-correction, in part because of optical thickness, and it will be much more difficult to observe than dust emission, especially for more normal galaxies at high redshift, where the high-J lines might not be excited.

Another point to note on Fig. 52, is that the maximum of CO emission is always at longer wavelengths than for the continuum (notice the different λ scales). The emission peaks at a frequency lower by in average a factor 5 than for the continuum. For $T = 90$ K, the dust peaks at $\sim 60\,\mu$m, and the CO at $\sim 600\,\mu$m. This comes from the fact that the lines always reflect the energy

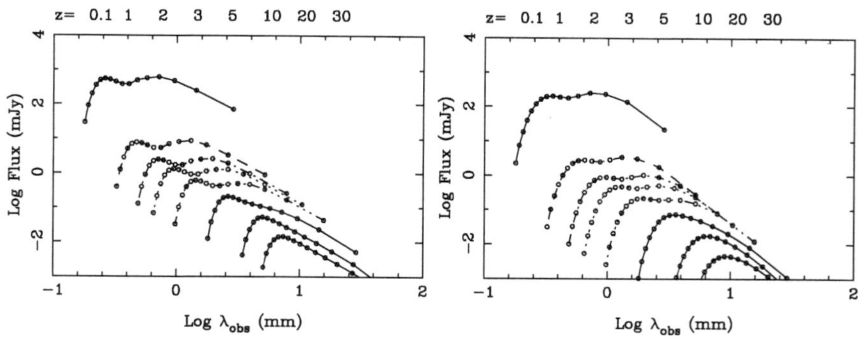

Fig. 53. (*Left*) Expected flux for the two-component cloud model, as in Fig. 52, but with a CO abundance 10 times lower, i.e., $CO/H_2 = 10^{-5}$. (*Right*) Same for $CO/H_2 = 10^{-6}$.

difference between two levels, and only the upper level correspond to the temperature of the gas. This is an advantage, given the largest atmospheric opacity at high frequencies.

The increase of the background temperature does not help in detecting the CO lines, in spite of the higher degree of excitation. Indeed, molecular gas at $T = T_{bg}$ is always undetectable, and the signal received is always obtained by comparison with the off-source position, which is at the T_{bg} temperature.

Given these flux predictions, it is now possible to choose the best strategy to observe a high-z galaxy, depends on excitation and redshift. It is not needed to go at very high frequency, and lower frequency instruments (cf. Table 3) would anyway be complementary to estimate the molecular masses (see the table of integration times with various instruments and frequencies in Combes et al. 1999).

Table 3. Panorama of mm and submm instruments.

Telescope	Area	λ_{min}	θ_{min}
IRAM-30 m	707 m^2	1 mm	10"
IRAM-PdB	6 ant 15 m=1060 m^2	1 mm	0.5"
NRO	6 ant 10 m=509 m^2	1 mm	0.5"
OVRO	6 ant 10 m= 509 m^2	1 mm	0.5"
BIMA	10 ant 6 m=282 m^2	1 mm	0.5"
CARMA*	791 m^2	1 mm	0.5"
SMA	7 ant 6 m=200 m^2	0.3 mm	0.1"
GBT	ant 100 m=7854 m^2	2.6 mm	7"
In project			
LMT	ant 50 m=1963 m^2	1 mm	6"
ALMA	64 ant 12 m=7238 m^2	0.3 mm	0.1 – 0.01"
EVLA	35 ant 25 m=17200 m^2	6 mm	0.004"

* project, combining OVRO and BIMA antennae

8.3 Prediction of Source Counts

How many sources will be detectable with the future instruments? In the frame of the hierarchical theory of galaxy formation, it is possible to estimate the number of starbursts as a function of z, such as to be consistent with the main observations: star formation history, level of cosmic infrared background radiation, and the submm continuum source counts. The cosmology adopted is an Einstein-de Sitter model, $\Omega = 1$, with no cosmological constant, and $H_0 = 75\,\mathrm{km/s/Mpc}$, $q_0 = 0.5$. The number of mergers as a function of redshift can be easily computed through the Press-Schechter formalism (Press & Schechter 1974), assuming self-similarity for the probability of dark halos merging, and an efficiency of mergers in terms of star-formation peaking at $z \sim 2$ (i.e. Blain & Longair 1993).

The main parameter use to fit the data is the rate of energy released in a merger (or the life-time of the event): this rate must increase strongly with redshift (cf. Blain et al. 1999). Once the counts are made compatible with the submm observations, the model yields the various relative contributions of the different redshifts. Figure 54 shows that the intermediate redshifts dominate the continuum source counts ($2 < z < 5$), if we allow the star formation to begin before $z = 6$. At higher dust temperature, the counts are dominated by the highest redshifts ($z > 5$).

The same curves can be obtained for the CO lines, as shown at the bottom of Fig. 54. The scale now indicates much lower numbers, as expected, and they are dominated by the high redshift sources. It is not useful to observe at λ below 1 mm for high-z protogalaxies, but instead to shift towards $\lambda = 1$ cm. At 2 mm wavelength, the dominant contribution is from $2 < z < 5$, and at 5 mm, the dominant sources are at $z > 5$.

8.4 Conclusions

Already some exceptional objects have been detected in the CO lines at high redshift, and the new generation of mm instruments will make a breakthrough in this domain, in detecting a few objects per square degree. The fraction of gas mass in galaxies is expected to increase with redshift as well as the fraction of starburst and their efficiency (shorter dynamical time).

The CO lines bring much more information than the mm continuum, in giving the mass of H_2 in the galaxy, the kinematics, the efficiency of star formation as a function of redshift. They will be used to find redshifts from the continuum-detected objects, when those are too obscured to be detected in the optical or NIR. In any case, they give complementary and unbiased information than in the optical, when the line width does not reflect the total mass, because of outflows and heavy extinction.

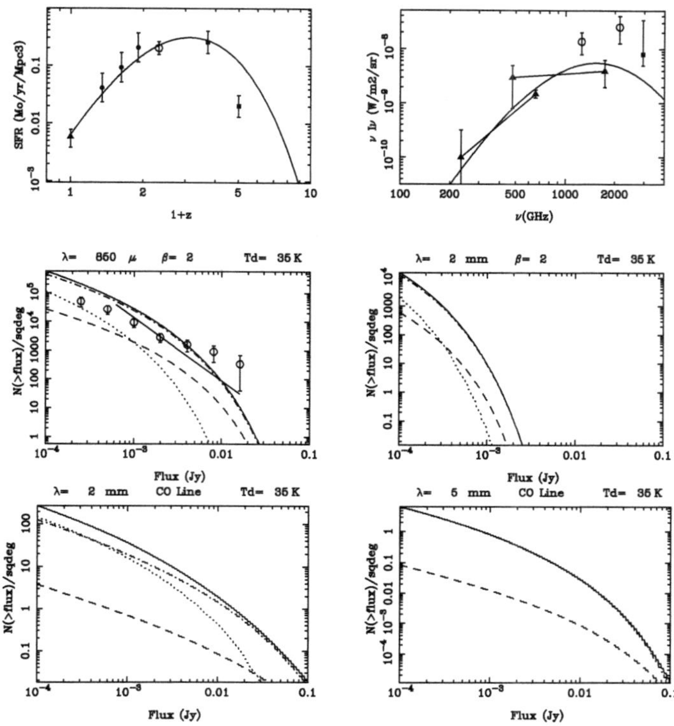

Fig. 54. (*Top left*) History of the star formation rate in the model (*curve*), compared with observations. The point at $z = 0$ is from Gallego et al. (1995), the 3 following ones (*full circles*) from Flores et al. (1999), the empty circle from Yan et al. (1999), and the two high redshift points (*filled squares*), from Pettini et al. (1998). (*Top right*) Model curve for the far-infrared cosmic background. The filled triangles with a line joining them are derived by Puget et al. (1996) from COBE, the filled circles are from Hauser et al. (1998), and the filled square from Dwek et al. (1998). (*Middle left and right*) Source counts predicted at 850 μm and 2 mm respectively: the solid line is the total. Dash is the lowest redshifts ($z < 2$); Dot-dash, intermediate ($2 < z < 5$); Dots are the highest redshifts ($z > 5$). The emission from the dusty starbursts have been assumed a grey-body at a temperature of 35 K, and an opacity varying as ν^2. The empty circles are data from Blain et al. (2000), and the straight line is a fit derived by Barger et al. (1999) at 850 μm. (*Bottom left*) Source counts for the CO lines at an observed frequency of 2 mm, assuming optically thick gas at $T_{\rm ex} = 35$ K. (*Bottom right*) Same for $\lambda = 5$ mm (from Combes 2000).

9 History of the Molecular Component

9.1 Formation of H_2

The formation of H_2 molecules is fundamental in the early universe. H_2 is the main coolant of the primordial gas, to allow the collapse of structures (taken over afterward by HD and LiH). The H_2 molecule may be formed in gas-phase reactions, involving H^- and/or H^+. It cannot be formed directly by radiative association

$$H + H \rightarrow H_2 + \gamma \;, \tag{56}$$

since the molecule has no dipole (on the contrary HD can be formed through radiative association). At very high densities (larger than $10^9\,\mathrm{cm}^{-3}$), the three-body reaction can also take place. It is essential that a third body takes away the formation energy of the molecule.

The formation mechanism is very different at $z = 0$, the dust acts as a catalyst, and helps to form H_2 much quicker. As soon as dust is present, and ions have decreased, H_2 is formed preferentially on dust.

The H_2 formation on grains may occur according to two processes (e.g. Herbst, 2000):

- The Langmuir-Hinshelwood process, in which H atoms land on a grain and diffuse on the surface, by either tunneling, or hopping.
- The Eley-Rideal process, where H atoms are fixed in position, chemisorbed, and a reaction occurs only when another H lands atop.

The number of binding sites for adsorbers is of the order of $N = 10^6$ on a 0.1 μm grain. The binding energy, or energy required for desorption (E_D) varies, according to the composition of the grain, it is for H on olivine of $E_D = 372\,\mathrm{K}$ (Katz et al. 1999), and H on amorphous carbon $E_D = 658\,\mathrm{K}$; for H_2 respectively 314 and 542 K (Katz et al. 1999).

Since the desorption energy is larger for heavier molecules, only H_2 can be formed like that, at 10 K. The grain, as a catalyst, absorbs the energy of formation, that can help to desorb another molecule.

At low temperature, H-atoms don't diffuse on the grain, and the second (Eley-Rideal) process is the only one. Recent experimental results have shown that the rate of diffusion of H-atoms on interstellar-like surfaces is significantly lower than estimated in the past.

The formation of H_2 on grain surfaces is still poorly understood, because the details characteristics of the surfaces are poorly constrained and time-dependent (Herbst 2000).

The interaction of H-atoms with the two kinds of interstellar grains, carbonaceous and silicate, have been studied theoretically (although more for the H-graphite interaction). The first kind include the PAH-like grains, for instance coronene ($C_{24}H_{12}$). For the second kind, the formation of H^- on MgO or forsterite (Mg_2SiO_4) by charge transfer would enable the formation of H_2 via associative detachment (Sidis et al. 2000). Experimental formation of H_2 on amorphous carbon, or olivine, in the conditions of cold temperature $5-20\,\mathrm{K}$ to simulate interstellar conditions, have also given some insight in the mobility of H on grains (Pirronello et al. 1999).

Formation in Gas Phase Reactions: Lepp & Shull (1984) have computed the formation of the interesting molecules H_2, HD, LiH and HeH^+ during the post-recombination epoch ($z = 300 - 30$). In standard nucleosynthesis, there exist only H, D, ^{4}He, ^{3}He, ^{7}Li, and molecules have to be formed without dust. The molecules dominate the cooling and are responsible for the first cloud collapse and star formation.

Cooling by Lyman-α is ineffective below $T = 8000\,\mathrm{K}$, and therefore after $z = 500$. H_2 is formed in small quantities from H^-, and provides sufficient cooling to trigger the cloud collapse. Cooling is necessary since a Jeans mass will increase in adiabatic collapse, and there would be no fragmentation.

The two well known processes that can form H_2 in gas-phase reactions, at low densities, are the H^- and the H_2^+ processes. The first process (cf rates by Bieniek & Dalgarno 1979) is:

$$\mathrm{H} + e^- \rightarrow \mathrm{H}^- + \gamma \,, \tag{57}$$

$$\mathrm{H}^- + \mathrm{H} \rightarrow \mathrm{H}_2 + e^- \,. \tag{58}$$

The second process (cf. Karpas et al. 1979) is:

$$\mathrm{H} + \mathrm{H}^+ \rightarrow \mathrm{H}_2^+ + \gamma \tag{59}$$

$$\mathrm{H}_2^+ + \mathrm{H} \rightarrow \mathrm{H}_2 + \mathrm{H}^+ \,. \tag{60}$$

At high density, when the clouds are already collapsed, a third process intervenes (e.g. Palla et al. 1983), the three-body process, for $n > 10^9\,\mathrm{cm}^{-3}$:

$$3\mathrm{H} \rightarrow \mathrm{H}_2 + \mathrm{H} \,. \tag{61}$$

Results of Chemistry: Lepp and Shull (1984) made their computations in a standard Friedman model (i.e. $\Lambda = 0$), and $\Omega_0 = \Omega_b = 0.1$ (with only baryonic dark matter). The resulting evolution of fractional abundances are plotted in Fig. 55, with all channels for H_2 formation indicated. Once Compton heating becomes less than expansion cooling, radiation and matter cool adiabatically:

$$T_\mathrm{r} \sim (1+z) \quad \text{and} \quad T_\mathrm{m} \sim (1+z)^2 \,. \tag{62}$$

H_2 reaches an abundance of 10^{-5} in the diffuse medium. As for cooling rates per H_2 molecule, the LiH dominates, for $HD/H_2 = 10^{-4}$ and $LiH/H_2 = 10^{-6}$ (but since then, the abundance of the LiH molecule is thought to be much less, due to revised radiative association formation rate of LiH, Dalgarno et al. 1996).

The cooling is enough for cloud to collapse, without being stopped by adiabatic regime. When a self-similar free-fall collapse is considered, at $z = 50$, with the Jeans length $\lambda_\mathrm{J} = 560\,\mathrm{pc}$, and corresponding mass $M_\mathrm{J} = 10^5\,\mathrm{M}_\odot$, much more molecules are formed. The most efficient cooling is by H_2, then HD, then LiH. The cooling rotational lines become optically thick at:

$N = 10^{24}\,\mathrm{cm}^{-2}\Delta v$ for H_2,

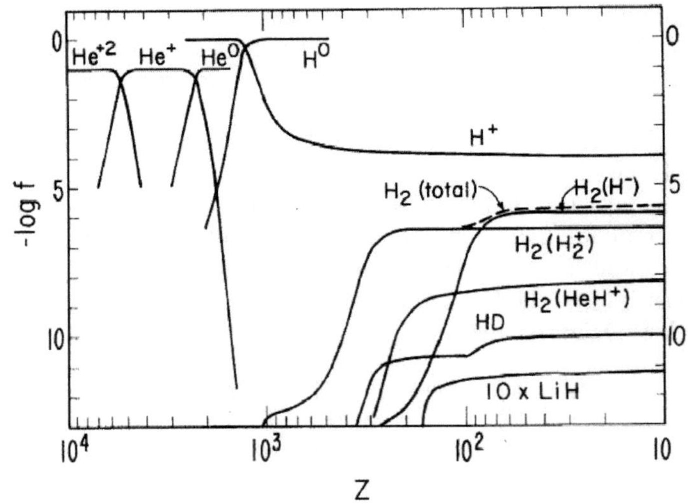

Fig. 55. Abundance fractions of atoms, ions and molecules versus redshift in a cosmological model where $\Omega = \Omega_b = 0.1$, $\Lambda = 0$ and $h = 0.5$. Various H_2 formation channels are indicated by intermediate species in parentheses (from Lepp & Shull, 1984).

$$N = 10^{20}\,\mathrm{cm}^{-2}\Delta v \text{ for HD,}$$
$$N = 10^{15}\,\mathrm{cm}^{-2}\Delta v \text{ for LiH,}$$

where Δ v in km/s is the velocity width of the line induced by cloud collapse, estimated at $\sim$ 100 by Lepp & Shull (1984). H_2 becomes optically thick at the same point as 3-body reaction sets in, then the H_2 formation by 3-body reaction begins to yield a significant heating. LiH remains thin and effective coolant until higher densities of LiH/H$= 10^{-10}$.

9.2 Chemistry of the Early Universe

A more detailed model, including 87 gas phase reactions with H, D, He, Li chemistry, has been devised by Galli & Palla (1998), who also compare their results with those of the various models in the literature (Lepp & Shull 1984; Black 1991; Puy et al. 1993; Palla et al. 1995; Giroux & Shapiro 1996). They take into account the smaller radiative association for LiH, which was overestimated by a factor 100 (Stancil et al. 1996), and also non equilibrium chemistry in radiative shocks during collapse, better H recombination, 2-3 less electrons, etc. They pointed out the minimal network of chemical reactions, which can reproduce the more complex network, it includes 13 reactions involving H and He, 6 for deuterium, and 14 for Li (Galli & Palla 1998).

At $z = 1$, they find $e^-/H = 3\cdot 10^{-4}$, $H_2/H \sim 10^{-6}$, in the diffuse medium. They improved the cooling functions of H_2 for $H-H_2$ collisions, as a function of density. The cooling function per molecule of H_2, HD, LiH and H_2^+ in the low density limit ($n < 100\,\mathrm{cm}^{-3}$) is plotted in Fig. 56. They also show the influence of the cosmological parameters: H_2 is almost independent of cosmology, while Li is more variable, LiH and HD can vary by $2-3$ orders of magnitude.

9.3 First Structures in the Dark Age

In the current best cosmological models, the first structures to form are the smallest one (bottom-up scenario); the structures form from small density fluctuations, that first expand with the Universe while growing linearly, then collapse non-linearly to bound systems. These fluctuations could be adiabatic, in which case they would be damped on scales below $3 \cdot 10^{13}(\Omega h^2)^{-5/4}\,\mathrm{M}_\odot$ before recombination (Silk 1968; where h is the Hubble constant in units of $100\,\mathrm{km\,s^{-1}\,Mpc^{-1}}$). Isothermal fluctuations can also exist (and, of course, also intermediate states), and in such cases, the first non-linear objects are much smaller.

The growth of adiabatic fluctuations at a scale of $10^{14}\,\mathrm{M}_\odot$ (8 Mpc) is described schematically in Fig. 57. Radiation fluctuations grow until their

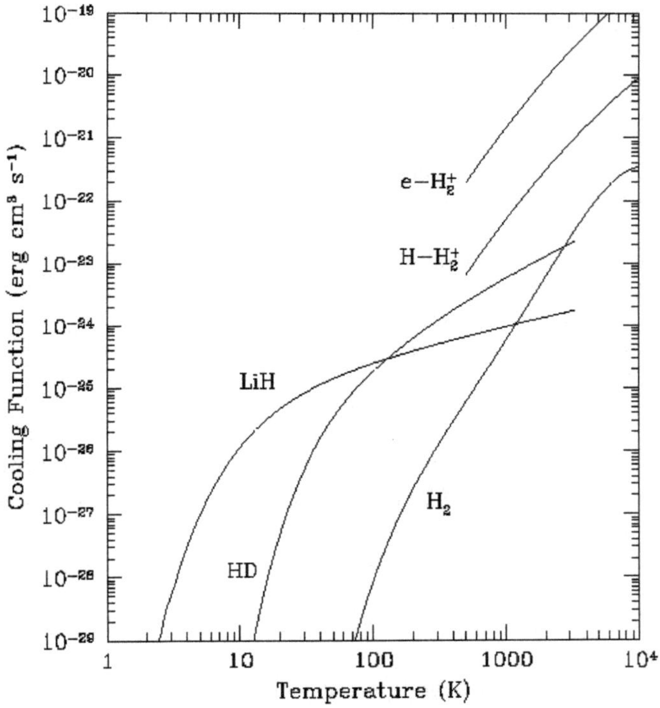

Fig. 56. Cooling function per molecule of H_2, HD, LiH and H_2^+ in the low density limit n(H) < 100 cm^{-3} (from Galli & Palla, 1998).

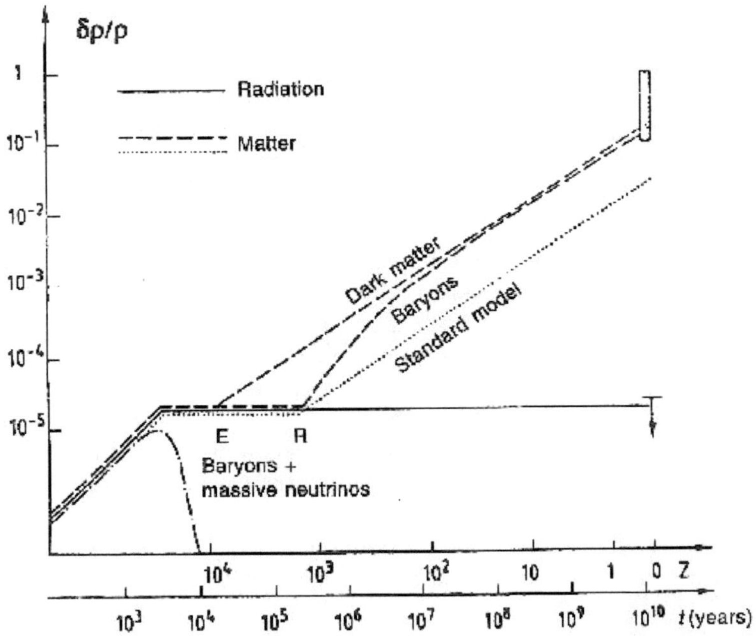

Fig. 57. The growth of adiabatic fluctuations at a scale of $10^{14}\,M_\odot$ (8 Mpc). The fluctuations of radiation (*solid line*) initially grow up to the point when the mass contained within the horizon (a sphere of radius *ct*) is equal to their own mass. They remain constant thereafter and their amplitude can be fitted so as to be compatible with current observations (*arrow*). The fluctuations of matter in the standard model (*dots*) follow the radiation until recombination (R). After they can grow, but too slowly to reach the amplitude observed (*hatched rectangle*). The fluctuations of non-baryonic dark matter also follow the radiation, but until the end of radiative era (E). The amplitude of the dark-matter fluctuations is therefore larger, and the baryons which are in this model only 10% of the total mass, fall in the dark potential wells. In the massive-neutrino scenario the fluctuations are destroyed by dissipative effects (*dot-dash*) (from Turner 1984).

mass equal that contained in the horizon *ct*; they remain constant after, and are calibrated with their present amplitude. The matter fluctuations follow radiation, if ionised. After recombination they grow too slowly in the standard model without non-baryonic matter, to reach the observed level today. Fluctuations of non-baryonic matter subject only to gravity, can begin to grow earlier, when the matter gravity dominates radiation. They trap baryons in their potential well, and the present amplitude of structures can be explained.

The power spectrum of the fluctuations, as a function of wave number k is assumed self-similar, as a power law n:

$$|\delta k|^2 = P(k) \sim k^n \;, \tag{63}$$

with $n = 1$ on large-scales (as confirmed by COBE), and becoming $n = -3$ on small scales, after a spectrum tilt, when the matter density equals the radiation density $\rho_r \sim \rho_m$ at the horizon scale (Peebles 1982). The density fluctuations can then be deduced:

$$\frac{\delta\rho}{\rho} \propto M^{-(n+3)/6} \tag{64}$$

then the largest masses to become non-linear after recombination, i.e., at $z \sim 1500$, are of the order of $M \sim 10^{6-8}\,\mathrm{M}_\odot$, depending on Ω, h and n, if the spectrum is calibrated from the presently observed correlation function (e.g., Peebles 1980). That means that all masses smaller than $M \sim 10^{6-8}\,\mathrm{M}_\odot$, and larger than the Jeans mass, which is $M \sim 10^5\,\mathrm{M}_\odot(\frac{\Omega_\mathrm{b}}{0.06})^{-1/2}(\frac{h}{0.5})^{-1}$ at $z = 1000$, will collapse, become bound, and decouple from expansion (at this epoch the Jeans length is of the order of $2\,\mathrm{pc}\,(\frac{\Omega_\mathrm{b}}{0.06})^{-1/2}(\frac{h}{0.5})^{-1})$. In fact, since in every hierarchical model with $n > -3$ the smallest masses have more density contrast, structures at precisely the Jeans mass collapse first. The masses correspond to typical $z = 0$ giant molecular clouds.

Nature of the Collapse: The nature of the collapse has been studied by many authors. If cooling is efficient enough ($\tau_\mathrm{cool} \sim \tau_\mathrm{ff}$), the collapse is quasi-isothermal, and fragmentation occurs, since the Jeans mass becomes smaller and smaller as the density increases (e.g., Hoyle 1953). This means that pressure forces are at most equal to gravity forces, and cannot stop the collapse. This fragmentation process occurs currently in the Galactic interstellar medium in a self-similar hierarchical structure (Larson 1981; Scalo 1985), and might even be pursued down to very low masses in particularly cold and quiet regions, such as the outer parts of galaxies (Pfenniger & Combes 1994). Fragmentation is limited by opacity, and the smallest fragments (or clumpuscules), which are at the transition of being pressure supported, are today of the order of $10^{-3}\,\mathrm{M}_\odot$, and their mass grows slowly, as $T^{1/4}$ or $(1+z)^{1/4}$, with redshift. This assumes a quasi-isothermal regime for the bottom of the hierarchy (containing most of the mass); therefore it is important that gas cooling remains efficient.

For gas just after recombination, above 10^4 K the main coolant is atomic hydrogen (by collisional excitation of Lyman-α), and below this temperature the vibration-rotation lines of molecular hydrogen take over until $T = 200$ K, and we have seen in the previous section that significant quantity of H_2 molecules forms through H^- and H_2^+. Below 200 K, HD is then more efficient (Palla & Zinnecker 1987; Puy & Signore 1996). Many groups have tackled the problem of computing the physico-chemistry of the primordial gas, to determine the size of the first forming bound structures (Yoneyama 1972; Hutchins 1976; Carlberg 1981; Palla et al. 1983; Lepp & Shull 1984). All of them have found that the cooling is indeed efficient, as soon as the redshift is below $z \sim 200$, and the calculated masses of fragments are in the wide range $0.1 - 100\,M_\odot$. Even if clouds of just the Jeans mass see their collapse somewhat delayed because of pressure forces, the latter are negligible for clouds of $M > 10^6\,M_\odot$ (Lahav 1986).

Most studies were interested in the first generation of stars. Already Peebles & Dicke (1968) concluded that the first objects to form were the globular clusters. Palla et al. (1983) in forming H_2 through the very efficient 3-body process, concluded that all the primordial gas is converted in H_2 at a density of $10^{12}\,cm^{-3}$, and that the Jeans mass eventually falls below $0.1\,M_\odot$, allowing the whole mass spectrum presently observed, for the first forming stars. Rotation could hinder the collapse (Kashlinsky & Rees 1983), and this could lead to Population III low-mass stars as well as super-massive objects.

Observations of the present ISM show that the efficiency of star formation is very low, and it is very likely that the bulk of the gas in this first collapse of GMC-size primordial clouds after recombination and decoupling form clumpuscules (Pfenniger & Combes 1994). Most of the fragments remain pressure supported, as soon as optically thick, and form a fractal of molecular clouds. All the hierarchical structure remains at the background temperature.

9.4 H_2 Formation and Cooling

To test this possibility, the H_2 formation and gas cooling has been computed, in the hypothesis of fractal formation (Combes & Pfenniger 1998). The simple scheme of chemistry reactions summarised by Tegmark et al. (1997), including the H^-, the H_2^+ and 3-H processes, was used. The cooling is essentially due to H_2 below 1000K, then HD below 100K.

The density is computed from the simple top-hat model before virialisation. Then assuming efficient cooling, fragmentation occurs, as a fractal structure, of dimension $D = 1.7$. The number of fragments at each scale is $N = 8$, compatible with the fractal of the interstellar medium in the Milky Way today (Scalo, 1985). The scale ratio between two levels of the hierarchy is then $N^{1/D} = 3.4$. The densities are in the ratio $r_d = N^{(3-D)/D} = 4.9$. In the hypothesis of efficient cooling, the density as a function of time can be computed by steps corresponding to the free-fall time $\tau_{\mathrm{ff},i}$ of a given level i of the hierarchy: $n(t) = n_i r_d^{(t-t_i)/\tau_{\mathrm{ff},i}}$. Fragmentation stops when the clumps become opaque to the H_2 rotational IR lines, i.e., when $n \sim 10^{10}\,cm^{-3}$.

The cooling is efficient, if all the energy of the collapse is radiated away, and the temperature of the gas in the collapsing structures never becomes higher than the virial temperature:

$$T_{\mathrm{vir}} \sim 1420\,\mathrm{K} \left(\frac{M}{10^5 M_\odot} \right)^{2/3} \frac{(1 + z_{\mathrm{vir}})}{100} , \tag{65}$$

where z_{vir} is the redshift at which the giant cloud virialises. Some results of the computations are displayed in Fig. 58, for $z_{\mathrm{vir}} = 100$. The horizontal full line indicates the virial temperature of the Jeans mass at recombination T_{vir}.

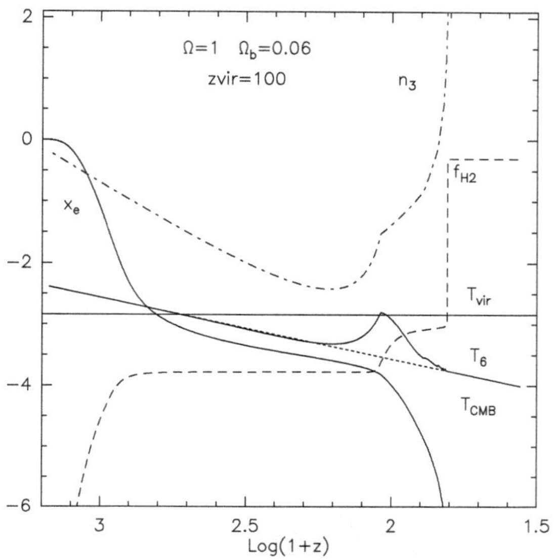

Fig. 58. Evolution of temperature (T_6 in 10^6 K), molecular fraction (f_{H_2}), ionization fraction (x_e) and density (n_3 in $10^3\,cm^{-3}$) for a gas cloud after recombination, collapsing at $z_{vir} = 100$. The Hubble constant is $h = 0.5$, $\Omega = 1$ and $\Omega_b = 0.06$ (from Combes & Pfenniger, 1998).

The density increases by steps (corresponding to each scale of the recursive fragmentation), their temperature stays always below T_{vir}, and the clumps are completely molecular at the end.

9.5 What Are the First Structures?

This old question has received many different answers, according to assumptions on the IMF and efficiency of star formation. The first objects could be gas clumps, brown dwarfs, $0.5\,M_\odot$ MACHOs, standard IMF stellar systems, or massive black holes. Numerically, often symmetries are assumed (Bodenheimer 1986) with hydrodynamics, chemistry, but following the fragmentation is difficult. There is the problem of the high dynamical range ($8 - 15$ orders of magnitude), of non-equilibrium chemistry, non-linear dynamics, and in particular problems when the structures become optically thick.

The highest resolution simulations have been carried out recently, with adaptive mesh refinement, to form the first star (see Fig. 59, Abel et al. 2000). While the virial radius of $5.6 \cdot 10^6 M_\odot$ halo is 100 pc, the cell size is 0.024 pc and the simulation box 6.4 kpc. In a hierarchical scenario, dense gas blobs form at filaments crossing (Abel et al. 1997, 1998).

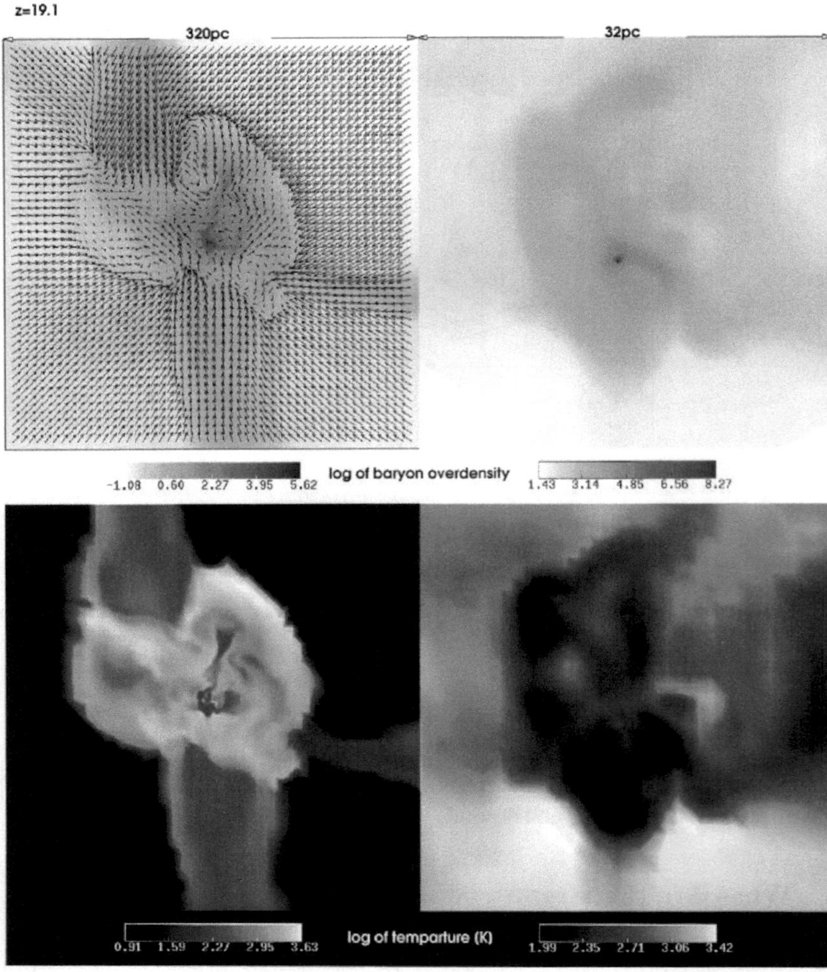

Fig. 59. Density and temperature in the first cosmological objects expected to form in hierarchical structure formation scenarios. The upper panels show the log of the baryonic overdensity in a slice through the point of highest gas density at a scale of 320 pc (*left*) and 32 pc (*right*). The lower panels give the corresponding plots of the log of the gas temperature. Additionally, the velocity field is also visualized in the upper left panel. Note that the computational volume simulated is 20 times larger than the left panels (from Abel et al. 2000).

The simulation reproduces a low star-forming efficiency: less than 1% of the gas will form stars. The first stars or quasars are likely to appear in clumps of $T_{\rm vir} \sim 10^4$ K at $z \sim 20$, and H_2 cooling plays a major role.

The direct H_2 radiation will not be detectable, but as soon as there is star formation with larger than 1% efficiency, or quasar radiating at Eddington limit, it will be detectable with NGST at $z = 15$ (Haiman & Loeb 1998). In case of mini-quasars, there will be reionisation by $z \sim 10$.

The average star formation efficiency can be probed through the enrichment of the Lyman-α forest. It is not larger than 2% in collapsed regions, not far from the 1% deduced from numerical simulations. But this deduction is based on the hypothesis of normal IMF, which is very uncertain.

Efficiency of Star Formation: Observations show that the efficiency of star formation is highly variable, and can be very low in some cases (e.g., in low surface brightness galaxies). Since the physical conditions of the gas at $z \sim 10 - 100$ is similar to the present outer parts of galaxies, we expect that the conditions for triggering global star formation are generally not met. The reasons for low star-forming efficiency are then:

- No deep potential wells, since galaxy-size structures have not yet virialized
- No critical surface density for star formation
- Collapsing mass always of the order of the Jeans mass, and not far from pressure support, since the Jeans mass decreases gradually as fragmentation proceeds.

When objects of galaxy-size collapse, $M \gg M_J$, then violent instabilities, and starburst can occur, unless rotation stabilises the perturbations.

Therefore, only sporadic star-formation (or even MACHOs formation) should occur at high redshift ($z > 10 - 100$), and that might be sufficient to initiate the reionization and the re-heating of the intergalactic medium (IGM), as is necessary to explain the high ionization fraction of the Lyman-α absorbers (Tegmark et al. 1984) and the omnipresent HeII gas at $z \sim 2$ (Jacobsen et al. 1994; Davidsen et al. 1996; Reimers et al. 1997). The first generation of stars could form today a small percentage of the dark halos around galaxies under the form of white or brown dwarfs (Alcock et al. 1997).

Once the bulk of the gas mass has condensed into molecular gas at $z \sim 150$, the clumps survive the reionization because of their high column density ($> 10^{25}\,\mathrm{cm}^{-2}$), and their fractal structure (Combes & Pfenniger 1997). At the interface with the extra-galactic background radiation, they will have an ionised and an atomic envelope. If the background comes essentially from the quasar UV light, and provides an ionization rate of $\xi \sim 2 \cdot 10^{-14}\,\mathrm{s}^{-1}$, corresponding to the study of low-redshift Lyman-α absorption lines (Madau 1992), then the column density at the interface is $N_{\mathrm{HI}} \sim 10^{18}\,\mathrm{cm}^{-2}$, as is observed in the outskirts of HI disks (Corbelli & Salpeter 1993).

Galaxy Formation: In this picture, a main peak of star formation will occur at galaxy formation, which could range between $z = 50$ and $z = 0$ (at $z = 200$, normal 10kpc galaxies would overlap). These galaxies would form through the merging of smaller entities. The protogalaxy collapse is violent

($M \gg M_J$), and starburst may occur in their centers. In the outer parts, there can remain cold self-gravitating gas, as a reservoir for quiescent star formation, or new starburst in hierarchical merging.

Cluster Formation: Another peak of star-formation is expected to occur when larger structures begin to collapse at $z = 2$, and galaxies merge in large quantities. Outer gaseous haloes are stripped, heated, and contribute to form the hot IGM, at T_{vir} of the galaxy cluster. A multiphase medium may survive (e.g. Ferland et al. 1994), in particular accounting for the cooling gas at the center (David et al. 1995).

Evolution of Gas Content: The evolution of the gas in galaxies can be traced through the damped Lyman-α absorbing systems (DLA), that are thought to come from external parts of galaxies.

The number of DLA traced by MgII between $z = 0$ and 1.65, then by Lyman-α are plotted in Fig. 60. They follow the power-law distribution:

$$n(DLA)(z) \propto (1 + z)^{2.27} \tag{66}$$

which is a strong evolution. However this result depends strongly on a few rare systems at the present time. Figure 60 shows that the gas density observed at $z = 3$ corresponds to the luminous density observed today (square in the figure).

9.6 Conclusion

After recombination, GMCs of $10^{5-6}\,M_\odot$ collapse and fragment down to $10^{-3}\,M_\odot$, due to efficient H_2 cooling. The bulk of the gas might not form stars but a fractal structure, in statistical equilibrium with the cosmic background temperature. Only a sporadic, low level, star formation is expected.

After the first stars (and the first quasars?) have formed, reheating and reionization of the universe begins. The cold gas survives and will be assembled in larger scale structures to form galaxies.

The fact that the bulk of the baryons condense without forming stars is a way to solve the cooling catastrophe (Blanchard et al. 1992). The H_2 fractal structure regulates the consumption of gas into stars, and play the role of a reservoir for later starbursts, and long-term quiescent star formation.

References

1. Abel T., Anninos P., Norman M.L., Zhang Y.: 1998, ApJ 508, 518
2. Abel T., Anninos P., Zhang Y., Norman M.L.: 1997, NewA 2 181
3. Abel T., Bryan G.L., Norman M.L.: 2000, ApJ 540, 39
4. Alard C.: 2001, A&A 379, L44

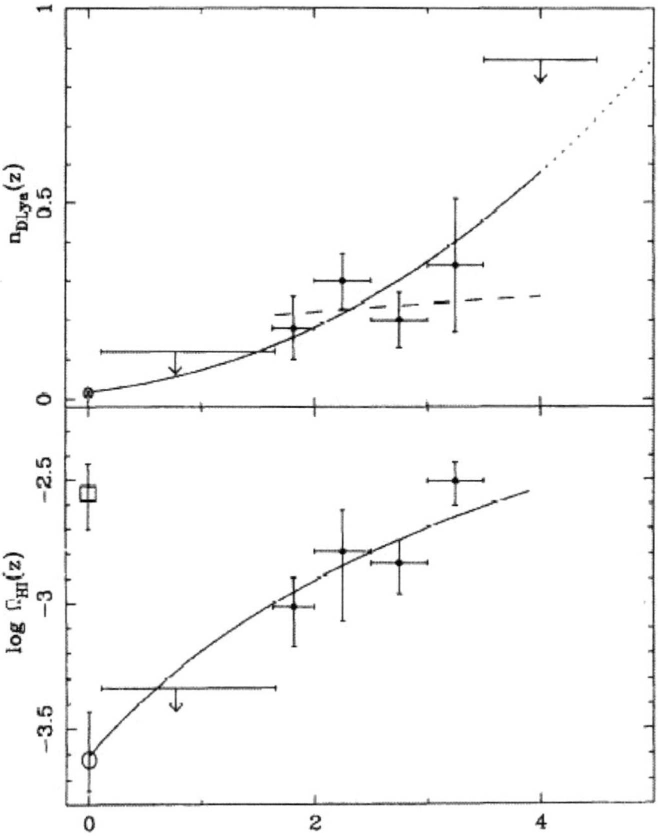

Fig. 60. (*Top*) Redshift distribution of the number density of Damped Lyman-α absorbers. At $z = 0$ the open circle is estimated from nearby galaxies. (*Bottom*) Evolution as a function of redshift of $\Omega_{\rm HI}$, the mass density of atomic hydrogen, for $q_0 = 0.5$. Power-law fits to the data are shown (from Rao et al. 1995).

5. Alcock C. et al.: 1997, ApJ 486 697
6. Andreani P., Cimatti A., Loinard L., Röttgering H.: 2000, A&A 354, L1
7. Arnault P., Kunth D., Casoli F., Combes F. 1988 A&A 205, 41
8. Bally J., Stark A. A., Wilson R. W., Henkel C.: 1987, ApJS 65, 13
9. Bally J., Stark A. A., Wilson R. W., Henkel C.: 1988, ApJ 324, 223
10. Bania T. M., Stark A. A., Heiligman G. M.: 1986, ApJ 307, 350
11. Barger A.J., Cowie L.L., Sanders D.B.: 1999, ApJ 518, L5
12. Barnes J., Hernquist L.: 1992, ARAA 30, 705
13. Barnes J., Hernquist L.: 1992, Nature 360, 715
14. Barone L. T., Heithausen A., Hüttemeister S., Fritz T., Klein U.: 2000, MNRAS 317, 649
15. Barvainis R., Alloin D., Guilloteau S., Antonucci R. 1998, ApJ 492, L13
16. Barvainis R., Tacconi L., Antonucci R., Coleman P.: 1994, Nature 371, 586
17. Bastien P.: 1981, A&A 93, 160

18. Bazell D., Desert F.X.: 1988, ApJ 333, 353
19. Bekki K. 1998 ApJ 499, 635
20. Bekki K., Freeman K.C.: 2002, ApJ in press (astro-ph/0206208)
21. Bieniek R.J., & Dalgarno A.: 1979, ApJ 228, 635
22. Binney J., Gerhard O. E., Stark A. A., et al.: 1991, MNRAS 252, 210
23. Black J.H.: 1991, in "Molecular Astrophysics", ed T.W. Hartquist, Cambridge University Press, p. 473
24. Black J.M., Dalgarno A.: 1976, ApJ 203, 132
25. Blain A.W., Ivison R.J., Kneib J-P., Smail I.: 2000, in Proceedings of "The Hy-redshift Universe", eds. A.J. Bunker, W.J.M. van Breugel, (astro-ph/9908024)
26. Blain A.W., Jameson A., Smail I. et al. : 1999, MNRAS 309, 715
27. Blain A.W., Longair M.S.: 1993, MNRAS 264, 509
28. Blanchard A., Valls-Gabaud D., Mamon G.A.: 1992, A&A 264 365
29. Blitz L., Williams J.P.: 1997 ApJ 488, L145
30. Bloemen J.B.G.M.: 1989, ARAA 27, 469
31. Bodenheimer P.: 1986, "Evolution of primordial gas clouds", Final Technical Report, 1984 -1986 California Univ., Santa Cruz.
32. Boissier S., Prantzos N.: 2000, MNRAS 312, 398
33. Borne K. D., Bushouse H., Lucas R. A., Colina L.: 2000, ApJ 529, L77
34. Bosma A.: 1981, AJ 86, 1825
35. Bothun G., Impey C., McGaugh S.: 1997, PASP 109, 745
36. Bottema R.: 1993, A&A 275, 16
37. Braine J., Combes F.: 1992, A&A 264, 433
38. Braine J., Combes F.: 1993, A&A 269, 7
39. Braine J., Duc P.-A., Lisenfeld U. et al.: 2001, A&A 378, 51
40. Braine J., Lisenfeld U., Duc P-A., Leon S.: 2000, Nature 403, 867
41. Brinks E., Burton W.B.: 1984, A&A 141, 195
42. Bronfman L., Cohen R.S., Alvarez H. et al.: 1988, A&A 230, 21
43. Brown R., Vanden Bout P.: 1992, ApJ 397, L19
44. Bureau M., Freeman K. C., Pfitzner D. W., Meurer G. R.: 1999, AJ 118, 2158
45. Burton W.B., Gordon M.A.: 1978, A&A 63, 7
46. Burton W.B.: 1992, in "The Galactic Interstellar Medium", Saas-Fee Lectures 21, p.1, Springer-Verlag
47. Buta R., Combes F.: 1996, Fundamental Cosmic Physics, 17, p. 95-282
48. Calzetti D., Armus L., Bohlin R.C. et al. 2000, ApJ, 533, 682.
49. Carilli C.L., Perlman E.S., Stocke J.T. 1992, ApJ 400, L13
50. Carilli C.L., Rupen M.P., Yanny B. 1993, ApJ 412, L59
51. Carlberg R.G.: 1981, MNRAS 197 1021
52. Casoli F., Dickey J., Kazes I. et al.: 1996, A&AS 116, 193
53. Casoli F., Dupraz C., Combes F.: 1992, A&A 264, 55
54. Casoli F., Dupraz C., Combes F.: 1992, A&A 264, 55
55. Casoli F., Sauty S., Gerin M. et al. 1998, A&A 331, 451
56. Cernicharo J., et al. 1997, A&A 323, L25
57. Charmandaris V., Combes F., van der Hulst J.M.: 2000, A&A 356, L1
58. Chin Y-N., Henkel C., Langer N., Mauersberger R.: 1999, 512, L143
59. Chin Y-N., Henkel C., Millar T. J., Whiteoak J. B., Mauersberger R.: 1996, A&A 312, L33
60. Cimatti A., Andreani P., Röttgering H., Tilanus R.: 1998, Nature 392, 895

61. Clarke D.A., Burns J.O., Norman M.L., 1992, ApJ 395, 444
62. Combes F. 2001, Advanced Lectures on the Starburst-AGN Connection, Eds Itziar Aretxaga, Daniel Kunth, and Rl Mjica: World Scientific, p.223 (astro-ph/0010570)
63. Combes F., Becquaert J-F.: 1997, A&A 326, 554
64. Combes F., Braine J., Casoli F., Gerin M., van Driel W.: 1992, A&A 259, L65
65. Combes F., Dupraz C., Casoli F., Pagani L.: 1988, A&A 203, L9
66. Combes F., Maoli R., Omont A.: 1999, A&A 345, 369
67. Combes F., Pfenniger D.: 1997, A&A 327 453
68. Combes F., Pfenniger D.: 1998, in Proceedings of Firenze-Arcetri conference, "H_2 in the Early Universe", Dec 4-6 1997, ed. E. Corbelli, D. Galli, F. Palla, Mem. S. A. It. Press, p. 413
69. Combes F., Wiklind T. 1995, A&A 303, L61
70. Combes F., Wiklind T. 1997, ApJ 486, L59
71. Combes F., Wiklind T. 1998, A&A 334, L81
72. Combes F., Wiklind T., 1996, in Cold Gas at High Redshift, eds. M.N. Bremer, P. van der Werf, H.J.A. Röttgering, C.L. Carilli, Kluwer Academic Pub., p. 215
73. Combes F., Wiklind T., Nakai N.: 1997, A&A 327, L17
74. Combes F.: 1991, ARAA 29, 195
75. Combes F.: 2000, in "Toward a New Millenium in Galaxy Morphology" ed. D. Block, Astr. and Space Sci. library, Kluwer, p. 405 (astro-ph/9909016)
76. Combes F.: 2001, in "The Central kpc of Starbursts and AGN: The La Palma Connection", ASP Conf Series, eds J.H. Knapen, J.E. Beckman, I. Shlosman and T.J. Mahoney (astro-ph/0106436)
77. Corbelli E., Salpeter E.E.: 1993, ApJ 419, 94
78. Cox P., Omont A., Djorgovski S. G. et al.: 2002, A&A 387, 406
79. Crutcher R. M., Troland T. H., Lazareff B., Paubert G., Kazès I.: 1999 ApJ 514, L121
80. Désert F.-X., Boulanger F., Puget J.-L. 1990, A&A, 237, 215.
81. Dalgarno A., Kirby K., Stancil P. C.: 1996, ApJ 458, 397
82. Dame T. M., Hartmann Dap, Thaddeus P.: 2001, ApJ 547, 792
83. Dame T. M., Ungerechts H., Cohen R. S. et al.: 1987, ApJ 322, 706
84. Dame T.M., Elmegreen B.G., Cohen R.S., Thaddeus P.: 1986, ApJ 305, 982
85. Das M., Jog C.J.: 1999, ApJ 527, 600
86. David L.P., Jones C., Forman W.: 1995, ApJ 445 578
87. Davidsen A.F., Kriss G.A., Zheng W.: 1996, Nature 380 47
88. Davis R.J., Diamond P.J., Goss W.M.: 1996, MNRAS 283, 1105
89. Dey A., Graham J. R., Ivison R. J. et al.: 1999, ApJ 519, 610
90. Diamond P.J., Goss W.M., Romney J.D. et al: 1989, ApJ 347, 302
91. Dickey J.M., Hanson M.M., Helou G.: 1990 ApJ 352, 522
92. Dixon D. D., Hartmann D. H., Kolaczyk E. D. et al.: 1998, NewA 3, 539
93. Downes D., Neri R., Wiklind T. et al. : 1999, ApJ 513, L1
94. Downes D., Solomon P. M.: 1998, ApJ 507, 615
95. Downes D., Solomon P.M., Radford S.J.E. 1995, ApJ 453, L65
96. Dubinski J.: 1994, ApJ 431, 617
97. Duc P.A., Mirabel I.F. 1994 A&A 289, 83
98. Dupraz C., Combes F., 1987, A&A 185, L1

99. Dwek E. et al. : 1998, ApJ 508, 106
100. Dwek E., Arendt R. G., Hauser M. G. et al.: 1995, ApJ 445, 716
101. Eckart A., Cameron M., Genzel R. et al.: 1990, ApJ 365, 522
102. Elbaz D., Arnaud M., Casse M., et al.: 1992, A&A 265, L29
103. Elmegreen B.G., Elmegreen D.M.: 2001 AJ 121, 1507
104. Elmegreen B.G., Kaufman M., Thomasson M.:1993, ApJ, 412, 90
105. Elmegreen B.G.: 1997, ApJ, 486, 944
106. Elmegreen B.G.: 1999, ApJ 515, 323
107. Elston R., Rieke G.H., Rieke M.J.: 1988, ApJ 331, L77
108. Emsellem E., Greusard D., Combes F. et al.: 2001, A&A 368, 52
109. Faison M.D., Goss W.M., Diamond P.J., Taylor G.B., 1998, AJ 116, 2916
110. Falconer K.J.:1990, Fractal geometry, Wiley, Chichester
111. Falgarone E., Phillips T.G., Walker C.K.: 1991, ApJ 378, 186
112. Falgarone E., Puget J-L., Perault M., 1992, A&A 257, 715
113. Federman S. R., Cardelli J. A., Sheffer Y., Lambert D. L., Morton D. C.: 1994, ApJ 432, L139
114. Ferland G.J., Fabian A.C., Johnstone R.M.: 1994, MNRAS 266 399
115. Fey A.L., Clegg A.W., Fiedler R.L., 1996, ApJ 468, 543
116. Fich M., Tremaine S.: 1991, ARAA 29, 409
117. Fiedler R.L., Dennison B., Johnston K., Hewish A.: 1987, Nature 326, 675
118. Fiedler R.L., Pauls T., Johnston K., Dennison B.: 1994, ApJ 430, 595
119. Fioc M., Rocca-Volmerange B.: 1997, A&A 326, 950
120. Fleck R.C.: 1981, ApJ 246, L151
121. Flores H., Hammer F., Thuan T. X. et al. : 1999, ApJ 517, 148
122. Frail D. A., Weisberg J. M., Cordes J. M., Mathers C.: 1994, ApJ 436, 144
123. Frayer D.T., Ivison R.J., Scoville N.Z., et al., 1998, ApJ 506, L7
124. Frayer D.T., Ivison R.J., Scoville N.Z., et al., 1999, ApJ 514, L13
125. Fux R.: 1999, A&A 345, 787
126. Gallego J., Zamorano J., Aragón-Salamanca A., Rego M.: 1995, ApJ 455, L1 (erratum ApJ 459, L43)
127. Galli D., Palla F.: 1998, A&A 335, 403
128. Garcia-Burillo S., Combes F., Neri R. 1999, A&A 343, 740
129. Garcia-Burillo S., Martin-Pintado J., Fuente A., Neri R.: 2001, ApJ 563, L27
130. Garcia-Burillo S., Sempere M.J., Bettoni D. 1998b ApJ 502, 235
131. Garcia-Burillo S., Sempere M.J., Combes F., Neri R. 1998a A&A, 333, 864
132. García-Burillo S., Sempere M. J., Combes F. et al.: 2000, A&A 363, 869
133. Gerin M., Phillips T.G.: 2000, ApJ 537, 644
134. Gerlich D.: 1990, J. Chem. Phys. 92, 2377
135. Giroux M.L., Shapiro P. R.: 1996, ApJS 102, 191
136. Gondhalekar P.M., Johansson L.E.B., Brosch N., Glass I.S., Brinks E. 1998 A&A 335, 152
137. Goodman A. A., Barranco J. A., Wilner D. J.,Heyer M. H.: 1998, ApJ 504, 223
138. Grabelsky D., Cohen R.S., Bronfman L. et al.: 1987, ApJ 315, 122
139. Guélin M., Zylka R., Mezger P. et al.: 1993 A&A 279, L37
140. Guilloteau S., Omont A., Cox P. et al. 1999, A&A 349, 363
141. Guilloteau S., Omont A., McMahon R.G. et al. 1997, A&A 328, L1
142. Hüttemeister S., Henkel C., Mauersberger R., Brouillet N., Wiklind T., Millar T. J.: 1995, A&A 295, 571

143. Hüttemeister S., Mauersberger R., Henkel C.: 1997, A&A 326, 59
144. Haiman Z., Loeb A.: 1998, Memorie della Societa Astronomia Italiana, Vol. 69, p.367
145. Hamilton A.J.S.: 1993, ApJ 417, 19
146. Hauser M. et al., 1998, ApJ 508, 25
147. Heiles C.: 1997, ApJ 481, 193
148. Heithausen A., Bensch F., Stutzki J., Fakgarone E., Panis J.F.: 1998, A&A 331, L65
149. Heithausen A.: 1996; A&A 314, 251
150. Henderson A.P.: 1979, A&A 75, 311
151. Henkel C., Chin Y.-N., Mauersberger R., Whiteoak J. B.: 1998, A&A 329, 443
152. Henkel C., Mauersberger R., Peck A. B., Falcke H., Hagiwara Y.: 2000, A&A 361, L45
153. Henkel C., Mauersberger R., Wiklind T. et al.: 1993, A&A 268, L17
154. Herbst E.: 2000, in "H_2 in Space", eds F. Combes & G. Pineau des Forêts, Cambridge University Press, p. 85
155. Hibbard J. E., Mihos J. C.: 1995, AJ, 110, 140
156. Hirashita H., Kamaya H.: 2000, AJ 120, 728
157. Hoekstra H., van Albada T.S., Sancisi R.: 2001, MNRAS, 323, 453
158. Houlahan P., Scalo J.: 1992, ApJ 393, 172
159. Hoyle F.: 1953, ApJ 118 513
160. Huber D., Pfenniger D.: 2001, A&A 374, 465
161. Hutchins J.B.: 1976, ApJ 205 103
162. Imamura K., Sofue Y.: 1997, A&A 319, 1
163. Impey C., Bothun G.: 1989, ApJ 341, 89
164. Jacobsen P. et al: 1994, Nature 370 35
165. Jacq T., Despois D., Baudry A.: 1988, A&A 195, 93
166. Jauncey D. L., Reynolds J. E., Tzioumis A. K. et al. 1991 Nature 352, 132
167. Jog C.J.: 1996, MNRAS 278, 209
168. Karpas Z., Anicich V., Huntress W.: 1979, J. Chem. Phys. 70, 2877
169. Kashlinsky A., Rees M.J.: 1983, MNRAS 205 955
170. Katz N., Furman I., Biham O. et al.: 1999, ApJ 522, 305
171. Katz N., Gunn J.E.: 1991, ApJ 377, 365
172. Kennicutt R. C., Jr., Edgar B. K., & Hodge P. W. 1989, ApJ, 337, 761
173. Kennicutt R.C., Roettiger K. A., Keel W. C., van der Hulst J. M., Hummel E.: 1987, AJ 93, 1011
174. Kennicutt R.C.: 1989, ApJ 344, 685
175. Kennicutt R.C.: 1998, ApJ 498, 541
176. Kim D.-C., Veilleux S., Sanders D. B.: 1998 ApJ 508, 627
177. Klessen R., 1997, MNRAS 292, 11
178. Klessen R.S., Burkert A., Bate M. R.: 1998 ApJ, 501, L205
179. Knapp G. R., Guhathakurta P., Kim D-W., Jura M. A.: 1989, ApJ 70, 329
180. Knapp G. R., Kerr F. J., Henderson A. P.: 1979, ApJ 234, 448
181. Kneib J-P., Prieur J-L., Ivison R., Smail I., Blain A.: 2002, A&A in prep
182. Kneib J.-P., Alloin D., Mellier Y. et al.: 1998, A&A 329, 827
183. Kolmogorov A.: 1941, in "Compt. Rend. Acad. Sci. URSS" 30, 301
184. Lahav O.: 1986, MNRAS 220 259
185. Lamer G., Newsam A. M., McHardy I. M.: 1999, MNRAS 309, 1085

186. Larson R.B., 1973, ARAA 11, 219
187. Larson R.B., 1978, MNRAS 184, 69
188. Larson R.B., 1981, MNRAS 194, 809
189. Larson R.B., 1982, MNRAS, 200, 159
190. Larson R.B., 1995, MNRAS 272, 213
191. Larson R.B.: 1981, MNRAS 194 809
192. Le Bourlot J., Pineau des Forets G., Roueff E., Schilke P.: 1993, ApJ 416, L87
193. Lees J.F., Knapp G.R., Rupen M.P. et al.: 1991, ApJ 379, 177
194. Lemme C., Walmsley C.M., Wilson T.L., Muders D., 1995, A&A 302, 509
195. Lepp S., Shull J.M.: 1984, ApJ 280 465
196. Lewis J.R., Freeman K.C.:1989 AJ, 97, 139
197. Lin D.N.C., Pringle J.E.: 1987, ApJ 320, L87
198. Lisenfeld U., Voelk H. J., Xu C.: 1996, A&A 306, 677
199. Liszt H. S., Lucas R.: 1996 A&A 314, 917
200. Liszt H. S., Lucas R.: 1998 A&A 339, 561
201. Liszt H. S., Lucas R.: 2000 A&A 355, 333
202. Liszt H. S., Lucas R.: 2001 A&A 370, 576
203. Liszt H. S., Wilson R. W.: 1993, ApJ 403, 663
204. Lucas R., Liszt H.S., 1994, A&A 282, L5
205. Lucas R., Liszt H.S., 1996, A&A 307, 237
206. Lucas R., Liszt H.S., 1998, A&A 337, 246
207. Lutz D., Spoon H. W. W., Rigopoulou D., Moorwood A. F. M., Genzel R.: 1998 ApJ 505, L103
208. Lynden-Bell D., Wood R.: 1968, MNRAS 138, 495
209. Madau P. 1992, ApJ 389 L1
210. Magnani L., Blitz L., Mundy L. 1985, ApJ 295, 402
211. Magnani L., Blitz L., Mundy L.: 1985, ApJ, 295, 402
212. Malhotra S.: 1994, ApJ 433, 687
213. Malhotra S.: 1995, ApJ 448, 138
214. Malin D. F., Carter D.: 1983, ApJ 274, 534
215. Maloney P., Black J.H. 1988, ApJ 325, 389
216. Mandelbrot B.B.: 1975, 'Les objets fractals', Paris, Flammarion
217. Mao R. Q., Henkel C., Schulz A. et al.: 2000, A&A 358, 433
218. Marscher A. P., Bania T. M., Wang Z.: 1991, ApJ 371, L77
219. Marscher A. P., Moore E. M., Bania T. M.; 1993, ApJ 419, L101
220. Marsh K.A., Helou G.: 1995 ApJ 445, 599
221. Masset F., Bureau M.: 2002, ApJ preprint
222. Matthews L. D., Gao Y.: 2001, ApJ 549, L191
223. Mauersberger R., Henkel C., Walsh W., Schulz A.: 1999, A&A 341, 256
224. Mauersberger R.,Henkel C., Chin Y.-N.: 1995, A&A 294, 23
225. Mayya Y. D., Regaradjan T. N.: 1997 AJ 114, 932
226. McGaugh S. S., Schombert J. M., Bothun G. D., de Blok W. J. G.: 2000 ApJ 533, L99
227. McGaugh S., de Blok W.J.G 1997 ApJ 481, 689
228. McHardy I. M., Merrifield M. R., Abraham R. G., Crawford C. S.: 1994, MNRAS 268, 681
229. Melchior A-L., Combes F., Guiderdoni B., Hatton S.: 2001, in "Semaine de l'Astrophysique Francaise", Lyon, France, May 28-June 1st, 2001, Eds.: F. Combes, D. Barret, F. Thvenin, EdP-Sciences, p.243

230. Merrifield M.: 1992, AJ 103, 1552
231. Mihos J.C., Hernquist L.:1994 ApJ, 437, 611
232. Mihos J.C., Hernquist L.:1996, ApJ, 464, 641
233. Motte F., André P., Neri R.: 1998, A&A 336, 150
234. Mulchaey J.S., Regan M.W.: 1997, ApJ 482, L135
235. Mulder W.A., Liem B.T.: 1986, A&A 157, 148
236. Murphy M. T., Webb J. K., Flambaum V. V., Drinkwater M. J., Combes F., Wiklind T.: 2001 MNRAS 327, 1244
237. Murray S.D., Lin D.N.C.: 1996, ApJ 467, 728
238. Myers P.C.: 1983, ApJ 270, 105
239. Myers P.C.: 2000, ApJ 530, L119
240. Narayan C.A., Jog C.J.: 2002, A&A 390, L35
241. Neininger N., Guelin M., Garcia-Burillo S., Zylka R., Wielebinski R.: 1996, A&A 310, 725
242. Neininger N., Guelin M., Ungerechts H. et al. 1998, Nature 395, 871
243. Neininger N.: 2000, in "Dynamics of Galaxies: from the Early Universe to the Present" Eds.: Francoise Combes, Gary A. Mamon, and Vassilis Charmandaris ASP Conference Series, Vol. 197, p. 355
244. Norberg P., Maeder A.: 2000, 359, 1025
245. Oey M. S., & Clarke C. J. 1998, AJ, 115, 1543
246. Ohta K., Yamada T., Nakanishi K., et al.:1996, Nature 382, 426
247. Omont A., Petitjean P., Guilloteau S. et al. 1996, Nature 382, 428
248. Oort J.H.: 1977, ARAA 15, 295
249. Ossenkopf V., Klessen R. S., Heitsch F.: 2001 A&A 379, 1005
250. Padmanabhan 1990, Phys. Rep. 188, 285
251. Palla F., Galli D., Silk J.: 1995, ApJ 451, 44
252. Palla F., Salpeter E.E., Stahler S.W.: 1983, ApJ 271 632
253. Palla F., Salpeter E.E., Stahler S.W.: 1983, ApJ 271, 632
254. Palla F., Zinnecker H.: 1987, in *Starbursts and Galaxy Evolution* XXIInd Rencontres de Moriond, ed. T.X. Thuan, T. Montmerle & J.T.T. Van Ed. Frontières, page 533
255. Papadopoulos P.P., Ivison R., Carilli C., Lewis G.: 2001, Nature 409, 58
256. Papadopoulos P.P., Rottgering H.J.A., van der Werf P.P., et al. 2000 ApJ. 528, 626
257. Patnaik A. R., Porcas R. W., Browne I. W. A.: 1995, MNRAS 274, L5
258. Peebles P.J.E., Dicke R.H.: 1968, ApJ 154 891
259. Peebles P.J.E.: 1980, in *The Large-Scale Structure of the Universe*, Princeton Univ. Press.
260. Peebles P.J.E.: 1982, ApJ 263, L1
261. Peters W.L.: 1975, ApJ 195, 617
262. Pettini M., Kellog M., Steidel C.C. et al.: 1998, ApJ 508, 539
263. Pfenniger D., Combes F., Martinet L.: 1994, A&A 285 79
264. Pfenniger D., Combes F.: 1994, A&A 285, 94
265. Pirronello V., Liu C., Roser J. E., Vidali G.: 1999, A&A 344, 681
266. Planesas P., Martin-Pintado J., Neri R., Colina L.: 1999 Science 286, 2493
267. Press W.H., Schechter P.: 1974, ApJ 187, 425
268. Puget J.-L., Abergel A., Bernard J.-P., et al. : 1996, A&A 308, L5
269. Puy D., Alecian G., Le Bourlot J., Leorat J., Pineau Des Forets G.: 1993, A&A 267, 337

270. Puy D., Signore M.: 1996, A&A 305 371
271. Quinn P. 1984, ApJ 279, 596
272. Rachford B.L., Snow T.P., Tumlinson J. et al.: 2001, ApJ 555, 839
273. Rao S.M., Turnshek D.A., Briggs F.H.: 1995, ApJ 449, 488
274. Rees M.J.: 1976, MNRAS 176, 483
275. Regan M. W., Thornley M. D., Helfer T. T. et al.: 2001, ApJ 561, 218
276. Reimers D., Kohler S., Wisotzki L. et al.: 1997, A&A 327 890
277. Reynaud D., Downes D. 1997, A&A, 319, 737
278. Reynaud D., Downes D. 1999, A&A, 347, 37
279. Richter P., Sembach K.R., Wakker B.P., Savage B. D.: 2001 ApJ 562, L181
280. Roberts M., Haynes M.: 1994, ARAA 32, 115
281. Rodríguez-Fernández N. J., Martín-Pintado J., de Vicente P. et al.: 2000, A&A 356, 695
282. Romeo A.B.: 1992, MNRAS 256, 307
283. Rousseau G., Chate H., Le Bourlot J.: 1998, MNRAS, 294, 373
284. Rownd B. K., Young J. S.: 1999 AJ 118, 670
285. Rubin V.C., Kenney J.D.P., Young J.S.:1997 AJ 113, 1250
286. Rubio M., Lequeux J., Boulanger F. 1993, A&A 271, 9
287. Sage L.J., Salzer J.J., Loose H.H., Henkel C. 1992 A&A 265, 19
288. Sakamoto K., Okumura S., Minezaki T. et al. 1995 AJ 110, 2075
289. Sakamoto S., Hasegawa T., Hayashi M., et al.: 1995, ApJS 100, 125
290. Sanders D,B., Mirabel I.F.: 1996, ARAA 34, 749
291. Sanders D. B., Soifer B. T., Elias J. H., Neugebauer G., Matthews K.: 1988, ApJ 328, L35
292. Sanders D.B., Clemens D.P., Scoville N.Z., Solomon P.M.: 1986, ApJS 60, 1
293. Sanders R.H., Bania T. M.: 1976, ApJ 204, 341
294. Savage B.D., Bohlin R.C., Drake J.F., Budich W.: 1977, ApJ 216, 291
295. Scalo J.M., 1987 in 'Interstellar Processes' D.J. Hollenbach and H.A. Thronson Eds., D. Reidel Pub. Co, p. 349
296. Scalo J.M.: 1985, in *Protostars and Planets II*, ed. D.C. Black & M.S. Matthews, Univ. of Arizona Press, Tucson, p. 201
297. Scalo J.M.: 1998, in The Stellar Initial Mass Function, ed. G. Gilmore, I. Parry, & S. Ryan (Cambridge: Cambridge Univ. Press), 201
298. Schiminovich D., van Gorkom J.H., van der Hulst J.M., Kasow S.: 1994, ApJ 432, L101
299. Schombert J.M., McGaugh S.S., Eder J.A.: 2001, AJ 121, 2420
300. Scoville N.Z., Padin S., Sanders D.B. et al. : 1993, ApJ 415, L75
301. Scoville N.Z., Yun M.S., Bryant P.M.: 1997a, ApJ 484, 702
302. Scoville N.Z., Yun M.S., Windhorst R.A. et al. 1997b, ApJ 485, L21
303. Semelin B., Combes F.: 1999, A&A sub
304. Semelin B., Combes F.: 2000, A&A 360, 1096
305. Sheth K., Vogel S.N., Regan M.W. et al.: 2002, AJ in press (astro-ph/0208018)
306. Shu F. H., Adams F. C., & Lizano S. 1987, ARAA, 25, 23
307. Shull J.M., Tumlinson J., Jenkins E.B. et al.: 2000, ApJ 538, L73
308. Sidis V., Jeloaica L., Borisov A.G., Deutscher S.A..: 2000, in "H_2 in Space", eds F. Combes & G. Pineau des Forêts, Cambridge University Press, p. 89
309. Silk J., Spaans M.: 1997, ApJ 488, L79
310. Silk J.: 1968, ApJ 151 459
311. Silk J.: 1977, ApJ 214, 718

312. Silk J.: 1997, ApJ 481, 703
313. Smail I., Ivison R.J., Kneib J-P., et al.: 1999, MNRAS 308, 1061
314. Sodroski T. J., Dwek E., Hauser M. G., Kerr F. J.: 1987, ApJ 322, 101
315. Solomon P. M., Downes D., Radford S. J. E., Barrett J. W.: 1997, ApJ 478, 144
316. Solomon P. M., Downes D., Radford S. J. E.: 1992, ApJ 387, L55
317. Solomon P.M., Downes D., Radford S.J.E., Barrett J.W.: 1997, ApJ 478, 144
318. Solomon P.M., Downes D., Radford S.J.E.: 1992, Nature 356, 318
319. Solomon P.M., Radford S.J.E., Downes D.: 1990, ApJ 348, L53
320. Solomon P.M., Rivolo A.R., Barrett J.W., Yahil A.: 1987, ApJ 319, 730
321. Solomon P.M., Sanders D.B., Scoville N.Z.: 1979, in "The Large-Scale Characteristics of the Galaxy", IAU 84, ed. W.B. Burton, p.35 Kluwer.
322. Sreenivasan K.R., & Méneveau C.: 1986, J. Fluid Mech. 173, 357
323. Srianand R., Petitjean P., Ledoux C.: 2000, Nature 408, 931
324. Stancil P.C., Lepp S., Dalgarno A.: 1996, ApJ 458 401
325. Stark A. A., Bania T. M.: 1986, ApJ 306, L17
326. Strong A.W., Bloemen J.B.G.M., Dame T.M. et al.: A&A 1988, 207, 1
327. Strong A.W., Mattox J.R.: 1996 A&A 308, L21
328. Strong A.W., Moskalenko I. V., Reimer O.: 1999, Proc. 26th ICRC (Salt Lake City), 1999, v.4, p.52-55 (astro-ph/9906229)
329. Stutzki J., Graf U. U., Haas S. et al.: 1997, ApJ 477, L33
330. Tacconi L., Young J.S.: 1986, ApJ 308, 600
331. Tamura N., Hirashita H., Takeuchi T. T.: 2001, ApJ 552, L113
332. Tan J.: 2000, ApJ 536, 173
333. Taniguchi Y., Sofue Y., Wakamatsu K-I. Nakai N.: 1990, AJ 100, 1086
334. Taylor C.L., Kobulnicky H.A., Skillman E.D. 1998, AJ 116, 2746
335. Tegmark M., Silk J., Blanchard A.: 1994, ApJ 420 484
336. Tegmark M., Silk J., Rees M.J., Blanchard A., Abel T., Palla F.: 1997, ApJ 474 1
337. Tilanus R. P. J., Allen R. J.: 1991, A&A 244, 8
338. Tohline J.E.: 1980, ApJ 235, 866
339. Toomre A.: 1964, ApJ, 139, 1217
340. Toomre A.: 1990, in "Dynamics and Interactions of Galaxies", ed. Roland Wielen, Springer-Verlag, p. 292
341. Troland T.H., Crutcher R.M., Goodman A.A., et al. : 1996, ApJ 471, 302
342. Truelove J. K., Klein R. I., McKee C. F., et al. : 1997, ApJ 489, L179
343. Tumlinson J., Shull J.M., Rachford B.L. et al.: 2002, ApJ 566, 857
344. Turner M.S.: 1984 in "Fundamental Interactions and Cosmology", ed. J. Audouze and J. Tran Thanh Van (Editions Frontières, Gif-sur-Yvette), p. 300.
345. Valentijn E. A., van der Werf P.: 1999, ApJ 522, L29
346. Vazquez-Semadeni E., Ballesteros-Paredes J., Rodriguez L.F.: 1997, ApJ 474, 292
347. Vazquez-Semadeni E., Canto J., Lizano S.: 1998, ApJ 492, 596
348. Vazquez-Semadeni E., Ostriker E.C., Passot T., Gammie C.F., Stone J.M.: 1999, in 'Protostars and Planets IV', eds. V. Mannings, A. Boss, S. Russell (astro-ph/9903066)
349. Vila-Costas M.B., Edmunds M.G. 1992, MNRAS 259, 121
350. Vogelaar M.G.R., Wakker B.P.: 1994, A&A 291, 557
351. Wada K., Norman C.: 1999, ApJ 516, L13

352. Wakker B. P., Howk J. C., Savage B. D. et al.: 1999, Nature 402, 388
353. Walker M., Wardle M., 1998, ApJ 498, L125
354. Wang Y., Evans N.J. II, Zhou S., Clemens D.P., 1995, ApJ 454, 217
355. Ward-Thompson D., Scott P.F., Hills R.E., André P., 1994, MNRAS 268, 276
356. Watson D. M., Guptill M. T., Buchholz L. M.: 1994, ApJ 420, L21
357. Webb J.K., Murphy M.T., Flambaum V.V. et al.: 2001, Phys. Rev. Lett. 87, p. 091301
358. Weil M. L., Hernquist L.: 1993, ApJ 405, 142
359. Weinberg M.D.: 1992 ApJ 384, 81
360. Wiklind T., Combes F. 1994, A&A 286, L9
361. Wiklind T., Combes F. 1995, A&A 299, 382
362. Wiklind T., Combes F. 1996a, Nature, 379, 139
363. Wiklind T., Combes F. 1996b, A&A 315, 86
364. Wiklind T., Combes F. 1997a, A&A 324, 51
365. Wiklind T., Combes F. 1997b, A&A 328, 48
366. Wiklind T., Combes F. 1998, ApJ 500, 129
367. Wiklind T., Combes F. 2001, in "Gravitational Lensing: Recent Progress and Future Goals", ASP Conference Proceedings, Vol. 237. Edited by Tereasa G. Brainerd and Christopher S. Kochanek. San Francisco: p.155
368. Wiklind T., Combes F., Henkel C.: 1995, A&A 297, 643
369. Williams J.P., de Geus E.J., Blitz L., 1994, ApJ 428, 693
370. Wilson C.D. 1995, ApJ 448, L97
371. Wink J.E., Guilloteau S., Wilson T.L., 1997, A&A 322, 427
372. Wolfendale A. W., Worrall D. M.: 1977 A&A 60, 165
373. Wouterloot J. G. A., Brand J., Burton W. B., Kwee K. K.: 1990, A&A 230, 21
374. Wyse R., Silk J.: 1989, ApJ 339, 700
375. Yan L., McCarthy P.J., Freudling W. et al.:1999, ApJ 519, L47
376. Yoshii Y., Arimoto N.: 1987 A&A 188, 13
377. Yoshii Y., Saio H.: 1985, ApJ 295, 521
378. Young J., Knezek M. 1989, ApJ, 347, L55
379. Young J., Scoville N.Z. 1991, A.R.A.A. 29, 581
380. Young L.M.: 2002, AJ 124, 788
381. Zaritsky D., Lorrimer S.J. 1993 in The Evolution of Galaxies and Their Environment, Proceedings NASA. Ames Research Center, p. 82-83
382. Zinnecker H.: 1984, MNRAS 210, 43
383. de Blok W.J.G., van der Hulst J.M. 1998, A&A 336, 49
384. de Grijs R., O'Connell R., Gallagher J.: 2001, AJ 121, 768
385. de Grijs R., Peletier R. F., van der Kruit P. C.:1997, A&A 327, 966
386. de Paolis F., Ingrosso G., Jetzer Ph., Roncadelli,M.: 1999, ApJ 510, L103
387. van Gorkom J., Schiminovich D.: 1997, in "The Nature of Elliptical Galaxies" 2nd Stromlo Symposium. ASP Conference Series; Vol. 116; ed. M. Arnaboldi; G. S. Da Costa; and P. Saha , p.310
388. van Zee L., Haynes M.P., Salzer J.J. 1997 AJ 114, 2497
389. van der Kruit P.C., Searle L.:1981 A&A 95, 105
390. van der Kruit P.C., Searle L.:1982 A&A 105, 351
391. van der Werf P.: 2001, in "Starburst galaxies - near and far", proceedings of the Ringberg workshop, eds. D. Lutz and L.J. Tacconi, Springer (astro-ph/0101185)

Astrophysics of Dust in Cold Clouds

Bruce T. Draine

Princeton University, Department of Astrophysical Sciences,
Peyton Hall 108, Princeton, NJ 08544-1011, USA

Dust plays an increasingly important role in astrophysics. Historically, dust was first recognized for its obscuring effects, and the need to correct observed intensities for attenuation by dust continues today. But with the increasing sensitivity of IR, FIR, and submm telescopes, dust is increasingly important as a diagnostic, with its emission spectrum providing an indicator of physical conditions, and its radiated power bearing witness to star populations of which we might otherwise be unaware. Finally, and most fundamentally, dust is now understood to play many critical roles in galactic evolution. By sequestering selected elements in the solid grains, and by catalyzing formation of the H_2 molecule, dust grains are central to the chemistry of interstellar gas. Photoelectrons from dust grains can dominate the heating of gas in regions where ultraviolet starlight is present, and in dense regions the infrared emission from dust can be an important cooling mechanism. Finally, dust grains can be important in interstellar gas dynamics, communicating radiation pressure from starlight to the gas, and providing coupling of the magnetic field to the gas in regions of low fractional ionization.

We would like to understand these effects of dust in the Milky Way, in other galaxies, and as a function of cosmic time. These lectures are organized around topics in the astrophysics of dust in the the Milky Way, as this is our best guide to understanding and modelling dust long ago and far away.

1 Introduction to Interstellar Dust

We begin with a brief review of some of the observational evidence which informs our study of interstellar dust.

1.1 Interstellar Extinction

Through study of open star clusters in the Galaxy, Trumpler (1930) found that distant stars were dimmed by something in addition to the inverse square law, and concluded that interstellar space in the galactic plane contained "fine cosmic dust particles of various sizes ... producing the observed selective absorption". Over the past 7 decades we have built on Trumpler's pioneering study, but many aspects of interstellar dust – including its chemical composition! – remain uncertain. Let us therefore begin by reviewing the different ways in which nature permits us to study interstellar dust.

Trumpler analyzed the interaction of light with interstellar dust, and this remains our most direct way to study interstellar dust. We use stars as "standard candles", and study the "selective extinction" – or "reddening" – of starlight by the dust. With the assumption that the extinction ($\equiv$ absorption + scattering) goes to zero at wavelengths $\lambda \to \infty$, and observing the star at sufficiently long wavelength (to, in effect, determine its distance) one can determine the attenuation of the starlight by dust as a function of wavelength. Because atomic hydrogen absorbs strongly for $h\nu > 13.6\,\mathrm{eV}$, it is only possible to measure the contribution of dust to the extinction at $h\nu < 13.6\,\mathrm{eV}$, or $\lambda > 912$Å. A typical "extinction curve" – the extinction as a function of wavelength or frequency — is shown in Fig. 1, showing the rapid rise in extinction in the vacuum ultraviolet. Observed extinction curves vary in shape from one line-of-sight to another, but appear to approximately form a one-parameter family (Cardelli et al. 1989); the parameter is often taken to be the ratio $R_V \equiv (A_B - A_V)/A_V$, where A_B and A_V are the extinctions measured in the B (4400Å) and V (5500Å) spectral bands. A parametrization of the extinction curve was provided by Cardelli et al.; the curves in Fig. 1 were

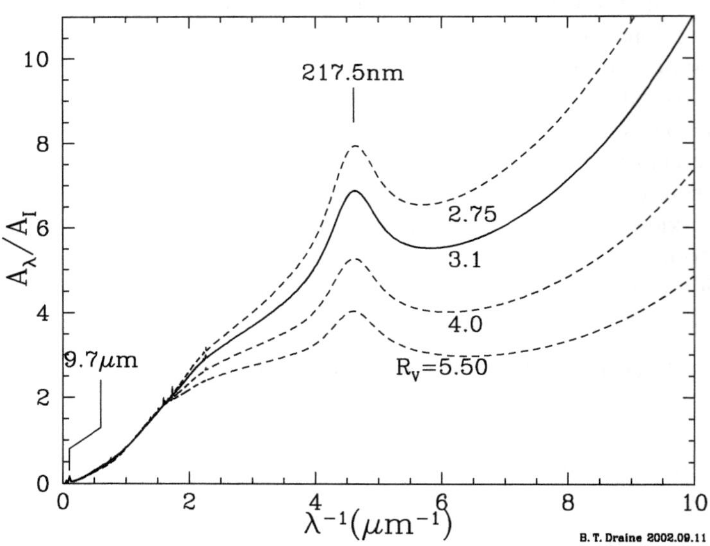

Fig. 1. Extinction at wavelength λ, relative to the extinction at $I = 900\,\mathrm{nm}$, as a function of inverse wavelength λ^{-1}, for Milky Way regions characterized by different values of $R_V \equiv A_V/E(B-V)$, where A_B is the extinction at $B = 4400\,\mu\mathrm{m}$, A_V is the extinction at $V = 5500\,\mu\mathrm{m}$, and the "reddening" $E(B-V) \equiv A_B - A_V$. Note the rapid rise in extinction in the vacuum ultraviolet ($\lambda \lesssim 0.2\,\mu\mathrm{m}$) for regions with $R_V \lesssim 4$. The normalization per H nucleon is approximately $A_I/N_\mathrm{H} \approx 2.6 \times 10^{-22}\,\mathrm{cm}^2/\mathrm{H}$. The silicate absorption feature (§1.5) at 9.7$\,\mu$m and the diffuse interstellar bands (§1.6) are barely visible.

calculated using a more recent prescription by Fitzpatrick (1999), with the extinction in the infrared following Draine (1989b).

We will discuss dust grain optics below, but it is clear that if the dust grains were large compared to the wavelength, we would be in the "geometric optics" limit and the extinction cross-section would be independent of wavelength. Therefore the tendency for the extinction to rise even at the shortest wavelengths where we can measure it tells us that grains smaller than the wavelength must be making an appreciable contribution to the extinction at all of observed wavelengths. As we will see below, "small" means (approximately) that $2\pi a|m-1|/\lambda \lesssim 1$, where $m(\lambda)$ is the complex refractive index. Thus if $|m-1| \approx 1$ at $\lambda = 0.1\,\mu$m,

- we must have large numbers of grains with $a \lesssim 0.015\,\mu$m.

1.2 Scattering of Starlight by Dust Grains

When an interstellar cloud happens to be unusually near one or more bright stars, we have a "reflection nebula", where we see the starlight photons which have been scattered by the dust in the cloud. The spectrum of the light coming from the cloud surface shows the stellar absorption lines, showing that scattering rather than some emission process is responsible. By comparing the observed scattered intensity with the estimated intensity of the starlight incident on the cloud, it is possible to infer the albedo of the dust – the ratio of scattering cross section to extinction cross section. The result is that in the optical the interstellar dust mixture has an albedo $\omega \approx 0.5$ – scattering is about as important as absorption – and the grains are somewhat forward scattering, with $\langle\cos\theta\rangle \approx 0.5$. Rayleigh scattering by particles small compared to the wavelength has $\langle\cos\theta\rangle \approx 0$, so this tells us that

- the particles dominating the scattering at $\lambda \approx 0.6\,\mu$m have $a \gtrsim \lambda/2\pi \approx 0.1\,\mu$m.

1.3 Polarization of Starlight

The polarization of starlight was discovered in 1949 (Hall 1949; Hall & Mikesell 1949; Hiltner 1949a,b). When it was realized that the degree of polarization tended to be larger for stars with greater reddening, and that stars in a given region of the sky tended to have similar polarization directions, it was obvious that the polarization is produced by the interstellar medium: light propagating through the interstellar medium becomes linearly polarized as a result of preferential extinction of one linear polarization mode relative to the other. The polarization percentage typically peaks near the V band (5500Å), and can be empirically described by the "Serkowski law" (Serkowski 1973):

$$p(\lambda) \approx p(\lambda_{\rm max}) \exp[-K \ln^2(\lambda/\lambda_{\rm max})] \quad , \qquad (1)$$

with $\lambda_{\max} \approx 5500$Å and $K \approx 1.15$. This "linear dichroism" of the interstellar medium is due to dust grains which are partially aligned by the interstellar magnetic field. The peak polarization $p_{\max}$ is found to fall within an envelope $0 < p_{\max} \leq 0.09(E(B-V)/\mathrm{mag})$, or $0 < p_V \lesssim 0.03\,\tau_V$; the maximum values are presumed to arise on sightlines where the magnetic field is uniform and perpendicular to the line-of-sight. While the "Serkowski law" was put forward as a fit to the observed polarization at $0.3\,\mu\mathrm{m} \lesssim \lambda \lesssim 1\,\mu\mathrm{m}$, it turns out to give a surprisingly good approximation to the measured linear polarization in the vacuum ultraviolet (Clayton et al 1992, Wolff et al 1997) as seen in Fig. 2.

The mechanism responsible for the grain alignment remains a fascinating puzzle, which we will discuss below. Independent of the grain alignment mechanism, however, we can infer the sizes of the interstellar grains responsible for this polarization by noting that the extinction rises rapidly into the UV whereas the polarization drops (Kim & Martin 1995). This can be understood if the grains responsible for the polarization have sizes a such that $a \approx (\lambda_{\max}/2\pi) \approx 0.1\,\mu\mathrm{m}$: then as one proceeds into the UV one moves toward

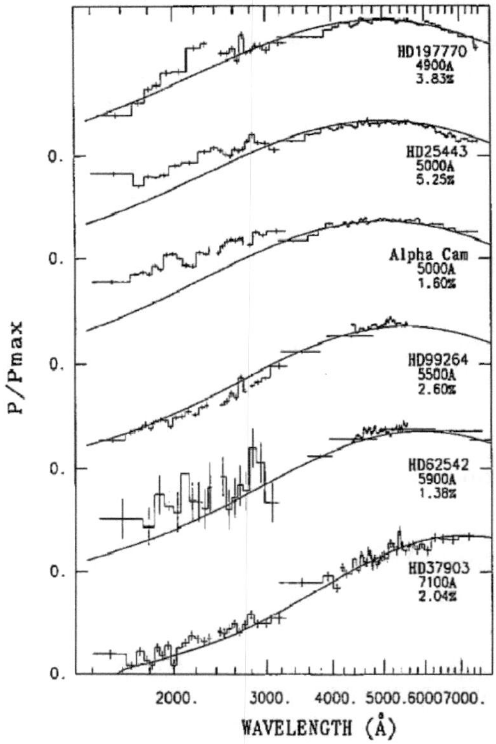

Fig. 2. Linear polarization in the ultraviolet measured by Clayton et al. (1992). The solid line is the "Serkowski law" fit to the data. Figure from Clayton et al. (1992).

the "geometric optics" limit where both polarization modes suffer the same extinction, so the polarization goes to zero:

- The extinction at $\lambda \approx 0.55\,\mu$m has an appreciable contribution from grains with sizes $a \approx 0.1\,\mu$m which are non-spherical and substantially aligned.
- The grains with $a \lesssim 0.05\,\mu$m which dominate the extinction at $\lambda \lesssim 0.3\,\mu$m are either spherical (which seems unlikely) or minimally aligned.

1.4 Spectroscopy of Dust: The 2175Å Feature

Of what is interstellar dust composed? One may look for spectroscopic clues in the extinction. The extinction curves in Fig. 1 show a conspicuous extinction feature at $\lambda^{-1} = 4.6\,\mu\text{m}^{-1}$, or $\lambda = 2175$Å. The feature is well-described by a Drude profile. The central wavelength is nearly identical on all sightlines, but the width varies significantly from one region to another (Fitzpatrick & Massa 1986).

The strength of this feature implies that the responsible material must be abundant (Draine 1989a): it must be made from H, C, N, O, Mg, Si, S, or Fe. Small graphite grains would have a strong absorption peak at about this frequency (Stecher & Donn 1965; Draine 1989a), due to $\pi \rightarrow \pi^*$ electronic excitations in the sp^2-bonded carbon sheets. Since the carbon skeleton of polycyclic aromatic hydrocarbon (PAH) molecules resembles a portion of a graphite sheet, such molecules also tend to have strong electronic transitions at about this frequency. It therefore seems likely that the 2175Å feature is due to some form of sp^2-bonded carbon material.

1.5 Spectroscopy of Dust: The Silicate Features

There is a conspicuous infrared absorption feature at $9.7\,\mu$m, shown in Fig. 3. Silicate minerals generally have strong absorption resonances due to the Si-O stretching mode near $10\,\mu$m, and it seems virtually certain that the interstellar $9.7\,\mu$m feature is due to silicates. This conclusion is strengthened by the fact that the $10\,\mu$m emission feature is seen in the outflows from oxygen-rich stars (which would be expected to condense silicate dust) but not in the outflows from carbon-rich stars. The interstellar $9.7\,\mu$m feature is seen both in emission (e.g., in the Trapezium region in Orion [Gillett, Forrest, et al. 1975]) or in extinction in the interstellar medium (Roche & Aitken 1984).

Crystalline silicate minerals generally have sharp features in their $10\,\mu$m absorption which are not seen in the broad interstellar $10\,\mu$m feature, leading to the conclusion that interstellar silicates are probably amorphous.

Near $18\,\mu$m warm interstellar dust shows another emission feature, which is attributable to the Si-O-Si bending mode in amorphous silicates.

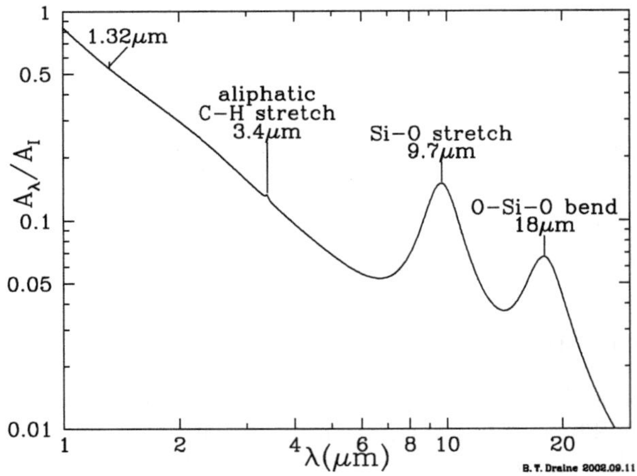

Fig. 3. Infrared extinction, relative to extinction at $I = 900\,\text{nm}$, showing the strong $9.7\,\mu\text{m}$ and $18\,\mu\text{m}$ silicate features, the $3.4\,\mu\text{m}$ aliphatic C-H stretch, and a weak unidentified DIB at $1.32\,\mu\text{m}$ (Joblin et al. 1990).

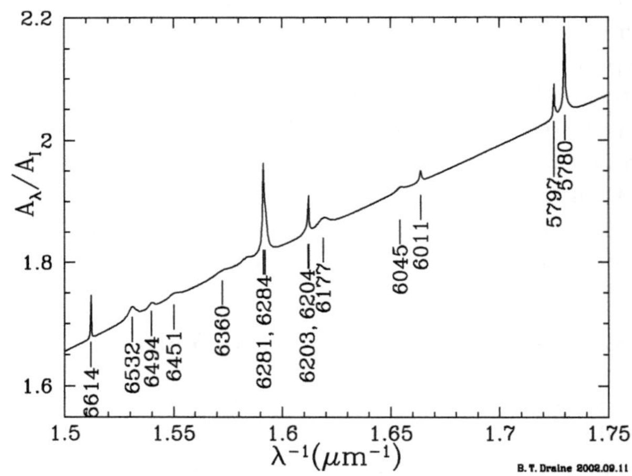

Fig. 4. Extinction at wavelength λ, relative to the extinction at $I = 900\,\text{nm}$, showing some of the diffuse interstellar bands.

1.6 Spectroscopy of Dust: Diffuse Interstellar Bands

The 3 features at $0.22\,\mu\text{m}$, $9.7\,\mu\text{m}$, and $18\,\mu\text{m}$ are by far the strongest features seen in diffuse interstellar dust. There are, in addition, numerous weaker features in the optical known as the "diffuse interstellar bands", or DIBs. Figure 4 shows the extinction for $1.5\,\mu\text{m}^{-1} < \lambda^{-1} < 1.75\,\mu\text{m}^{-1}$, with several conspicuous DIBs present, most notably the DIB at $0.5780\,\mu\text{m}$. The strongest DIB falls at 443.0nm. Jenniskens & Desert (1994) report a total of 154 "certain" DIBs in the interval $0.38 - 0.868\,\mu\text{m}$, plus another 52 "prob-

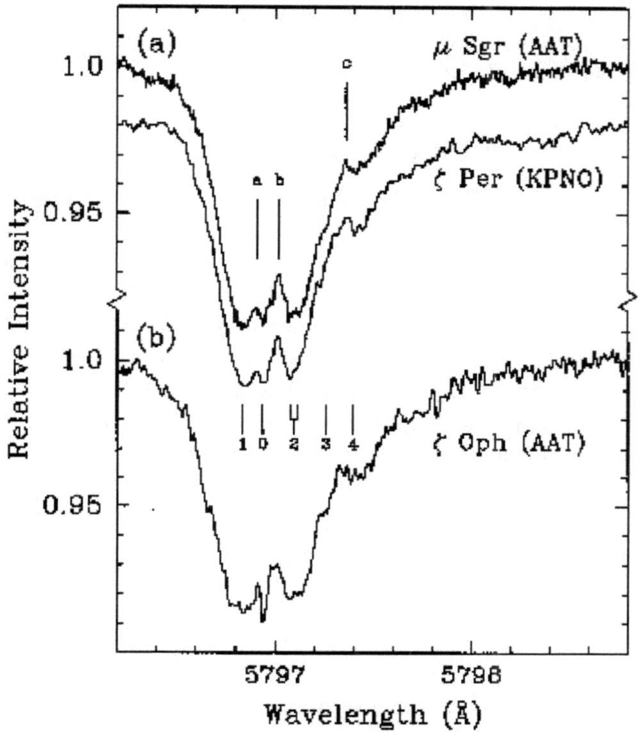

Fig. 5. Fine structure of the 5797Å DIB toward ζ Oph, ζ Per, and μ Sgr (from Kerr et al. 1998). Note in particular the very narrow feature labelled 0.

able" detections. DIBs were discovered 80 years ago (Heger 1922) and their interstellar nature was established 68 years ago (Merrill 1934).

It is embarrassing that Nature has provided astrophysicists with this wealth of spectroscopic clues, yet as of this writing not a single one of the DIBs has been convincingly identified! It seems likely that some of the DIBs may be due to free-flying large molecules; this hypothesis has received support from high resolution spectra of the 5797Å feature (see Fig. 5) showing intrinsic ultra-fine structure (Sarre et al. 1995; Kerr et al. 1998).

1.7 Spectroscopy of Dust: The 3.4μm Feature

There is a broad absorption feature at 3.4 μm which is almost certainly due to the C-H stretching mode in hydrocarbons. A recent study by Pendleton & Allamandola (2002) concluded that hydrocarbons with a mixed aromatic (ring) and aliphatic (chain) character provided a good fit to the observed interstellar absorption, including the $3.35 - 3.53\,\mu$m region. This included hydrocarbon films deposited following laser-ablation of amorphous carbon in

Ar, followed by exposure to atomic H (Mennella et al. 1999) or from a weakly-ionized plasma produced by laser-ablation of graphite in hydrogen (Scott & Duley 1996; Duley et al. 1998).

1.8 Spectroscopy of Dust: Ice Features

In dark molecular clouds a number of additional absorption features are seen, most notably a strong band at 3.1 μm which is attributed to the O-H stretching mode in H_2O ice. However, the 3.1 μm feature is *not* seen on sightlines which are confined to diffuse interstellar clouds (Gillett, Jones et al. 1975).

When a strong 3.1 μm feature appears in absorption, a number of other absorption features are also seen, including features due to CO (4.67 μm), CH_3OH (3.53 μm), and CO_2 (15.2 μm). The shape of the 3.1 μm H_2O feature is indicative of the type of ice and the impurities present in it. The relative strengths of the various features indicate that H_2O is the dominant "ice" species, with NH_3, CO, CH_3OH, and CO_2 as secondary constituents.

1.9 Spectroscopy of Dust: PAH Emission Features

A wide variety of galactic objects, including planetary nebulae, H II regions, photo-dissociation fronts, and reflection nebulae, have strong infrared emission in the 3-13 μm region. Most of the radiated power emerges in 5 broad infrared bands, at 3.3, 6.2, 7.7, 8.6, and 11.3 μm. In Fig. 6 we show the 4-15 μm emission observed from the reflection nebula NGC 7023 (Cesarsky et al. 1996). These features are seen in many other objects as well (see, e.g., Boulanger et al. 1998). The emission is quite strong: a significant fraction of the starlight energy incident on the reflection nebula is re-radiated in these

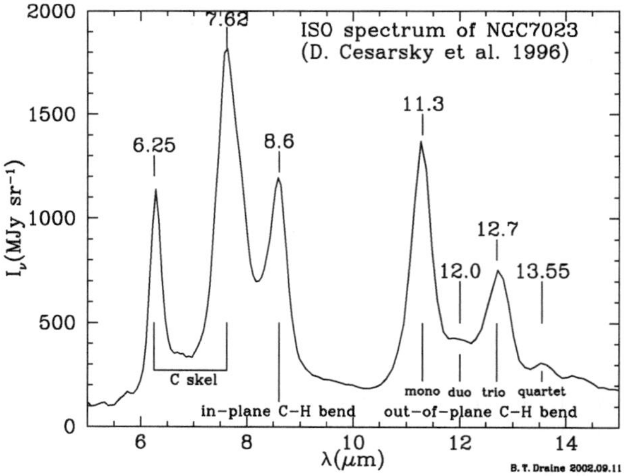

Fig. 6. Emission spectrum of the reflection nebula NGC 7023 measured by ISO-CAM (Cesarsky et al. 1996).

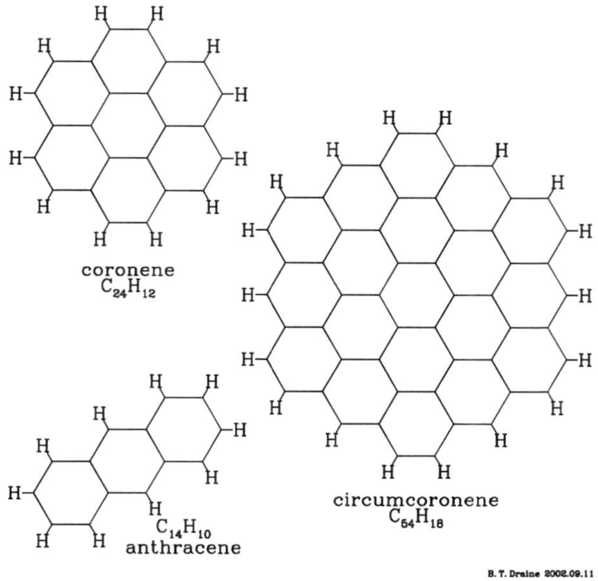

Fig. 7. Three examples of PAH molecules.

infrared emission bands – the fraction can be as large as 10-20% depending on the spectral type of the star – so the particles responsible for this emission must be quite abundant, as we will discuss further below.

The emission is thought to result from vibrational modes of polycyclic aromatic hydrocarbon (PAH) molecules. PAHs can be thought of as a section of a graphite "sheet" with the C atoms in a 2-dimensional hexagonal lattice, with the H atoms attached to C atoms at the edges of the lattice. Three examples of PAHs are shown in Fig. 7. When H atoms are attached to the edge of an aromatic ring,

- the C-H stretching mode tends to be at $3.3\,\mu$m,
- the C-H in-plane bending mode is near $8.6\,\mu$m
- the C-H out-of-plane bending mode tends to fall at 11.3, 11.9, 12.7, or $13.6\,\mu$m, depending on whether there are one, two, three, or four adjacent C-H bonds ("solo-", "duo-", "trio-", or "quartet-" C-H). The observed spectra suggest approximately equal fractions of solo, duo, or trio C-H, and little quartet C-H, consistent with what one would expect for a mixture of large symmetric compact PAHs (Stein & Brown 1991).
- The carbon skeleton has C-C-C bending modes near 6.2 and $7.7\,\mu$m

The precise position of the bands, and their relative strengths, vary from one individual PAH to another, and upon their state of ionization, but the observed spectra are in good agreement with what one might obtain from a mixture of neutral and ionized PAHs (Allamandola, Hudgins, & Sandford 1999).

1.10 IR and FIR Emission

The energy which grains absorb from starlight is re-radiated at longer wavelengths, mainly into the far-infrared. We can estimate the emission spectrum of the Galaxy using data from all-sky surveys by the IRAS and COBE satellites, and pointed observations by the Japanese IRTS satellite. At high galactic latitudes we know that most of the dust is illuminated by the local average starlight background. Theoretical studies of dust grain heating by starlight suggested that $a \gtrsim 0.01\,\mu\mathrm{m}$ interstellar grains would be heated to temperatures $15\,\mathrm{K} \lesssim T \lesssim 20\,\mathrm{K}$ by the diffuse starlight (see, e.g., Draine & Lee 1984).

Because there is little dust at high galactic latitudes, the infrared surface brightness is low, but from the correlation of infrared surface brightness with 21 cm emission one can extract the infrared emission per H atom. This has been done using photometry in the 4 IRAS photometric bands, the 9 photometric bands of the DIRBE instrument on COBE, and using the spectrophotometry by the FIRAS instrument on COBE.

The Japanese IRTS satellite obtained the 3-13 μm spectrum of the galactic plane ($l \approx 45°$, $b \approx 0$). If we assume that the 3-13 μm surface brightness is proportional to the 100 μm surface brightness, we can estimate the 3-13 μm emission per H atom. The results are shown in Fig. 8. The $3-13\,\mu$m emission appears to be dominated by the same infrared emission features observed in

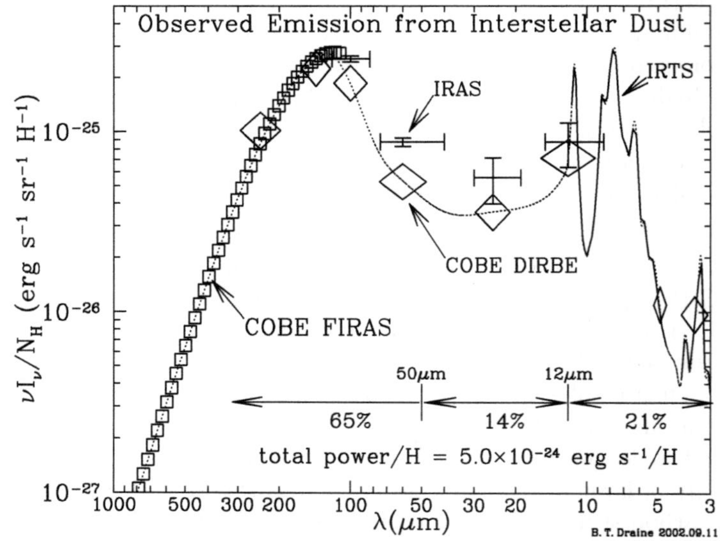

Fig. 8. Observed infrared emission per H nucleon from dust heated by the average starlight background (see text). Crosses: IRAS (Boulanger & Perault 1988); Squares: COBE-FIRAS (Wright et al. 1991); Diamonds: COBE-DIRBE (Arendt et al. 1998); Heavy Curve: IRTS (Onaka et al. 1996, Tanaka et al. 1996). The interpolated dotted line is used to estimate the total power.

reflection nebulae and planetary nebulae, and widely attributed to polycyclic aromatic hydrocarbons.

If we interpolate to obtain the dotted line in Fig. 8, we can numerically integrate to obtain the total infrared emission power per H nucleon = $5.0 \times 10^{-24}\,\mathrm{erg\,s^{-1}H^{-1}}$. About 2/3 of this power is radiated at wavelengths $\lambda > 50\,\mu\mathrm{m}$, which is where interstellar dust at $T \lesssim 20\,\mathrm{K}$ was expected to radiate. However, about 1/3 of the power is radiated at $\lambda < 50\,\mu\mathrm{m}$, where the emission from $T < 20\,\mathrm{K}$ dust should be negligible. Furthermore, about 1/6 of the total power is radiated at $\lambda < 10\,\mu\mathrm{m}$, primarily in 6.6, 7.7, and 8.6 μm PAH features.

1.11 Interstellar Depletions: Atoms Missing from the Gas

Narrow optical and ultraviolet absorption lines seen in absorption in stellar spectra can be used to determine interstellar gas phase abundances of many elements. While some species, such as N and S, have gas phase abundances (relative to H) which are approximately solar, certain others, such as Mg, Al, Si, Ti, Ca, Fe, Ni, Cr show abundances which are far below solar. Since we presume that the interstellar abundances are approximately solar (Sofia & Meyer 2001), the atoms missing from the gas must be locked up in dust grains. In Fig. 9 the abundances have been plotted as a function of "condensation temperature" – the temperature at which cooling, solar abundance gas at LTE would begin to condense a particular element into a solid or liquid phase. Elements with $T_{\mathrm{cond}} \gtrsim 1200\,\mathrm{K}$ are able to form "refractory" solids with large binding energies, and it is therefore not surprising that the strongly depleted

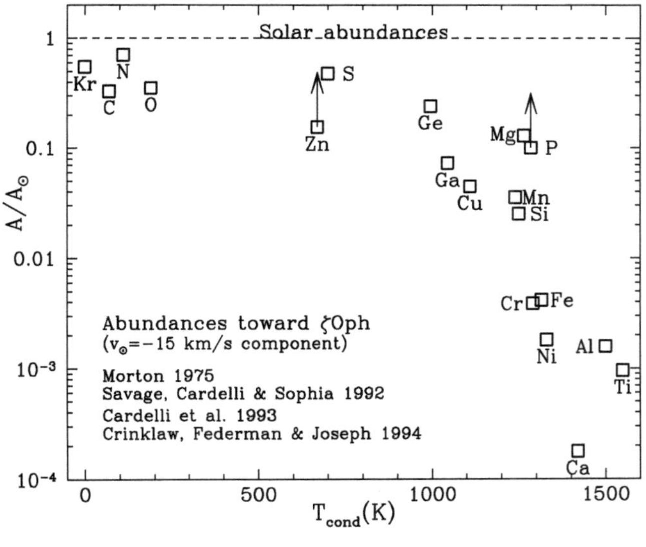

Fig. 9. Depletions in the diffuse molecular cloud toward ζOph.

elements tend to be those with high values of $T_{\rm cond}$. The observed elemental depletions provide a clue to the composition of interstellar dust. Restricting ourselves to abundant elements, we see that the "electron donor" species in interstellar dust material must be predominantly C, Mg, Si, and Fe. The grains could contain a significant amount of O, and perhaps H.

1.12 A Provisional Grain Model

Below we will be discussing the astrophysics of interstellar dust. It is helpful to have a specific grain model in mind so that one can carry out calculations. Accordingly, here we describe briefly a grain model which appears to be consistent with observations. The model consists of a mixture of carbonaceous particles and silicate particles. The carbonaceous particles have the physical and optical properties of polycyclic aromatic hydrocarbon molecules when they are small $a < 50$Å, or $N_{\rm C} \lesssim 6 \times 10^4$ carbon atoms. When they are larger, the carbonaceous grains are assumed to have the optical properties of randomly-oriented graphite spheres. The silicate grains are assumed to be amorphous silicates.

The emission in the $3 - 12\,\mu$m region from the diffuse ISM is attributed to dust grains small enough that a single starlight photon can raise the vibrational temperature high enough for thermal emission in the observed vibrational modes. In order for this emission to amount for $\sim 21\%$ of the total radiated power, very large numbers of very small dust grains are required. Li & Draine (2001) estimate that $\sim$15% of the total carbon abundance must be in carbonaceous particle containing $< 10^3$ C atoms. Since these particles must account for $\sim$20% of the total absorption of starlight, it is obvious that they must make a significant contribution to the interstellar extinction curve.

Weingartner & Draine (2001a, hereafter WD01) have shown that there are dust size distributions which include the required large numbers of ultra-small carbonaceous grains and which are compatible with the observed extinction curves for various regions in the Milky Way, LMC, and SMC galaxies. The size distributions are by no means unique – the observed extinction curve cannot be "inverted" – but they are nevertheless strongly constrained, so the WD01 size distributions are probably not too far from the truth. Figure 10 shows a grain size distribution appropriate for diffuse clouds in the local Milky Way. The total mass in grains with $a \lesssim 100$Å is constrained by the observed extinction in the ultraviolet, but since these grains are in the "Rayleigh limit" the extinction is virtually independent of the actual sizes and numbers of these particles, provided only that the constraint on the total volume of these grains is satisfied.

The size distribution for $a \lesssim 15$Å carbonaceous grains is adjusted to make the predicted infrared emission agree with observations. The bimodal nature of the carbonaceous size distribution for $a \lesssim 100$Å is probably an artifact of the fitting procedure.

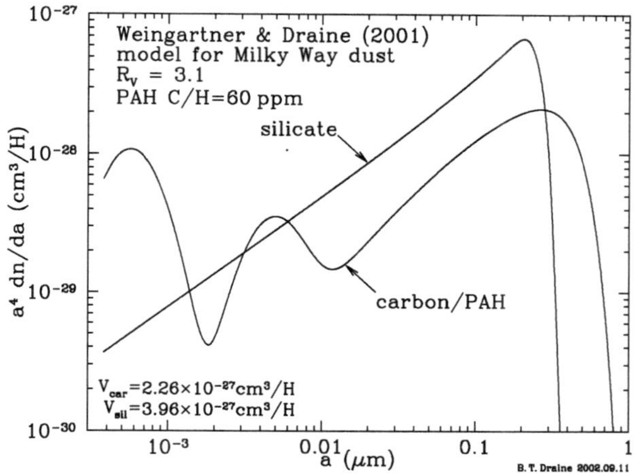

Fig. 10. WD01 size distribution for dust consistent with an $R_V = 3.1$ extinction curve for the local Milky Way. The quantity plotted – $a^4 dn/da$ – is proportional to the dust volume per logarithmic size interval.

1.13 Far-Infrared and Submm Opacities

If dust grains in a cloud have a temperature T_d, the emergent intensity is

$$I_\nu = B_\nu(T_\mathrm{d})\left[1 - e^{-\tau_\mathrm{d}}\right] \quad , \tag{2}$$

$$\tau_\mathrm{d}(\lambda) = \kappa(\lambda)\int \rho\, ds \quad , \tag{3}$$

where the dust opacity

$$\kappa(\lambda) = \frac{1}{1.4 n_\mathrm{H} m_\mathrm{H}} \int da \frac{dn_\mathrm{d}}{da} C_\mathrm{abs}(a,\lambda) \quad , \tag{4}$$

ρ is the total mass density (gas + dust), and we have assumed $n_\mathrm{He}/n_\mathrm{H} = 0.1$. Dust opacities κ are given in Fig. 12; these opacities are also available at `http://www.astro.princeton.edu/~draine/dust/dust.html`.

At long wavelengths the optical depth $\tau_\mathrm{d} \ll 1$. If we can estimate the dust temperature, the cloud mass per area can be obtained from the observed intensity I_ν by

$$\int \rho\, ds \approx \frac{1}{\kappa(\lambda)} \frac{I_\nu}{B_\nu(T_\mathrm{d})} \quad . \tag{5}$$

The derived mass density depends sensitively on the value of T_d unless one is in the Rayleigh-Jeans limit, $\lambda \gtrsim hc/kT_\mathrm{d}$, in which case $\int \rho\, ds \propto T_\mathrm{d}^{-1}$.

2 Optics of Interstellar Dust Grains

Our knowledge of interstellar dust is based in large part on the interaction of dust grains with electromagnetic radiation: absorption, scattering, and

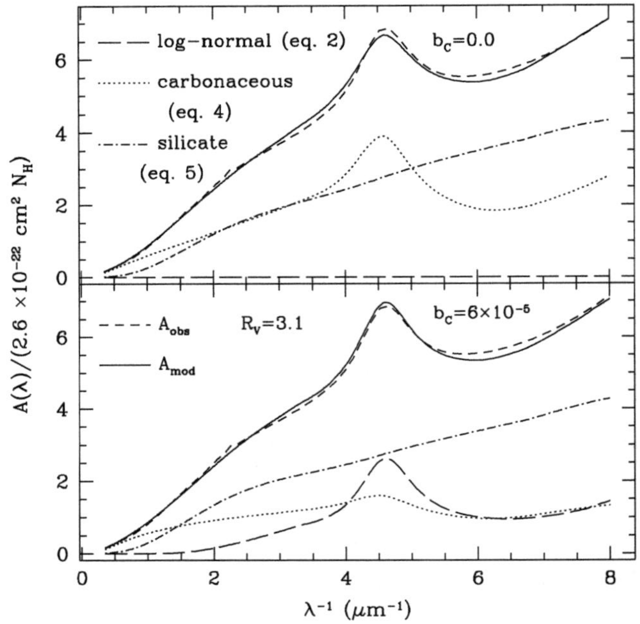

Fig. 11. *Dashed line*: average diffuse cloud extinction curve. *Solid lines*: model extinction curve for two extreme values of the ultra-small carbonaceous grain abundance: $b_C = 0$ (no PAHs) and $b_C = 60\,\mathrm{ppm}$ C in PAHs. For each case the size distribution of carbonaceous and silicate grains was adjusted to obtain good overall agreement with the observed extinction. From Weingartner & Draine (2001a).

emission. The interaction of electromagnetic radiation with a "target" will depend on the composition of the target (which determines the "response" of the target material to applied electric and magnetic fields) and the geometry of the target – both the shape and the size. Lacking definite knowledge of both the composition and the grain geometry, we must make some assumptions to proceed.

2.1 Cross Sections, Scattering Matrices, and Efficiency Factors

We are generally interested in calculating the following quantities for unpolarized plane waves incident on the target:

- $C_{\rm abs}$ = the total absorption cross section.
- $C_{\rm sca}$ = the total scattering cross section.
- $dC_{\rm sca}/d\Omega$ = differential scattering cross section. This is related to the Muller matrix element S_{11} by $dC_{\rm sca}/d\Omega = S_{11}/k^2$, where $k = 2\pi/\lambda$.
- polarization P for the radiation scattered in a particular direction.

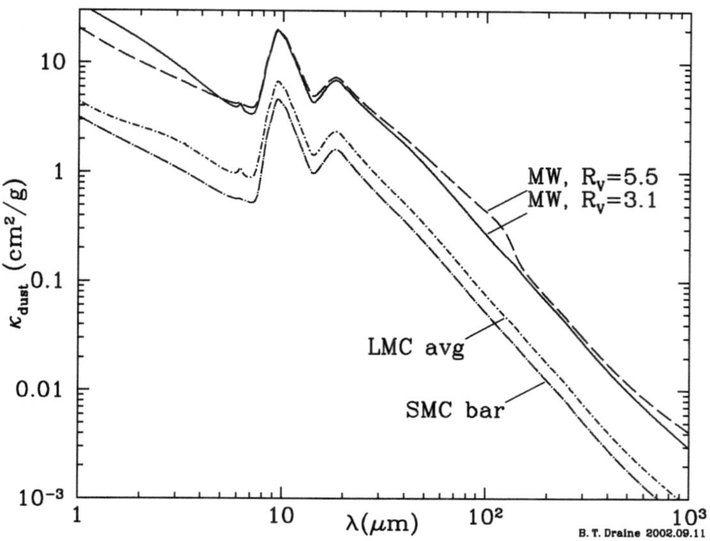

Fig. 12. Opacities $\kappa(\lambda)$ = (dust absorption cross section)/(mass of gas + dust) for Milky Way dust with $R_V = 3.1$ and $R_V = 5.5$, LMC average dust, and dust in the SMC bar, computed for the carbonaceous/silicate dust model (Weingartner & Draine 2001a; Li & Draine 2001).

In some cases we wish to explicitly consider polarized light. For this case the Stokes vector for the scattered radiation is obtained by multiplying the Stokes vector of the incident radiation by the 4×4 Mueller scattering matrix S_{ij}. The 4×4 scattering matrix (which, for a fixed grain, is a function of direction of incidence and direction of scattering) fully characterizes the scattering properties of a grain. See Bohren & Huffman (1983) or Mishchencko, Hovenier, & Travis (2000) for discussions of scattering concepts.

It is convenient to normalize scattering and absorption cross sections by dividing them by a geometric cross section. In the case of spherical target of radius a, the standard convention is to simply divide by the geometric cross section πa^2. In the case of non-spherical targets, there is more than one convention in use. Some authors divide by the geometric cross section of the target as seen from the direction of the illumination. Other authors divide by the average geometric cross section for random orientations.

My preference is to normalize by dividing by the geometric cross section of an "equal volume sphere". Thus for a target of mass M, and made of a material with density ρ, the volume is $V = M/\rho$, and we define "efficiency factors" $Q_{\rm sca}$, $Q_{\rm abs}$, $Q_{\rm ext} \equiv Q_{\rm sca} + Q_{\rm abs}$ by

$$Q = \frac{C}{\pi a_{\rm eff}^2} \quad \text{where} \quad a_{\rm eff} \equiv \left(\frac{3V}{4\pi}\right)^{1/3} \quad . \tag{6}$$

2.2 Grain Geometry?

Since interstellar dust is able to polarize starlight, we know that at least some interstellar dust grains must be appreciably non-spherical. Nevertheless, for many purposes we will approximate dust grains as spheres for the simple reason that we can use "Mie theory" to calculate scattering and absorption of light by spherically-symmetric targets with sizes comparable to the wavelength. Exact series solutions have been found for spheroids (Asano & Yamamoto 1975, Asano & Sato 1980; Voshchinnikov & Farafonov 1993) but numerical calculations are much more challenging than for spheres. Solutions for other shapes (other than the unphysical case of infinite cylinders) are not available.

2.3 Dielectric Functions

First, a few words about conventions: We will follow c.g.s. electromagnetism, so that $\epsilon = \mu = 1$ for the vacuum. We will represent plane waves as proportional to $e^{ikx-i\omega t}$; with this convention, absorption corresponds to $\mathrm{Im}(\epsilon) > 0$.

When an electric field is applied to a material, there are two distinct types of response possible: the response of the *bound* charge and the *free* charge. The *bound* charge (e.g., electrons and nucleus within an atom) undergoes a finite displacement when the electric field is applied, with the individual atoms or molecules acquiring an electric dipole moment in response to the applied electric field. The response of the bound charge is characterized by a complex dielectric function $\epsilon^{(\mathrm{bound})}(\omega)$.

The *free* charge responds in the form of an electric current density,

$$J = \sigma E \quad , \tag{7}$$

where $\sigma(\omega)$ is the electrical conductivity.

When dealing with a monochromatic electric field it is convenient to define a dielectric function ϵ describing the response of both bound and free charge:

$$\epsilon = \epsilon^{(\mathrm{bound})}(\omega) + \frac{4\pi i \sigma(\omega)}{\omega} \quad . \tag{8}$$

With this definition, the free charge current J is now absorbed into the displacement current $(1/4\pi)\partial D/\partial t$, so that Maxwell's equation becomes

$$\nabla \times B = \frac{1}{c}\frac{\partial D}{\partial t} \quad , \tag{9}$$

where $D = \epsilon E$.

2.3.1 A Model Insulator

It is instructive to consider a simple classical model for the dielectric response of an insulator. Suppose the insulator consists of "molecules", with no permanent dipole moment, but which can be polarized in response to the local applied electric field plus the electric field from all of the other polarized molecules. The dipole moment of a molecule is qx, where x is the displacement of the bound charge q. Suppose the displacement x behaves like a driven, damped harmonic oscillator,

$$m\ddot{x} = qE - m\omega_0^2 x - m\frac{\dot{x}}{\tau_0} \quad , \tag{10}$$

where ω_0 is the resonant frequency, and τ_0 is a damping time. For a periodic electric field $E \propto e^{-i\omega t}$ we can readily solve for the molecular polarizability $\alpha = qx/E$:

$$\alpha(\omega) = \frac{\alpha_0}{1 - (\omega/\omega_0)^2 - i\omega/(\omega_0^2\tau_0)} \quad , \tag{11}$$

where

$$\alpha_0 \equiv \frac{q^2}{m\omega_0^2} \tag{12}$$

is the zero-frequency polarizability of the molecule. We now consider a medium consisting of a number density n of such molecules. Each now responds to the externally applied electric field plus the electric field due to all of the other polarized molecules. In the low frequency limit, the effective dielectric function ϵ of the medium is related to the molecular polarizability through the famous Clausius-Mossotti relation (see, e.g., Jackson 1975)

$$\epsilon = 1 + \frac{4\pi n\alpha}{1 - (4\pi/3)n\alpha} \quad . \tag{13}$$

For molecules on a cubic lattice, the Clausius-Mossotti relation (13) is exact in the limit $|m|kd \ll 1$, where $d \equiv n^{-1/3}$ is the intermolecular spacing and $k \equiv \omega/c$. For finite kd, corrections to $O[(kd)^3]$ have been derived by Draine & Goodman (1993). Substituting (11) into (13) we obtain

$$\epsilon = 1 + \frac{4\pi n\alpha_0}{1 - (4\pi/3)n\alpha_0 - (\omega/\omega_0)^2 - i\omega/(\omega_0^2\tau_0)} \quad . \tag{14}$$

In the low frequency limit $\omega \to 0$, we have

$$\mathrm{Re}(\epsilon^{(\mathrm{bound})}) \to \epsilon_0^{(\mathrm{bound})} \equiv 1 + \frac{4\pi\alpha_0}{1 - (4\pi/3)\alpha_0} \quad , \tag{15}$$

from which we note that

$$4\pi\alpha_0 = 3\left[\frac{\epsilon_0^{(\mathrm{bound})} - 1}{\epsilon_0^{(\mathrm{bound})} + 2}\right] \quad . \tag{16}$$

At low frequencies, the imaginary part of $\epsilon^{(\mathrm{bound})}$ varies linearly with frequency:

$$\mathrm{Im}(\epsilon^{(\mathrm{bound})}) \approx A\omega \quad , \tag{17}$$

$$A = \frac{3}{\omega_0^2 \tau} \frac{\epsilon_0^{(\text{bound})} - 1}{\epsilon_0^{(\text{bound})} + 2} \quad . \tag{18}$$

An insulator has no mobile charges ($\sigma = 0$), and for our model at low frequencies we have

$$\text{Re}\,(\epsilon) \to \epsilon_0^{(\text{bound})} = \text{const} \quad , \tag{19}$$

$$\text{Im}\,(\epsilon) \to A\omega \quad . \tag{20}$$

2.3.2 A Model Conductor

For a material with a density n_e of free electrons, a simple classical model for the electron dynamics would be

$$m_e \ddot{x} = eE - m_e \frac{\dot{x}}{\tau_e} \quad , \tag{21}$$

where τ_e is the electron collision time. With this equation of motion, the conductivity $\sigma(\omega) \equiv n_e e \dot{x}/E$ is just

$$\sigma(\omega) = \frac{\sigma_0}{1 - i\omega\tau_e} \quad , \tag{22}$$

where

$$\sigma_0 = \frac{n_e e^2 \tau_e}{m_e} = \frac{\omega_p^2 \tau_e}{4\pi} \quad , \tag{23}$$

where σ_0 is the d.c. conductivity, and $\omega_p \equiv (4\pi n_e e^2/m_e)^{1/2}$ is the plasma frequency. For this case, the low-frequency behavior of ϵ is

$$\text{Re}\,(\epsilon) \to \epsilon_0^{(\text{bound})} - 4\pi\sigma_0\tau_e = \text{const} \quad , \tag{24}$$

$$\text{Im}\,(\epsilon) \to A\omega + \frac{4\pi\sigma_0}{\omega} \quad . \tag{25}$$

Comparing (20) and (25), we see that the low-frequency behavior of Im(ϵ) is qualitatively different for an insulator or for a conductor. Below we will see what this implies for the frequency-dependence of the absorption by small particles.

2.4 Calculational Techniques

In many astrophysical applications we are concerned with particles which are neither very small nor very large compared to the wavelength of the incident radiation. There are several different methods which can be used to calculate scattering and absorption cross-sections by targets which are neither very large nor very small compared to the wavelength:

- Mie theory solution for spherical targets (Bohren & Huffman 1983). Mie theory codes are highly-developed and readily available. They break down due to numerical roundoff error when the target becomes too large compared to the wavelength. A modified version of the Bohren & Huffman code is available at `http://www.astro.princeton.edu/~draine/scattering.html`
- Series solution for homogeneous spheroid or layered spheroid (boundaries must be confocal spheroids) (Asano & Yamamoto 1975; Asano & Sato 1980; Voshchinnikov & Farafonov 1993). Codes for spheroids have been developed but have not been widely used. Codes based on the Asano & Yamamoto treatment had a reputation for being somewhat numerically delicate (Rogers & Martin 1979). Voshchinnikov & Farafanov (1993) have developed a treatment based upon separation of variables which is reported to be more robust. They have recently generalized this to core-mantle spheroidal targets (Farafanov, Voshchinnikov, & Somsikov 1996).
- Extended Boundary Condition Method (EBCM), often referred to as the "T-matrix method" (Mishchenko, Travis & Macke 2000). The EBCM method (Mischchenko, Travis & Macke 2000) is used to construct the "T matrix" which gives the coupling between vector spherical harmonic components of the incoming wave and vector spherical harmonic components of the scattered wave. Once the T-matrix has been obtained, orientational averages can be efficiently calculated. The EBCM method appears to be well-suited to targets with rotational symmetry which are not extremely elongated. EBCM T-matrix codes have been made available by M. Mishchenko at `http://www.giss.nasa.gov/~crmim/t_matrix.html`
- Discrete Dipole Approximation (Draine & Flatau 1994; Draine 2000). The Discrete Dipole Approximation can be quite readily applied to complex geometries. It has been used to calculate the optical properties of graphite particles, with a highly anisotropic dielectric tensor (Draine 1988; Draine & Malhotra 1993). In recent years it has been accelerated by use of FFT techniques (Goodman, Draine, & Flatau 1991), and its accuracy has been improved by refinements in the assignment of dipole polarizabilities (Draine & Goodman 1993). The Fortran code DDSCAT is available from `http://www.astro.princeton.edu/~draine/DDSCAT.html` , and a detailed User Guide is available (Draine & Flatau 2000).

2.5 Scattering by Homogeneous Isotropic Spheres

The solution to Maxwell's equations for a plane wave incident on a homogeneous and isotropic sphere was independently obtained by Mie (1908) and Debye (1909). The solution – now generally referred to as "Mie theory" – is given in terms of a series expansion in powers of the "size parameter" $x \equiv 2\pi a/\lambda$. The expansion converges, but the number of terms which must be retained is of $O(x)$. The details of the Mie theory solution are nicely described in the excellent monograph by Bohren & Huffman (1983). There are a number of computer codes to evaluate the absorption cross section and differential scattering cross sections given by Mie scattering theory. When x is large, care must be taken to avoid errors due to finite precision arithmetic in numerical evaluation of the series solution.

For an incident monochromatic plane wave, the full Mie theory solution depends on the dielectric function $\epsilon(\omega)$ – describing the electric polarization of the material in response to an applied electric field oscillating at angular frequency ω – and the magnetic permeability function $\mu(\omega)$ – characterizing the magnetization of the material in response to an applied oscillating magnetic fields. Because the magnetic response is generally negligible (i.e., $|\mu - 1| \ll |\epsilon - 1|$) at frequencies above 100 GHz) it is customary to neglect the magnetization of the grain material except insofar as it is due to eddy currents.

2.6 Discrete Dipole Approximation

The discrete dipole approximation consists in replacing the (continuum) target of specified dielectric function ϵ with an array of polarizable points (referred to as "dipoles"). The array geometry is chosen to mimic the target geometry. The polarizabilities α of the points are chosen so that an infinite lattice of such polarizable points would have the same dispersion relation as the material of dielectric function ϵ (Draine & Goodman 1993). With FFT techniques employed to speed the calculation, it is now feasible to calculate scattering and absorption by targets represented by more than 100,000 polarizable dipoles on a workstation with 256 MB of RAM. There are great efficiencies if the dipoles are situated on a cubic lattice, so DDSCAT requires that this be the case.

The DDA can be applied to inhomogeneous targets and targets with complex geometries. As an example of the type of problem which can be solved, in Fig. 13 we show a discrete-dipole array of 61432 dipoles intended to approximate a tetrahedral target. In Fig. 14 we show the calculated scattered intensity in the x-y plane as a function of scattering angle Θ. Notice the pronounced scattering peak at $\Theta = 240°$ – this is the direction where one would have a specular reflection peak in the geometric optics limit.

For a fixed target size D, complex refractive index m, and wavelength λ, the DDA converges to the exact answer in the limit where the inter-dipole separation $d \to 0$: the target structure is well-resolved and the dipole separation is small compared to the wavelength in the target material. The criterion for separation small compared to λ can be written $|m|kd \lesssim 1$. If $|m|$ is not too large, we find that the overall scattering and absorption cross section is calculated fairly accurately provided $|m|kd \lesssim 1$; the detailed scattering pattern will be calculated accurately if $|m|kd \lesssim 0.5$. For example, the two calculations in Fig. 14, with $|m|kd = 0.82$ and 0.68, are already in fairly good agreement.

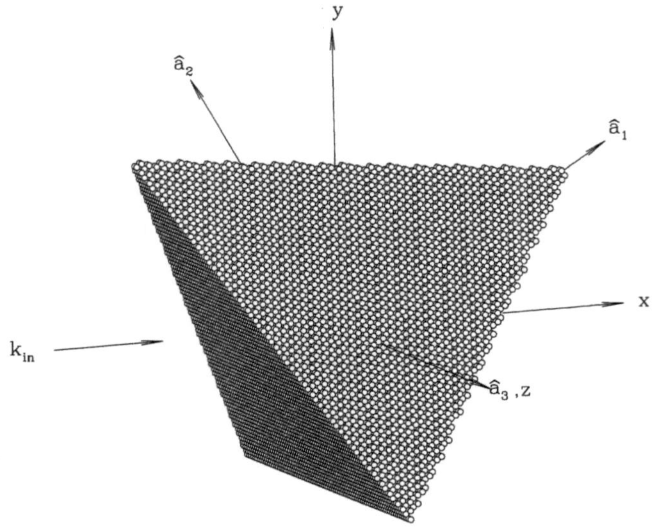

Fig. 13. Representation of a tetrahedron with a dipole array. The dipoles are actually point-like – here they are represented by small spheres for purposes of visualization. We consider incident radiation propagating along the direction of the vector $\mathbf{k}_{\rm in}$ illustrated here. Taken from Draine (2000).

2.7 Infrared and Far-Infrared

When the wavelength is long compared to the target size, we may use the "dipole approximation" (Draine & Lee 1984) and characterize the response of the target to the incident electromagnetic field purely in terms of the (oscillating) electric and magnetic dipole moments of the target. We may write

$$C_{\rm abs} = \frac{4\pi\omega}{c} \left[{\rm Im}(\alpha_E^e) + {\rm Im}(\alpha_H^m) \right] \quad , \tag{26}$$

$$C_{\rm sca} = \frac{8\pi}{3} \left(\frac{\omega}{c}\right)^4 \left[|\alpha_E^e|^2 + |\alpha_H^m|^2 \right] \quad , \tag{27}$$

where α_E^e is the (complex) electric polarizability along the direction of the incident electric field vector, and α_H^m is the (complex) magnetic polarizability along the direction of the incident magnetic field vector.

Consider a homogeneous ellipsoidal particle composed of an isotropic material, with semi-axes a, b, c. If the applied electric field is along one of its principal axes, the electric polarizability is

$$\alpha_E^e = \frac{V}{4\pi} \frac{\epsilon - 1}{(\epsilon - 1)L_E + 1} \quad , \tag{28}$$

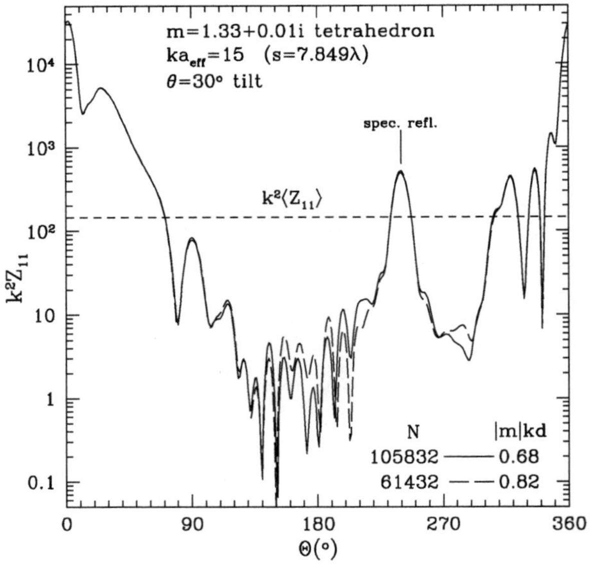

Fig. 14. $S_{11} = k^2 dC_{\rm sca}/d\Omega$ for scattering of unpolarized light by a tetrahedron, for scattering directions in the x-y plane (see Fig. 13), as a function of scattering angle Θ. The tetrahedron has refractive index $m = 1.33 + 0.01i$, has edges of length $s = 7.849\lambda$, where λ is the wavelength of the incident light in vacuo, and is oriented with an angle $\theta = 30°$ between $\mathbf{k}_{\rm in}$ and $\hat{\mathbf{a}}_1$. The peak at $\Theta = 240°$ corresponds to the direction of specular reflection for geometric optics. Results are shown for a tetrahedron represented by $N = 61432$ and $N = 105832$ dipoles. The scattered intensities are in excellent agreement in all directions where the scattering is at all strong. The dashed line shows $k^2\langle Z_{11}\rangle = C_{\rm sca}/4\pi$. Taken from Draine (2000).

where L_E is the "shape factor" along the direction of the applied E field, and $V = 4\pi abc/3$ is the volume of the ellipsoid. The shape factors along the three principal axes satisfy the sum rule $L_1 + L_2 + L_3 = 1$ and may be obtained by numerical quadrature (Bohren & Huffman 1983).

Consider now a spheroid with semi-axes a, b, b (prolate if $a/b > 1$, oblate if $a/b < 1$). For this case we have analytic expressions for the shape factors L_a and $L_b = (1 - L_a)/2$. For a prolate spheroid ($a < b$) we have (van de Hulst 1957)

$$L_a = \frac{1-e^2}{e^2}\left[\frac{1}{2e}\ln\left(\frac{1+e}{1-e}\right) - 1\right] \quad , \tag{29}$$

and for an oblate spheroid ($a < b$) we have

$$L_a = \frac{1+e^2}{e^2}\left[1 - \frac{1}{e}\arctan(e)\right] \quad , \tag{30}$$

where

$$e^2 \equiv |1 - (b/a)^2| \quad . \tag{31}$$

Exact solutions for α_E^e in the limit $\lambda \gg a_{\rm eff}$ are also available for layered grains provided the interfaces are confocal spheroids (Gilra 1972; Draine & Lee 1984).

For spheres we have $L_a = L_b = 1/3$, and

$$C_{\rm abs} = \frac{9\omega V}{c} \frac{\epsilon_2}{(\epsilon_1 + 2)^2 + \epsilon_2^2} \quad , \tag{32}$$

where $\epsilon_1 \equiv {\rm Re}(\epsilon)$ and $\epsilon_2 \equiv {\rm Im}(\epsilon)$.

Equation (32) contains an important result: the absorption cross section of a grain with size $\ll \lambda$ is simply proportional to the grain volume! The far-infrared opacity therefore depends only on the total volume of grain material present, but not on the sizes of the particles, provided only that they are small compared to the wavelength λ.

How do we expect the absorption cross section to depend on frequency ω at low frequencies? For an insulator, from (20) we see that

$$C_{\rm abs} \to \frac{9V}{c} \frac{A}{(\epsilon_0 + 2)^2} \; \omega^2 \quad , \tag{33}$$

while for a conductor, we see from (25) that

$$C_{\rm abs} \to \frac{9V}{c} \left(\frac{1}{4\pi\sigma_0} \right) \; \omega^2 \quad . \tag{34}$$

Thus we see that for both insulating grains and conducting grains, a simple physical model leads to $C_{\rm abs} \propto \omega^2$ at low frequencies! This is the basis for the expectation that dust opacities should vary as λ^{-2} in the far-infrared, as in the dust models of Draine & Lee (1984).

Now the above discussion has been based on simple classical models for the response of charge to applied electric fields. Real materials obey quantum mechanics, so results may differ. Furthermore, the above discussion assumed that $\omega \ll \omega_0$ in discussion of the response of the bound charge, but an amorphous material could have very low frequency vibrational modes, so that the assumption that $\omega \ll \omega_0$ might not be valid in the far-infrared. This might explain why some laboratory studies of amorphous materials find different behavior: for example, Agladze et al. (1996) find that the absorption coefficient for amorphous $MgO{\cdot}2SiO_2$ varies as $\omega^{1.2}$ between 800 μm and 4 mm.

Nature must be our guide, but it is worth keeping in mind that opacities varying as ω^2 emerge naturally from simple models.

2.8 Kramers-Kronig Relations

Purcell (1969) pointed out that the Kramers-Kronig relations can provide useful constraints in dust modelling. The Kramers-Kronig relations are general relations which apply to "linear response functions", such as a dielectric function, which specifies the response (e.g., the electric polarization) to an applied stress (e.g., an applied electric field). The *only* assumptions are that (1) the response is linear, and (2) the system is causal – the response can depend on the stress applied in the past, but cannot depend upon the future. With these very simple assumptions, it is possible to derive the Kramers-Kronig relations (see Landau & Lifshitz 1960 for a nice derivation). In the case of the dielectric function, the Kramers-Kronig relations are

$$\epsilon_1(\omega_0) = 1 + \frac{2}{\pi} P \int_0^\infty d\omega \, \frac{\omega \, \epsilon_2(\omega)}{\omega^2 - \omega_0^2} \quad , \tag{35}$$

$$\epsilon_2(\omega_0) = \frac{2}{\pi} \omega_0 \, P \int_0^\infty d\omega \, \frac{\epsilon_1(\omega)}{\omega_0^2 - \omega^2} \quad , \tag{36}$$

where P indicates that the principal value is to be taken.

Thus the real and imaginary parts of $\epsilon(\omega)$ are by no means independent – if either one is known at all frequencies, the other is fully determined. In order to have a physically acceptable dielectric function, one approach (see, e.g., Draine & Lee 1984) is to specify ϵ_2 at all frequencies, and then construct $\epsilon_1(\omega)$ using (35).

2.9 Kramers-Kronig Relations for the ISM

Purcell showed that one could apply the Kramers-Kronig relations directly to the interstellar medium. Plane waves propagate through the interstellar medium but are attenuated by scattering and absorption. One can describe this attenuation by an imaginary component of the dielectric constant $\tilde{\epsilon}$ of the interstellar medium, where $|\tilde{\epsilon} - 1| \ll 1$.

Recall that for an electromagnetic plane wave of frequency ω, the wave vector is $k = m(\omega)\omega/c$, where $m(\omega) = \sqrt{\epsilon}$ is the complex refractive index. The electric field intensity decays as

$$E \propto \exp\left[-\,\mathrm{Im}(k)x\right] = \exp\left[-\frac{1}{2}\,\mathrm{Im}(\tilde{\epsilon})\frac{\omega}{c}\right] \quad . \tag{37}$$

The energy in the wave is proportional to $|E|^2$ and therefore decays twice as rapidly as E. Thus

$$n_{\mathrm{gr}} \, C_{\mathrm{ext}}(\lambda) = \frac{\omega}{c} \tilde{\epsilon}_2 \quad . \tag{38}$$

We can now apply the Kramers-Kronig relation (35) to obtain the real part of the dielectric function $\tilde{\epsilon}$ of the ISM at zero frequency:

$$\tilde{\epsilon}_1(0) - 1 = \frac{2}{\pi} \int_0^\infty \frac{d\omega}{\omega} \frac{c}{\omega} \, n_{\mathrm{gr}} \, C_{\mathrm{ext}}(\omega) \quad . \tag{39}$$

The static dielectric function of the ISM is directly related to the electric polarizability α_{gr} of the grains in it:

$$\tilde{\epsilon}_1(0) - 1 = 4\pi n_{\mathrm{gr}} \, \alpha_{\mathrm{gr}}(0) \quad . \tag{40}$$

Thus

$$4\pi n_{\rm gr}\,\alpha_{\rm gr}(0) = \frac{2}{\pi} c n_{\rm gr} \int_0^\infty d\omega \,\frac{1}{\omega^2}\, C_{\rm ext}(\omega) \tag{41}$$

$$= \frac{1}{\pi^2} n_{\rm gr} \int_0^\infty d\lambda\, C_{\rm ext}(\lambda) \quad . \tag{42}$$

We have seen above how the static polarizability of a spheroidal grain depends on its volume, shape, and dielectric function. Averaged over random orientation, we have (for a dielectric grain)

$$4\pi\alpha_{\rm gr}(0) = 3VF(a/b,\epsilon_0) \quad , \tag{43}$$

$$F(a/b,\epsilon_0) \equiv \frac{(\epsilon_0-1)}{3}\left[\frac{1}{(\epsilon_0-1)3L_a+3} + \frac{2}{(\epsilon_0-1)3L_b+3}\right] \quad . \tag{44}$$

F is just the orientationally-averaged polarizability relative to the polarizability of an equal-volume conducting sphere. Using (43), we obtain the grain volume per H atom in terms of an integral over the extinction per H atom:

$$\frac{n_{\rm gr}V}{n_{\rm H}} = \frac{1}{3\pi^2 F}\int_0^\infty d\lambda\, \frac{n_{\rm gr}}{n_{\rm H}} C_{\rm ext}(\lambda) \tag{45}$$

Now consider a conducting material. From (24-25) we see that $\epsilon \to i\times\infty$, so that

$$F \to \frac{1}{9L_a} + \frac{2}{9L_b} \tag{46}$$

so that a conducting sphere ($L_a = L_b = 1/3$) has $F(a/b = 1, \epsilon_0 = \infty) = 1$.

We do not know the grain shape, and we do not know the static dielectric constant ϵ_0. However, from Fig. 15 we see that $F < 1.5$ unless the grain is *extremely* elongated ($a/b > 20$ or $b/a > 20$) *and* the static dielectric function is very large ($\epsilon_0 > 10$) – as for a metal. For a reasonable grain shape (e.g., $0.5 < a/b < 2$) and dielectric function $\epsilon_0 \approx 3$ we have $F \approx 0.4$.

Purcell's analysis is a delight, and has two important consequences:

2.9.1 Grain Volume per H

It is of course not possible to measure the extinction per H atom at wavelengths from 0 to ∞. However, since $C_{\rm ext} > 0$, measurements over a finite wavelength range can be used to obtain a lower bound on $FVn_{\rm gr}/n_{\rm H}$. The extinction per H nucleon is fairly well-known from 0.1 μm to 30 μm; a numerical evaluation gives

$$\int_{0.1\,\mu{\rm m}}^{30\,\mu{\rm m}} \frac{\tau_{\rm ext}(\lambda)}{N_{\rm H}}\, d\lambda \approx 1.1\times 10^{-25}\,{\rm cm}^3/{\rm H} \quad . \tag{47}$$

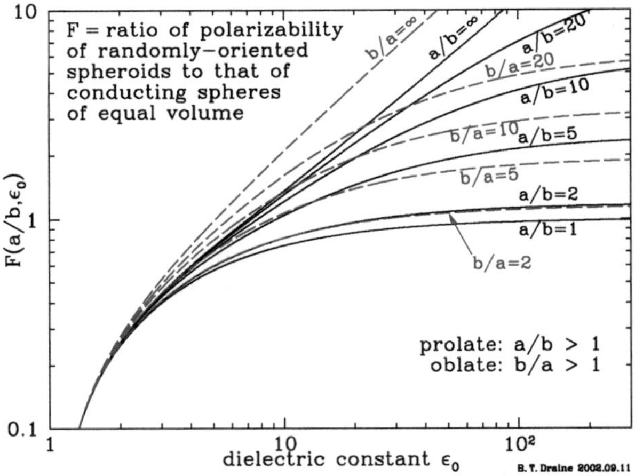

Fig. 15. Function $F(a/b, \epsilon_g)$ as a function of the static dielectric constant ϵ_0, for selected values of axial ratio a/b. After Purcell (1969).

Approximately 50% of the integral is contributed by wavelengths $0.1 < \lambda < 1\,\mu$m. If we estimate $F \approx 0.4$ this gives

$$\frac{n_{gr}V}{n_H} \gtrsim 9.3 \times 10^{-27}\,\mathrm{cm}^3/\mathrm{H} \quad , \tag{48}$$

or, for an assumed grain mass density $\rho = 2.5\,\mathrm{g\,cm}^{-3}$ (intermediate between graphite and silicate) we have a lower bound on the ratio of grain mass to H mass:

$$\frac{M_{gr}}{M_H} \gtrsim 0.014 \quad . \tag{49}$$

2.9.2 Asymptotic Behavior at Long Wavelengths

Equation (45) tells us that $\int_0^\infty C_{ext}(\lambda)\,d\lambda$ must be convergent, and therefore that $C_{ext}(\lambda)$ *must* decline more rapidly than $1/\lambda$: grain models in which $C_{ext} \propto 1/\lambda$ for $\lambda \to \infty$ are unphysical. It is of course possible to have $d\ln C_{ext}/d\ln\lambda \approx -1$ over a limited range of wavelengths, but this cannot be the asymptotic behavior.

2.10 Microwave

Dust grains would be expected to radiate thermally at microwave frequencies, but this emission was expected to be quite weak, based on an extrapolation from the thermal emission peak at $\sim 100\,\mu$m. It was therefore a surprise when sensitive maps of the microwave sky brightness (for the purpose of studying angular structure in the cosmic background radiation) showed relatively strong emission correlated with the Galactic $100\,\mu$m emission, and

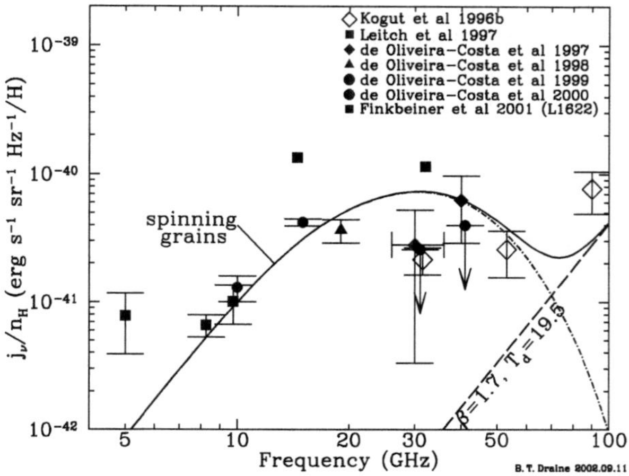

Fig. 16. Microwave sky brightness correlated with $100\,\mu$m emission. Shown is the estimated microwave emissivity per unit H nucleon.

therefore with interstellar dust (Kogut et al. 1996a,b). Figure 16 shows the inferred microwave emissivity of per H nucleon. The dashed line labelled "$\beta = 1.7, T_{\rm d} = 19.5$" is the contribution expected from thermal emission from dust grains if the dust opacity $\propto \lambda^{-1.7}$ for $\lambda \gtrsim 100\,\mu$m. We see that at 30 GHz ($\lambda = 1\,$cm the observed microwave emission is two orders of magnitude above the value expected from classical dust grains.

The power radiated in the microwave is a tiny fraction of the total, so this microwave emission is not an important source of cooling. Nevertheless, we would like to know what process is responsible. Possible mechanisms include synchrotron emission from relativistic electrons, or free-free emission from a thermal plasma, but these seem unable to account for the observed $15-50\,$GHz emission (see Draine & Lazarian 1999b). We are left with interstellar dust.

The possibility of radio emission from rotating dust grains dates back to Erickson (1957). As discussed above, the strong infrared emission in the $3-12\,\mu$m range appears to require a very large population of very small dust grains. Ferrara & Dettmar (1994) pointed out that these dust grains would produce detectable radio emission if they were undergoing Brownian rotation at the gas temperature.

Draine & Lazarian (1998a,b) analyzed the grain rotational dynamics, and showed that the population of ultra-small grains required to understand the infrared emission would be expected to produce microwave emission, with a predicted spectrum (and intensity) as shown in Fig. 16. We will discuss the grain dynamics in §6, but this rotational emission still seems likely to contribute a substantial fraction of the observed dust-correlated microwave emission. In principle, these rapidly spinning dust grains could be aligned, leading to polarized emission, but the polarization fraction has been estimated to be small (Lazarian & Draine 2000).

However, if dust grains contain magnetic materials – which does not seem implausible, considering the large amount of Fe in interstellar grains – there could also be *thermal* magnetic dipole emission from dust. Consider a ferromagnetic domain in a dust grain. This domain is spontaneously-magnetized: it is energetically favorable to have alignment of electron orbital angular momentum and electron spins. The lowest energy state has some specific magnetization, but there are nearby energy states where the domain has uniform magnetization with the same net magnetic moment but in a slightly different direction. In thermal equilibrium, it is possible for these higher energy states to be excited, so that the magnetization direction will fluctuate. This results in a time-varying magnetic dipole moment, and hence magnetic dipole radiation.

The optics of magnetic dust grains with size small compared to the wavelength has been developed by Draine & Lazarian (1999a; hereafter DL99a), and the magnetic susceptibility $\mu(\omega)$ was estimated for several magnetic materials, including metallic Fe and magnetite Fe_3O_4. The spectrum and intensity of this radiation was estimated by DL99a, and it was found that *if* a substantial fraction of the Fe in interstellar grains is incorporated into magnetic materials (e.g., magnetite), the resulting thermal emission could account for an appreciable fraction of the observed microwave emission!

How will we be able to distinguish between spinning dust grains and magnetic dust grains as sources of microwave emission? DL99a point out that small grains are believed to be under-abundant in dense clouds (they presumably coagulate to form bigger grains), so that the rotational emissivity per H nucleon would be reduced. The magnetic dipole emission from magnetic grain materials does not, however, depend on the grain size. Hence if we see strong microwave emission from dust in dense clouds, this would suggest that magnetic dust grains may be responsible. Future pointed observations of dense clouds may answer this question.

2.11 X-rays

The scattering of X-rays by dust grains was first discussed by Overbeck (1965), Slysh (1969), and Hayakawa (1970). Because the refractive index of grain materials is close to unity at X-ray energies, the scattering is through small angles. As a result, images of X-ray point sources show a "halo" of dust-scattered X-rays. The observable halo can extend for tens of arcminutes from the source.

For sufficiently small grains, this scattering can be calculated using the Rayleigh-Gans approximation (see Bohren & Huffman 1983), which is valid if $|m-1|ka \ll 1$, where a is the grain radius. Because X-rays have large k, this condition may not be satisfied for the larger interstellar grains, in which case one should resort to full Mie theory. This can be numerically challenging because the number of terms which must be retained is of order $ka \approx 507(a/0.1\,\mu\text{m})(E/\text{keV})$, so high numerical accuracy is required. Smith & Dwek (1998) have compared scattering halos estimated using the Rayleigh-Gans approximation with Mie theory, and show that the Rayleigh-Gans approximation fails for energies below 1 keV. Scattering by grains with $|m-1|ka \ll 1$ and $ka \gg 1$ can be treated using "anomalous diffraction theory" (van de Hulst 1957).

Nova Cygni 1992, a bright X-ray nova, was observed by ROSAT. The observed X-ray halo has been compared to predictions for different grain models by Mathis et al. (1995), Smith & Dwek (1998), and Witt, Smith & Dwek (2001), with differing conclusions. The most recent study (Draine & Tan 2002) concludes that our standard dust model is in reasonable agreement with the observed X-ray halo. Future observations of X-ray scattering halos will be valuable to test and constrain grain models.

3 IR and Far-IR Emission from Interstellar Dust

3.1 Heating of Interstellar Dust

While dust grains can be heated by collisions with gas atoms and molecules, starlight usually is the dominant heating process. The rate of energy deposition, the rate of photon absorptions, and the mean energy per absorbed photon are

$$\left\langle\frac{dE}{dt}\right\rangle_{\text{abs}} = \int_0^\infty d\nu\, u_\nu\, c\, C_{\text{abs}}(\nu) \quad , \tag{50}$$

$$\dot{N}_{\text{abs}} = \int_0^\infty d\nu\, \frac{u_\nu}{h\nu}\, c\, C_{\text{abs}}(\nu) \quad , \tag{51}$$

$$\langle h\nu\rangle_{\text{abs}} = \frac{1}{\dot{N}_{\text{abs}}}\left\langle\frac{dE}{dt}\right\rangle_{\text{abs}} \quad , \tag{52}$$

where $u_\nu\, d\nu$ is the photon energy density in $(\nu, \nu + d\nu)$, and $C_{\text{abs}}(\nu)$ is the photo-absorption cross section. We assume that all of the energy of an absorbed photon is converted to heat; this is not exactly correct, since energetic photons may eject a photo-electron, or excite fluorescence, but these processes take away only a small fraction of the total absorbed power.

Now suppose that the dust grain has energy E distributed among its vibrational degrees of freedom. Except when the energy E is very small, the number of vibrational modes which can be excited is very large. The density of states has been discussed by Draine & Li (2001): even for a small molecule like coronene $C_{24}H_{12}$, there are $\sim 10^{20}$ different states with total energy $E < 1\,\text{eV}$; for $C_{4000}H_{1000}$ this number becomes $\sim 10^{135}$. As a result, the statistical notion of "temperature" can be used, even for quite small grains: a grain with vibrational energy content E is assumed to be characterized by a temperature $T(E)$ such that if the grain were in contact with a heat bath at temperature T, the expectation value for its energy would be E.

At temperature T, the average power radiated by a grain is

$$\left(\frac{dE}{dt}\right)_{\rm rad} = \int_0^\infty d\nu\, C_{\rm abs}(\nu)\, 4\pi B_\nu(T) \quad , \tag{53}$$

where $C_{\rm abs}(\nu)$ is the angle-averaged absorption cross section for photons of frequency ν, and

$$B_\nu(T) \equiv \frac{2h\nu^3}{c^2}\frac{1}{e^{h\nu/kT}-1} \tag{54}$$

is the Planck function.

It is now natural to determine the "steady-state" temperature $T_{\rm ss}$ at which the power radiated equals the power absorbed:

$$\int_0^\infty d\nu\, C_{\rm abs}(\nu)\, 4\pi B_\nu(T_{\rm ss}) = \left\langle\frac{dE}{dt}\right\rangle_{\rm abs} \quad . \tag{55}$$

Once we solve for $T_{\rm ss}$, we calculate the thermal energy content of the grain at this temperature, $E(T_{\rm ss})$. There are two regimes:

- If $E(T_{\rm ss}) \gg \langle h\nu\rangle_{\rm abs}$, then individual photon absorption events do not substantially change the energy content of the grain, and we may assume that the grain temperature $T = T_{\rm ss}$. This is the "steady heating" regime, where we can approximate the discrete heating events as a continuous process.
- If $E(T_{\rm ss}) \ll \langle h\nu\rangle_{\rm abs}$, then individual photon absorption events heat the grain up to peak temperatures $T_{\rm max} \gg T_{\rm ss}$, and the grain will usually cool to a temperature $T_{\rm min} < T_{\rm ss}$ before the next photon absorption event. This is the "stochastic heating" regime, where the discrete nature of the heating has important consequences. As we will see below, stochastic heating is often important.

Figure 17 shows the temperature history of 4 grains of different sizes in the diffuse interstellar medium over a $\sim 10^5$ s interval – about one day. The temperature of the $a = 200$Å grain fluctuates in a small range around ~ 20 K. However, as the grain size is reduced, the fluctuations become increasingly extreme, with the peaks being higher and the low points being lower. The smallest grain shown ($a = 25$Å) absorbs about 1 starlight photon per day; this grain spends most of its time quite cool, but immediately following absorption of a photon the grain temperature reaches ~ 50 K. The grain cools by infrared emission; it is obvious that most of the infrared emission must take place during the brief interval while the grain is "hot". If we wish to calculate the time-averaged emission spectrum, we cannot use the "average" grain temperature – we need to integrate over a distribution of grain temperatures. This phenomenon is even more pronounced for smaller grains. A grain with

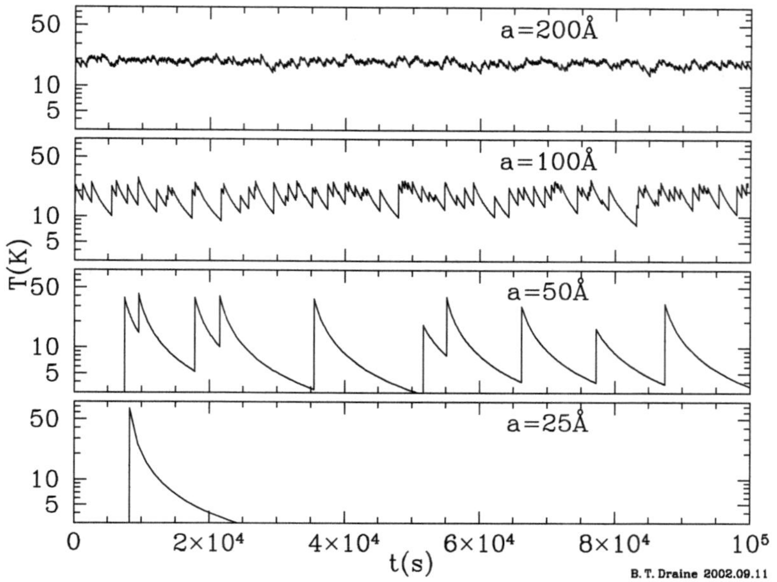

Fig. 17. A day in the life of an interstellar grain: grain temperature vs. time for 4 grain sizes, for grains heated by the average interstellar radiation field. Grains with $a \gtrsim 200$Å have a nearly constant temperature, but $a \lesssim 100$Å grains show conspicuous increases in temperature following each photon absorption, with gradual cooling between photon absorption events.

a radius of 5Å has a volume only 1/125 that of the 25Å shown in Fig. 17; the mean time between photon absorptions is of order $\sim 10^7$ s, but a 10 eV starlight photon can raise the grain temperature to $\sim 10^3$ K.

Let $P(E)$ be the probability of the grain having vibrational energy $E' \geq E$. In the steady heating regime, we approximate $P(E)$ as a step function, and $|dP/dE|$ as a delta function $\delta(E - E_{ss})$.

In the stochastic heating regime, we must solve for $P(E)$. Draine & Li (2001a) show examples of energy distribution functions. If the energy states are grouped into bins $j = 0, 1, \ldots$ (where $j = 0$ is the ground state), then we can calculate the state-to-state transition rates T_{ji} for transitions $i \to j$ due to both photon absorptions and photon emissions. Once $T_{i\to j}$ is known, we define the diagonal elements $T_{ii} \equiv -\sum_{j\neq i} T_{ji}$. The steady state solution P_j for the probability of being in state j then satisfies the N coupled linear equations

$$0 = \sum_{j=0}^{M} T_{ij} P_j \quad \text{for } i = 0, \ldots, M \quad . \tag{56}$$

Using the normalization condition $\sum_{j=0}^{M} P_j = 1$, we obtain a set of M linear equation for the first M elements of P_j:

$$\sum_{j=0}^{M-1} (T_{ij} - T_{iM})P_j = -T_{iM} \quad \text{for } i = 0, \ldots, M-1 \quad , \tag{57}$$

which we solve using standard techniques. In practice, we take $M \approx 500$, in which case iterative methods are required to efficiently solve for P_j. Once the P_j are determined for $j = 0, \ldots, M-1$ we could obtain P_M by subtraction $P_M = 1 - \sum_{j=0}^{M-1}$ but this is inaccurate; it is better to instead use

$$P_M = \frac{-1}{T_{MM}} \sum_{j=0}^{M-1} T_{Mj} P_j \quad . \tag{58}$$

Figure 18 shows $dP/d\ln E$ for carbonaceous grains heated by the average interstellar radiation field (ISRF) due to starlight in an HI region, as estimated by Mathis, Mezger, & Panagia (1983, hereafter MMP). For $a \lesssim 25$Å grains $dP/d\ln E$ becomes very small for $E/hc > (1/911\text{Å}) = 1.1 \times 10^5\,\text{cm}^{-1}$: this reflects the fact that grains cool essentially completely between photon absorption events, so that the energy content virtually never rises above the maximum energy (13.6 eV) of the illuminating photons. As the grain size increases, the time between photo-absorptions goes down, and the cooling time at fixed energy goes up; when the grain size exceeds ~ 30Å there is a significant probability of a photo-absorption taking place before the energy of the previous photo-absorption has been radiated away. However, stochastic heating effects are noticeable even for grains as large as 300Å – the energy distribution function has narrowed considerably, but is still appreciably broad.

3.2 IR and Far-IR Emission Spectrum

With the energy distribution function calculated as discussed above, we can now calculate the time-averaged emission spectrum for a carbonaceous grain:

$$F_\lambda = 4\pi \int dE \, \frac{dP}{dE} \, C_{\text{abs}}(\lambda) \, B_\lambda(T(E)) \quad . \tag{59}$$

In Fig. 19 we show the emission spectrum of PAH^+ molecules of various sizes heated by the ISRF, and in Fig. 20 we compare the emission spectrum calculated using the energy distribution functions $dP/d\ln E$ with emission spectra calculated for dust with steady temperature T_{ss}. We see that stochastic heating is important even for grains as large as ~ 100Å in the ISRF.

In Fig. 21 we show the emission spectrum calculated for a mixture of carbonaceous and silicate grains with size distributions consistent with the average ($R_V = 3.1$) interstellar extinction curve, illuminated by the local

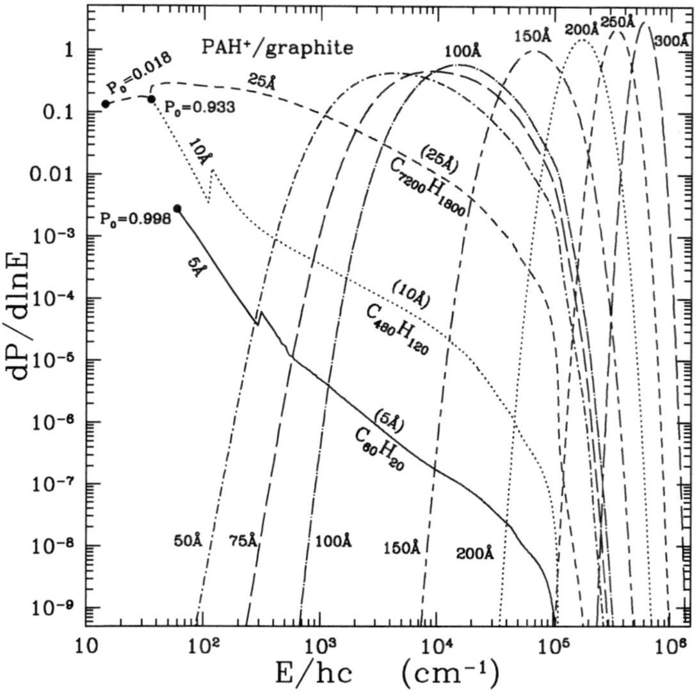

Fig. 18. Energy distribution functions for charged carbonaceous grains with radii $a = 5, 10, 25, 50, 75, 100, 150, 200, 250, 300\text{Å}$ in the interstellar radiation field. The discontinuity in the 5, 10, and 25Å curves is an artifact due to a change in the method of estimating the cooling when the energy is equal to the 20th vibrational mode. For 5, 10, and 25Å, a dot indicates the first excited state, and the probability P_0 of being in the vibrational ground state is given. Taken from Li & Draine (2001).

ISRF. This should be the emission spectrum of the "cirrus" clouds. Also shown in Fig. 21 are observations of the emissivity per H atom of the dust at high galactic latitudes.

Diamonds show DIRBE photometry; the small triangles show our calculated spectrum convolved with the DIRBE filters, for comparison with the diamonds. The squares show the FIRAS determination of the emissivity per H nucleon. The far-infrared emission is in excellent agreement with the predictions of our dust model. The model is in very good agreement with DIRBE photometry at 3.5 μm and 25 μm; the model appears too low by about a factor 2 at 5 and 12 μm. The observations are difficult; it is hoped that SIRTF will be able to measure the spectrum of selected cirrus clouds for comparison with our model.

In Fig. 22 we compare our model with observations taken on the galactic plane, where the FIR surface brightness is much higher. In addition to

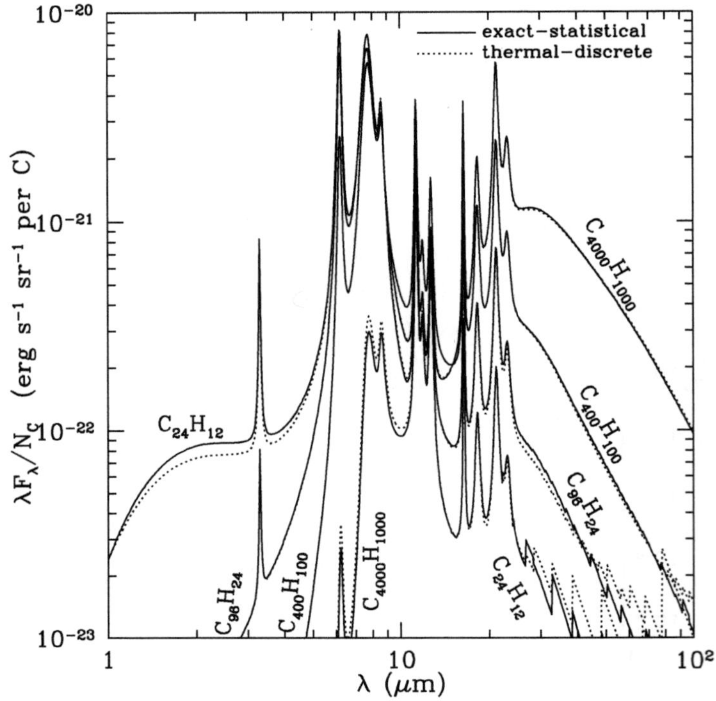

Fig. 19. IR emissivity per C atom for PAH^+ molecules of various sizes illuminated by the average ISRF. *Solid line*: "exact-statistical" calculation, which uses transition probabilities which do not involve the thermal approximation. *Dotted line*: "thermal-discrete" calculation where the spontaneous emission rates are calculated using a thermal approximation, as discussed in the text. The two methods are in excellent agreement, indicating that the thermal approximation can be used to calculate the transition rates. Taken from Draine & Li (2001).

DIRBE photometry, spectroscopic observations made with the IRTS satellite are available.

These observations are looking in the galactic plane at the "tangent" point of gas in the 6 kpc ring, at a distance of 6 kpc from us. At this distance, the $40'$ DIRBE beam subtends 70 pc, so the observations sum the emission from dust in a region $\sim 70 \times 70$ pc on a side, and many kpc long. We compare the observations to a model which is obviously oversimplified: a uniform slab of dust heated by a uniform radiation field with the spectrum of the local ISRF but whose intensity we scale so that the dust emission best reproduces the observed spectrum: the best fit is obtained with the starlight intensity equal to twice the ISRF. The DIRBE beam obviously includes emission from stars, which we believe dominate at $\lambda < 4\,\mu$m, so our model includes light from stars assumed to be mixed uniformly with the dust. This simple model is in good agreement with the observations. In particular, we successfully

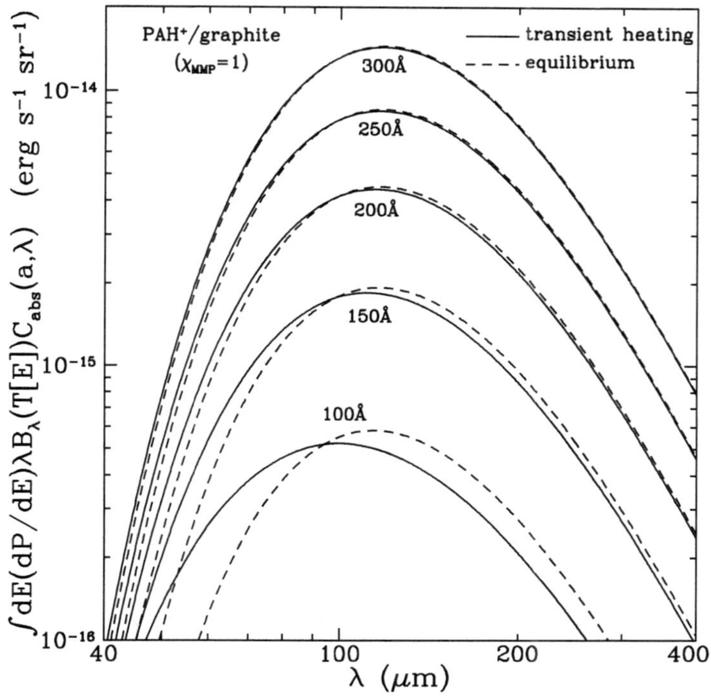

Fig. 20. IR emission per grain for carbonaceous grains of various sizes illuminated by the average ISRF. *Solid line*: full stochastic heating calculation. *Broken line*: assuming steady-state temperature T_{ss}. For $a \gtrsim 200$Å in the ISRF we see that stochastic heating has little effect on the emission spectrum, but for $a \lesssim 150$Å it significantly modifies the emission spectrum. Taken from Li & Draine (2001).

reproduce the very strong emission in the $5-12\,\mu$m region. The discrepancy at $100\,\mu$m is somewhat surprising, but may in part be due to oversimplification of our model, and perhaps in part to uncertainties in the DIRBE absolute calibration.

3.3 The Small Magellanic Cloud (SMC)

How successful is this dust model in reproducing the emission observed from other galaxies? The SMC, with metallicity $\sim$ 10% of solar (Dufour 1984), and dust-to-gas ratio $\sim$ 10% of the Milky Way (Bouchet et al. 1985) is a good test case. Li & Draine (2002) sum over both quiescent medium and regions of active star formation in the SMC, find that to match the observed spectrum they require a distribution of starlight intensities. Following Dale et al. (2001) they adopt a power-law distribution of starlight intensities.

The dust extinction in the SMC is known to differ from Milky Way dust, primarily in the absence of a 2175Å extinction bump on most (but not all)

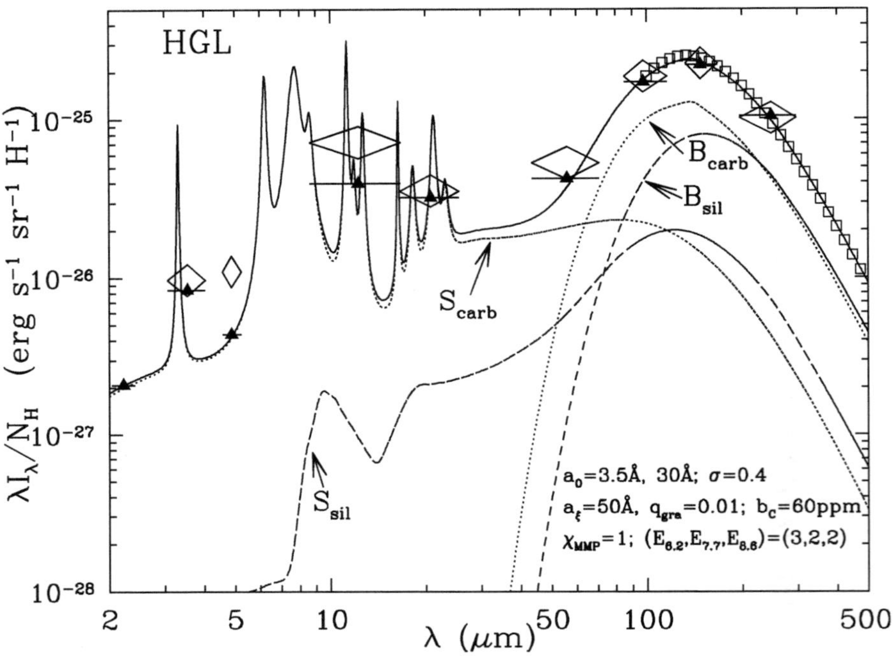

Fig. 21. IR emission per H for dust mixture illuminated by the ISRF. *Diamonds*: DIRBE (Arendt et al 1998). *Squares*: FIRAS (Finkbeiner et al. 1999). From Li & Draine (2001).

SMC sightlines (Gordon & Clayton 1998). Since we take the view that the 2175Å feature is due to PAHs, the absence of this feature implies a low PAH abundance. We adopt the dust mixture of Weingartner & Draine (2001a), which reproduces the measured SMC bar extinction curve.

With suitable choice of illuminating radiation field, the dust model is able to reproduce the observed IR emission from the SMC. It appears that the carbonaceous/silicate grain model (Weingartner & Draine 2001a; Li & Draine 2001) can reproduce both the observed interstellar extinction by dust and the observed IR/FIR emission from dust.

4 Charging of Interstellar Dust

Why do we care about the charging of dust grains? There are a number of reasons:

- Charged dust grains are coupled to the magnetic field, whereas neutral grains are not. This is important not only for the motions of the dust grains, but also as a mechanism for coupling magnetic fields to neutral gas.

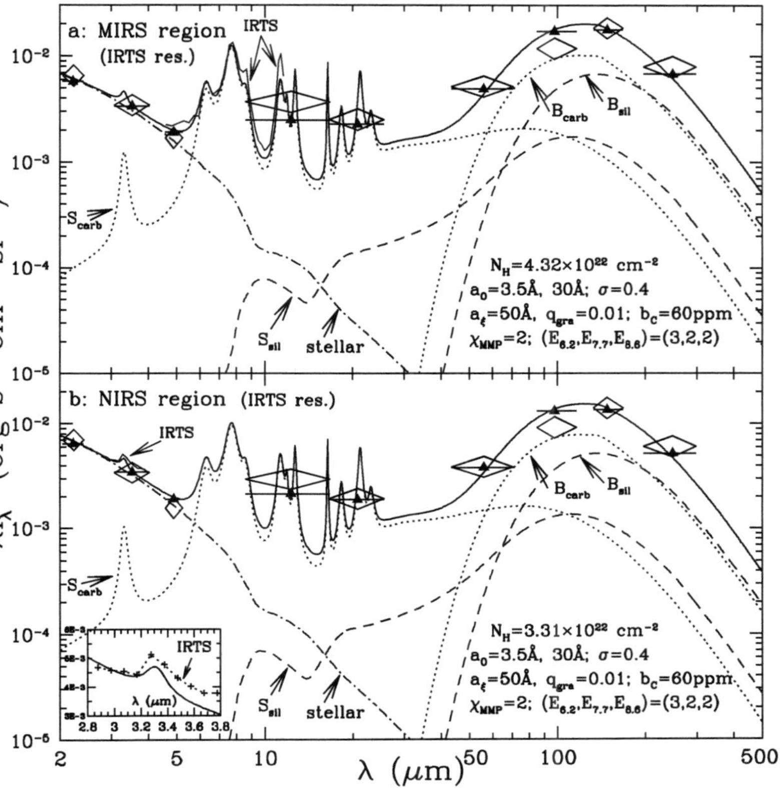

Fig. 22. IR surface brightness toward (a) $l = 44°20'$, $b = -0°20'$; (b) $l = 47°45'$. *Diamonds*: DIRBE photometry. *Thin line, upper panel*: $5-12\,\mu$m spectrum measured by MIRS on IRTS (Onaka et al. 1996). *Thin line, lower panel*: $2.8-3.9\,\mu$m spectrum measured by NIRS on IRTS (Tanaka et al. 1996). *Solid line*: carbonaceous-silicate dust model illuminated by 2×ISRF, plus reddened starlight. From Li & Draine (2001).

- Charged dust grains undergo stronger gas drag due to Coulomb interaction with ions in the gas.
- The photoelectric charging process injects energetic photoelectrons into the gas, which is a major mechanism for heating interstellar gas.
- Neutral and negatively-charged dust grains can play an important role in neutralization of ions.
- In cold, dense regions, the dust grains may "lock up" a substantial fraction of the electrons.

4.1 Collisions with Electrons and Ions

Consider a spherical grain of radius a and charge $Z_g e$, and a charged particle of charge ze, kinetic energy E, on a trajectory with impact parameter b. If the

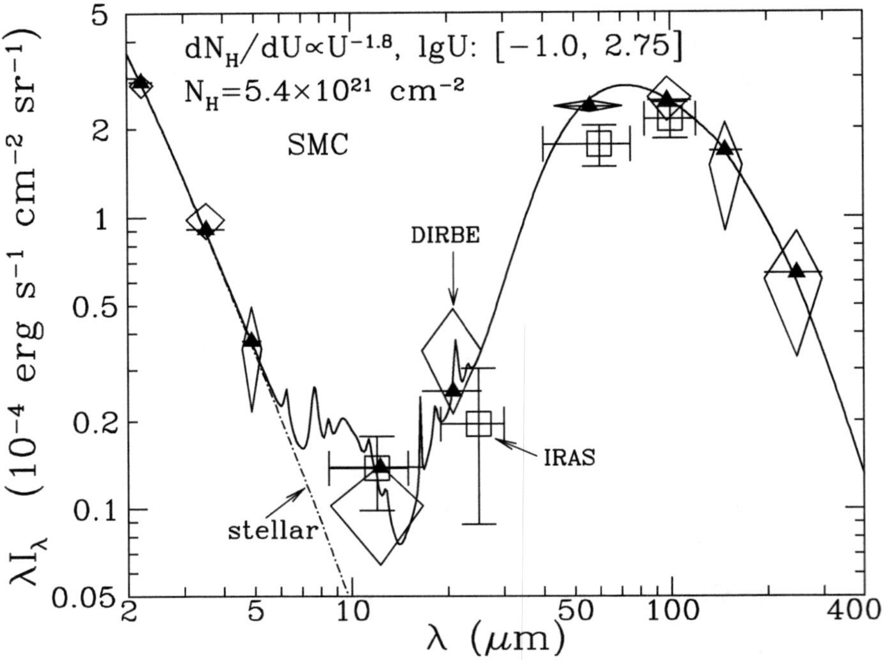

Fig. 23. Comparison between dust model (*solid line*) and the observed spectrum of the SMC obtained by COBE/DIRBE (*diamonds*) and IRAS (*squares*) averaged over a 6.25 $\deg^2$ region including the optical bar and the Eastern Wing (Stanimirovic et al 2000). Model spectrum includes infrared emission from dust heated by a range of starlight intensities. plus starlight, which dominates at $\lambda \lesssim 7\,\mu$m. Triangles show the model spectrum convolved with the DIRBE filters. Taken from Li & Draine (2002).

interaction potential is taken to be a Coulomb potential, then conservation of energy and angular momentum allow one to find the critical impact parameter $b_{\rm crit}$ for the particle to just graze the grain surface. Trajectories with $b < b_{\rm crit}$ will collide with the grain, and trajectories with $b > b_{\rm crit}$ will miss. The collision cross section is simply

$$C = \pi b_{\rm crit}^2 = \pi a^2 \left[1 - \frac{zeU}{E}\right] \quad \text{for} \quad E > zeU \quad , \tag{60}$$

$$= 0 \quad \text{for} \quad E \leq zeU \quad , \tag{61}$$

where $U \equiv Z_{\rm g}e/a$ is the electrostatic potential at the grain surface. If the velocity distribution is thermal, then the rate at which the charged particles reach the grain surface is just

$$\frac{dN}{dt} = n\pi a^2 \left(\frac{8kT}{\pi m}\right)^{1/2} \times \begin{cases} \exp\left(-\frac{zeU}{kT}\right) & \text{for} \quad zeU > 0 \\ \left[1 - \frac{zeU}{kT}\right] & \text{for} \quad zeU < 0 \end{cases} \quad . \tag{62}$$

This is a wonderfully simple result. If the sphere is located in a plasma of electrons and ions, and the only charging process is collisional, then the sphere will become negatively charged since the electrons move more rapidly than the ions. The steady-state charge is such that suppression of the electron arrival rate, and enhancement of the ion arrival rate, make them equal:

$$n_e s_e \left(\frac{8kT}{\pi m_e}\right)^{1/2} \exp\left(\frac{eU}{kT}\right) = n_I s_I \left(\frac{8kT}{\pi m_I}\right)^{1/2} \left[1 - \frac{eU}{kT}\right] \quad , \qquad (63)$$

where we have assumed $Z_I = +1$. Taking logarithms of both sides we obtain

$$\frac{eU}{kT} = -\frac{1}{2} \ln\left(\frac{m_I}{m_e}\right) + \ln\left(\frac{s_I}{s_e}\right) + \ln\left(1 - \frac{eU}{kT}\right) \quad , \qquad (64)$$

with the solution $eU/kT = -2.51$ for a hydrogen plasma ($m_I/m_e = 1836$) with $s_I = s_e$.

However, the Coulomb potential does not fully describe the interaction between a charged particle and a grain. In the case of a neutral grain, the electric field of the approaching charged particle will polarize the grain, resulting in an attractive potential – the same effect that causes ion-neutral scattering rates to be large, even at low temperatures.[1] Draine & Sutin (1987) have calculated collision rates including the image potential. Image potential effects are important for neutral grains when $e^2/a \gtrsim kT$, or $aT \lesssim 1.67\times10^{-3}$ cm K. Thus it is an important correction for small grains in cold gas (e.g., $a \lesssim 1.67\times10^{-5}$ cm for $T = 100$ K).

The "sticking coefficients" for electrons and small ions are not well known. In the case of an ion with an ionization potential larger than the work function for the grain material, it seems likely that the ion, upon arrival at the grain surface, will either remain stuck to the grain or will strip an electron from the grain and depart; in either case the ion "sticking coefficient" $s_I = 1$. The case of impinging electrons is less clear. The electron might be elastically reflected from the grain surface, or the electron might pass through a small grain and out the other side without energy loss, in which case the electron would have enough energy to return to infinity.

Experimental sticking coefficients for small neutral molecules or molecular ions can be estimated by dividing the measured electron capture rate coefficient by the estimated rate at which the electrons would reach the surface of the ion. Figures 24 and 25 show the electron sticking coefficients for small neutral and charged carbonaceous clusters containing $10 < N_C < 100$ carbon atoms.

In hot gas, it is possible for impinging electrons to eject an electron from the grain. For molecules, this is called "ionization"; for larger objects, it is called "secondary electron emission". It is therefore possible for grains to

[1] This is just the familiar "image potential" from electrostatics for a charge near a conductor.

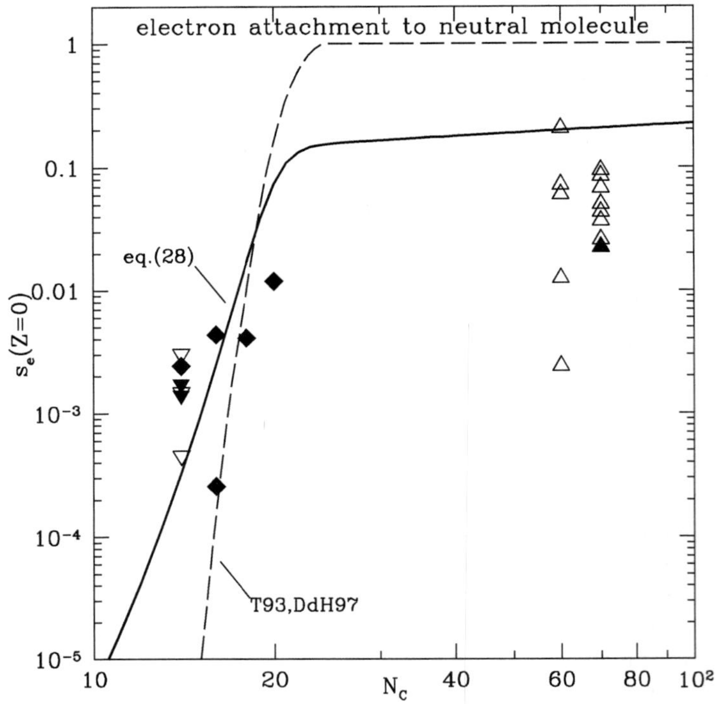

Fig. 24. Electron sticking coefficient for small neutral grains. From Weingartner & Draine (2001c).

become positively charged by collisions in a hot ($T \gtrsim 10^6$ K) plasma (Draine & Salpeter 1979a).

4.2 Photoelectric Emission

The electrons in a neutral grain can be thought of a being confined within a potential well, as shown in Fig. 26. The potential well is produced by a "double layer" of charge at the boundary of the grain (the electron charge density extends beyond the ion charge density, resulting in a region near the surface with a radial electric field – see standard texts on solid state physics, e.g., Ashcroft & Mermin 1976).

Three cases are shown:

- For a grain with $Z \geq 0$, the excited electron will be subject to a long-range Coulomb attraction by the rest of the charge. In this case the electron which physically gets outside of the grain must still have additional energy in order to reach ∞.
- For a grain with charge $Z = -1$, the excited electron sees a system with zero net charge, so once the electron is physically outside the double layer, it will continue to infinity.

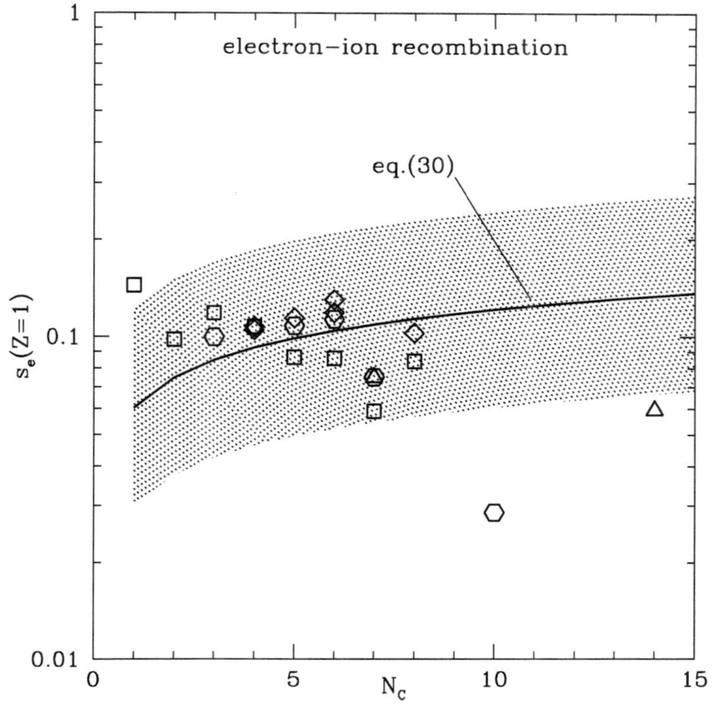

Fig. 25. Electron sticking coefficient for small grains with charge $Z = +1$. From Weingartner & Draine (2001c).

- For a grain with charge $Z < -1$, an electron which gets outside the double layer will be accelerated away from the grain. An excited electron can in principle penetrate the double layer by tunneling, so that it is not actually necessary for the electron to have an energy higher than the top of the double layer.

An insulating grain has all the "valence band" energy levels occupied when it is neutral. The difference in energy between the top of the valence band and the first available vacant energy level (in the "conduction band") is referred to as the "band gap". If an insulating grain is to be negatively charged, the excess electrons must occupy available energy levels in the conduction band. Less energy is required to photo-eject one of these "excess" electrons (a process known as "photo-detachment") than to eject an electron from the top of the valence band.

When an ultraviolet photon is absorbed in a dust grain, it raises an electron to an excited state. If the electron has an energy high enough to reach infinity, and does not lose this energy to inelastic scattering, it will escape and be counted as a "photo-electron". To calculate the rate of photoelectric charging of a grain, we require the "photoelectric yield" $Y(h\nu, Z, a)$, which is

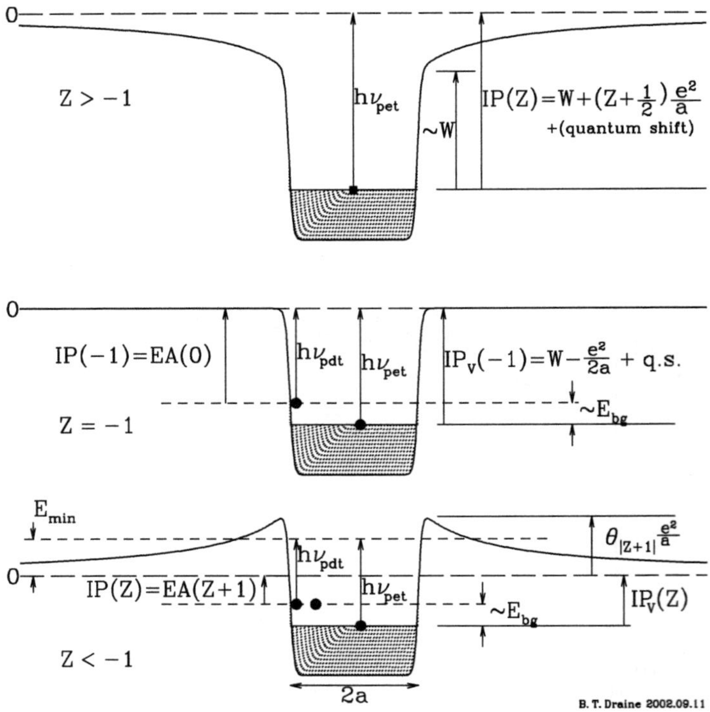

Fig. 26. Model for the potential confining electrons in a grain with charge Ze. Shaded regions show occupied energy levels. W is the work function for bulk material; $IP(Z)$ is the ionization potential; $EA(Z)$ is the electron affinity; $h\nu_{\rm pet}$ is the photoelectric threshold energy; $h\nu_{\rm pdt}$ is the photo-detachment threshold energy (the minimum photon energy required to detach an electron from a negatively charged grain); $IP_V(Z)$ is the energy difference between infinity and the top of the valence band; $E_{\rm bg}$ is the bandgap energy for the material ($E_{\rm bg} = 0$ for a metal). Taken from Weingartner & Draine (2001c).

the probability that absorption of a photon $h\nu$ will produce a photo-electron. To calculate the rate of photoelectric heating of the gas, we also need to know the energy distribution function $f(E, h\nu, Z, a)$ of the energy E of the photoelectrons leaving the grain.

The photoelectric ejection process consists of four stages:

1. Excitation of an electron of appropriate energy.
2. Motion of the electron from the point of excitation to the grain surface,
3. Penetration of the surface layer and overcoming of the image potential.
4. Once outside the grain surface, the electron must overcome the Coulomb potential (if the grain is now positively charged) to reach infinity.

Weingartner & Draine (2001c) have estimated the photoelectric yield for small grains, both neutral and charged, taking into account the above effects.

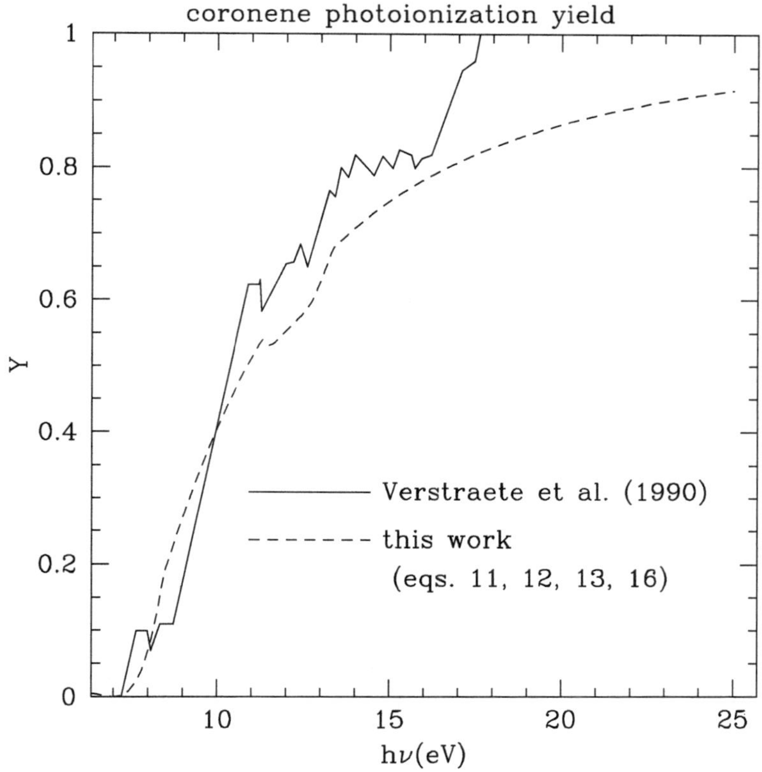

Fig. 27. Photoelectric yield for neutral coronene, as measured by Verstraete et al. (1990) and as estimated in the approach of Weingartner & Draine (2001c). Equations 11, 12, 13, 16 refer to equations in Weingartner & Draine (2001c), from which this figure is taken.

Allowance is made for dependence of the image potential on the grain radius a, and the possibility of tunneling through the double layer in the case of a negatively charged grain.

Watson (1973) pointed out that very small particles should have enhanced photoelectric yields because of step 2 above: in a small particle, photo-excitations will on average be closer to the grain surface, increasing the probability that the excited electron will reach the grain surface without energy loss. Draine (1978) proposed a simple function to estimate the magnitude of this yield enhancement, which we use:

$$y_1 = \left(\frac{\beta}{\alpha}\right)^2 \frac{\alpha^2 - 2\alpha + 2 - 2\exp(-\alpha)}{\beta^2 - 2\beta + 2 - 2\exp(-\beta)} \quad , \tag{65}$$

$$\beta = \frac{a}{l_a} \ , \quad \alpha = \frac{a}{l_a} + \frac{a}{l_e} \quad , \tag{66}$$

where $l_a = \lambda/[4\pi\,\mathrm{Im}(m)]$ is the photon attenuation length in the material, and l_e, the "electron escape length", is essentially the mean free path against energy loss to inelastic scattering for an excited electron. In the limit $\alpha \ll 1$ we have

$$y_1 \to \frac{\alpha}{\beta} = \frac{l_a + l_e}{l_e} \quad . \tag{67}$$

Photon attenuation lengths $l_a \approx 300$Å are typical in the vacuum ultraviolet. Martin et al. (1987) report $l_e \approx 9$Å for 6 eV electrons in thin carbon films, and McFeely et al. (1990) report $l_e \approx 6$Å for 8 eV electrons in SiO_2. Weingartner & Draine (2001c) adopt $l_e = 10$Å, independent of energy, for both graphite and silicate grains. Since $l_e \ll l_a$, it is clear that very large yield enhancement factors y_1 are possible for very small grains.

Weingartner & Draine write

$$Y(h\nu, Z, a) = \min\left[y_0 \times y_1(h\nu, a)\right] \times y_2(h\nu, Z, a) \tag{68}$$

where y_2 is the fraction of electrons which, having crossed the grain surface, have sufficient energy to overcome the long-range coulomb attraction if $Z \geq 0$ (for $Z < 0$, $y_2 = 1$). For PAHs and larger carbonaceous grains, y_0 is chosen so that $Y = y_0 y_1 y_2$ approximates the photoionization yield measured for coronene.

Figure 28 shows the photoelectric yields estimated for neutral carbonaceous and silicate grains of various radii.

4.3 Charge Distribution Functions

Figure 29 shows grain charge distribution functions calculated for dust in a molecular cloud region. The H nucleon density is taken to be $n_{\mathrm{H}} = 300\,\mathrm{cm}^{-3}$, the gas temperature $T = 25$ K, and the fractional ionization $n_e/n_{\mathrm{H}} = 10^{-4}$. The radiation field is assumed to have the spectrum of the MMP ISRF, but with intensity a factor of 10 below the MMP value. Under these conditions small grains ($a \lesssim 200$Å) remain negatively charged. For larger grains, the reduced importance of "image charge" effects (as well as a slight reduction in the work function) tilts the balance in favor of photoelectric charging, but the grain potentials remain very small.

Figure 30 shows grain charge distribution functions calculated for a diffuse region with starlight equal to the MMP ISRF. The electron density $n_e = 0.03\,\mathrm{cm}^{-3}$ is taken to be the same as for the molecular cloud region of Fig. 29, but the increased starlight intensity allows photoelectric emission to dominate the charging for grains with $a \gtrsim 25$Å. function The "CNM" conditions are appropriate to diffuse HI at 100 K. Despite the presence of ultraviolet radiation producing photoelectric emission, note that an appreciable fraction of the small grains can be neutral or even negatively charged.

We see that the competition between collisional charging and photoelectric charging can go either way, depending on the grain composition, the

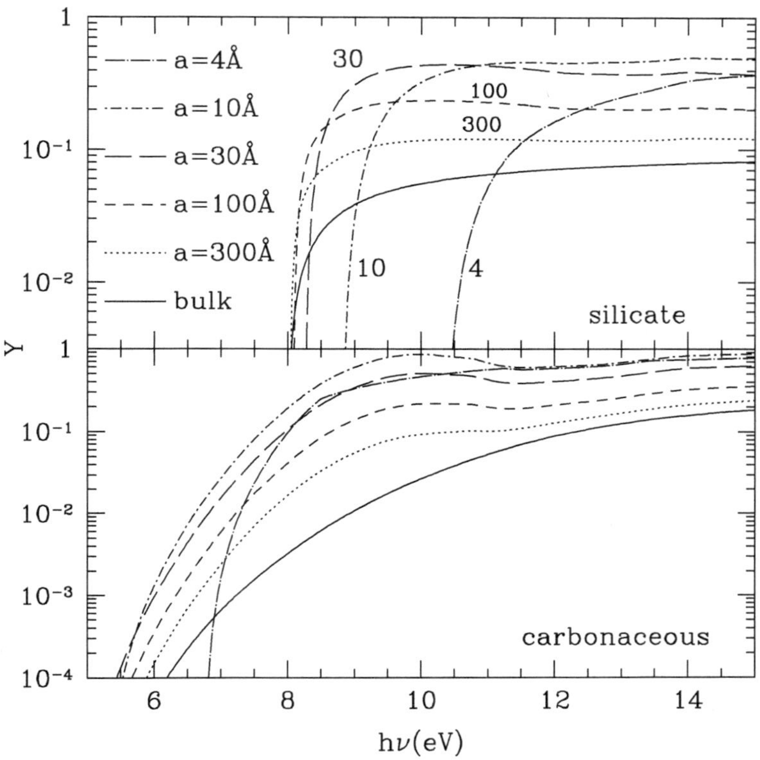

Fig. 28. Photoelectric yields (photoelectrons per absorbed photon) for neutral graphite and silicate grains as a function of incident photon energy $h\nu$, for selected values of the grain radius a. From Weingartner & Draine (2001c).

grain size a, the electron density n_e and temperature T, and the spectrum and intensity of the ultraviolet background due to starlight.

5 Dynamics of Interstellar Dust

What are the velocities of interstellar grains? To answer this question, we must understand the forces acting to accelerate or decelerate the grains.

5.1 Gas Drag

If the grain is moving relative to the gas, it will experience a drag force as momentum is transferred from the grain to the gas. This drag force is approximately $F_{\rm drag} = M_{\rm gr} v_{\rm gr}/\tau_{\rm gr}$ where $M_{\rm gr}$ is the grain mass, $v_{\rm gr}$ is the grain speed relative to the gas, and $\tau_{\rm gr}$ is the time for the grain to collide with its own mass in gas particles. For the case of an uncharged spherical

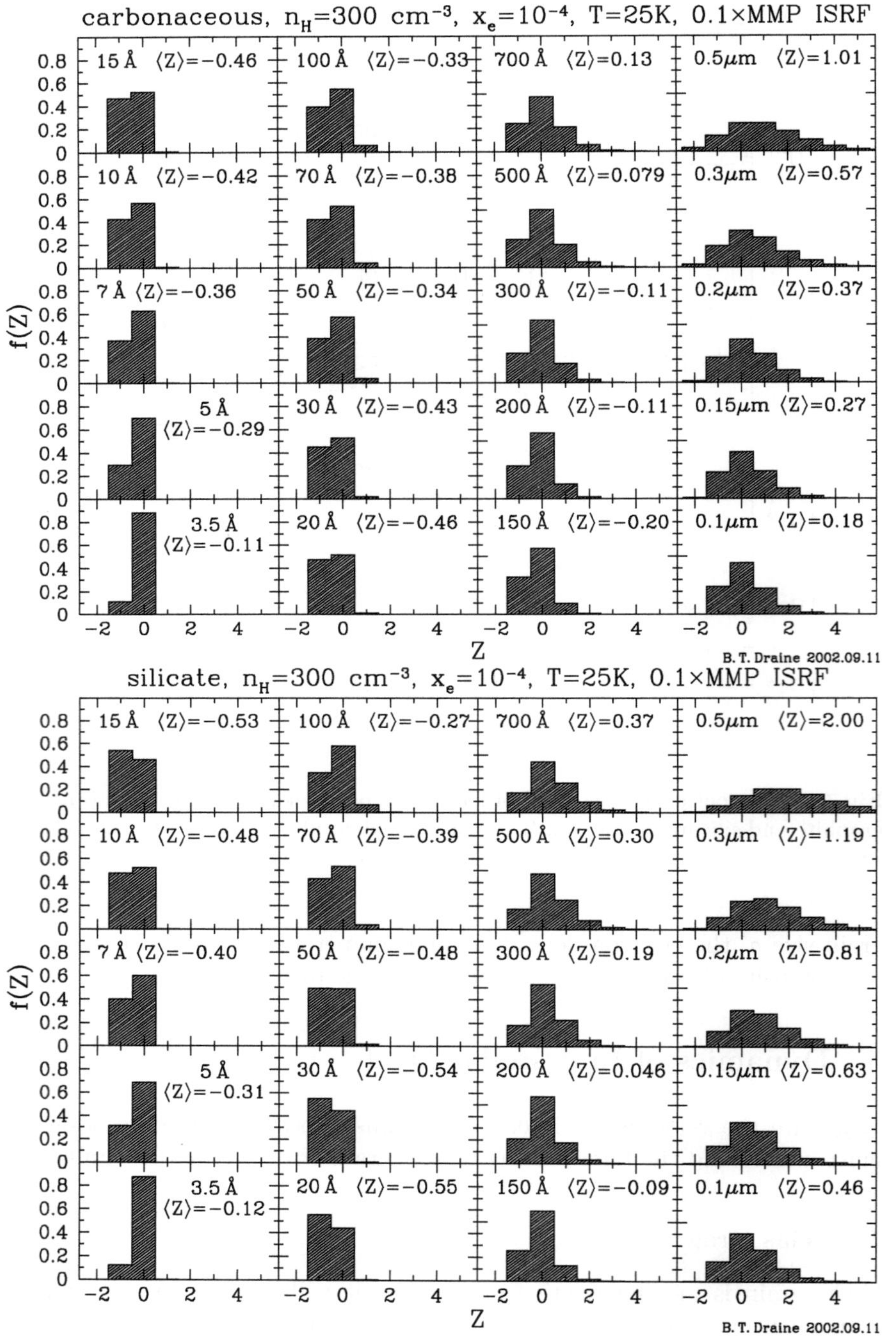

Fig. 29. Charge distribution functions for carbonaceous grains and silicate grains in cold molecular gas.

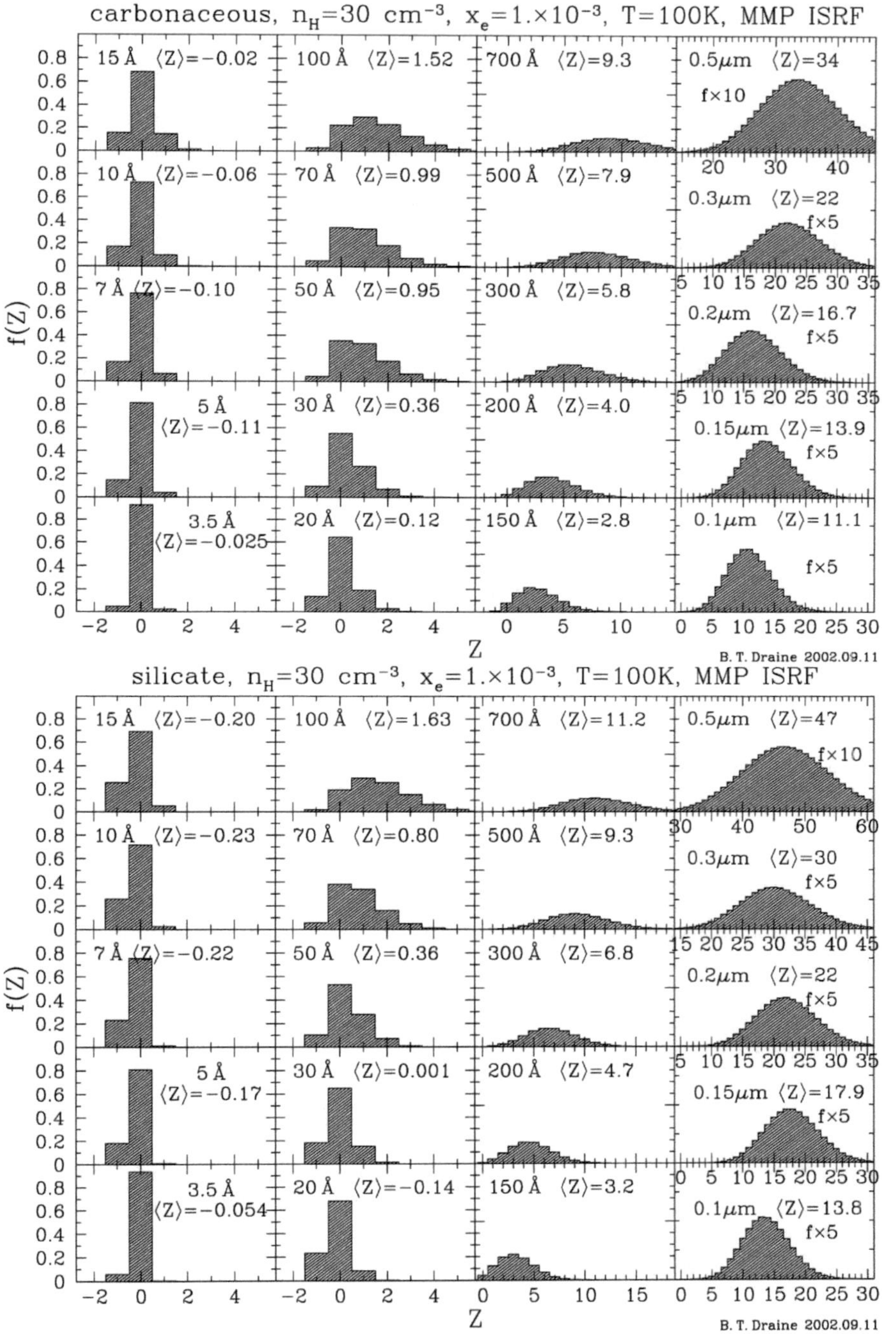

Fig. 30. Charge distribution functions for carbonaceous grains and silicate grains in "cold neutral medium" diffuse HI.

grain in neutral gas, the problem can be solved exactly, and approximated to within 1% by

$$F_{\rm drag} = 2\pi a^2 kT \sum_i n_i \frac{8 s_i}{3\sqrt{\pi}} \left[1 + \frac{9\pi}{64} s_i^2\right]^{1/2} \qquad s_i^2 \equiv \frac{m_i v^2}{2kT} \quad , \tag{69}$$

where the sum is over the species (H, H_2, He) in the gas (Draine & Salpeter 1979a). If the grain is charged and ions are present, there is an additional "Coulomb drag" term (Draine & Salpeter 1979a), but this is numerically unimportant in regions of fractional ionization $\lesssim 10^{-2}$. Equation (69) is obtained for either perfectly inelastic collision (where the gas atom is assumed to "stick") to the grain or perfectly elastic specular reflections. For subsonic motion, the drag time is

$$\tau_{\rm drag} = \frac{M_{\rm gr} v}{F_{\rm drag}} = \frac{\rho a \sqrt{2\pi/kT}}{\sum_i n_i \sqrt{m_i}} \quad . \tag{70}$$

For a cool region where the hydrogen is primarily molecular,

$$\tau_{\rm drag} = 1.2 \times 10^5\,{\rm yr} \left(\frac{a}{0.1\,\mu{\rm m}}\right) \left(\frac{\rho}{3\,{\rm g\,cm^{-3}}}\right) \left(\frac{300\,{\rm cm^{-3}}}{n_{\rm H}}\right) \left(\frac{25\,{\rm K}}{T}\right)^{1/2} \quad , \tag{71}$$

while in a more diffuse region where the hydrogen is atomic,

$$\tau_{\rm drag} = 4.4 \times 10^5\,{\rm yr} \left(\frac{a}{0.1\,\mu{\rm m}}\right) \left(\frac{\rho}{3\,{\rm g\,cm^{-3}}}\right) \left(\frac{30\,{\rm cm^{-3}}}{n_{\rm H}}\right) \left(\frac{100\,{\rm K}}{T}\right)^{1/2} \quad . \tag{72}$$

5.2 Lorentz Force

Dust grains are charged, and there are magnetic fields in clouds. The cyclotron period is

$$\frac{2\pi}{\omega_c} = \frac{2\pi M_{\rm gr} c}{|\langle Z\rangle| eB} \tag{73}$$

$$= \frac{5.2 \times 10^4\,{\rm yr}}{|\langle Z\rangle|} \left(\frac{a}{10^{-5}\,{\rm cm}}\right)^3 \left(\frac{\rho}{3\,{\rm g\,cm^{-3}}}\right) \left(\frac{3\mu{\rm G}}{B}\right) \quad , \tag{74}$$

where we take $\langle Z\rangle$ because the grain charge fluctuates on a time short compared to the cyclotron period. Values of $\langle Z\rangle$ are indicated in Figs. 29 and 30. The character of the grain dynamics depends on the ratio of the drag force to the Lorentz force, or

$$\omega_c \tau_{\rm drag} = \frac{3}{4\pi} \frac{|\langle Z\rangle| eB}{a^2 c} \frac{\sqrt{2\pi/kT}}{\sum_i n_i \sqrt{m_i}} \tag{75}$$

$$= 53\,|\langle Z\rangle| \left(\frac{B}{3\mu{\rm G}}\right) \left(\frac{10^{-5}\,{\rm cm}}{a}\right)^2 \left(\frac{100\,{\rm K}}{T}\right)^{1/2} \left(\frac{30\,{\rm cm^{-3}}}{n_{\rm H}}\right) \quad , \tag{76}$$

where we have assumed the hydrogen to be atomic. For grains in cold clouds, we see that we usually have $\omega_c \tau_{\rm drag} \gg 1$: grains are strongly coupled to the magnetic field, except for grain sizes where collisional charging and photoelectric charging happen to just balance so that $|\langle Z \rangle| \lesssim 0.02(a/10^{-5}\,{\rm cm})^2$.

5.3 Radiation Pressure

The interstellar radiation field is in general anisotropic. As a result, dust grains are subject to "radiation pressure" forces which can cause drift of the grain relative to the gas. This drift can be important as a means of transporting dust through gas, but also because the drifting grains transfer momentum to the magnetic field (via the Lorentz force) and to the gas (via gas drag).

Consider a unidirectional beam of photons of energy $h\nu$, with energy density $u_\nu d\nu$ in interval $(\nu, \nu + d\nu)$. The photons have momentum $h\nu/c$, and this momentum can be transferred to the grain by absorption or by scattering. It is easy to see that the rate of momentum transfer is

$$F_{\rm rad} = \int \frac{u_\nu d\nu}{h\nu} c \left[C_{\rm abs} + C_{\rm sca}(1 - \langle \cos\theta \rangle)\right] \frac{h\nu}{c} \tag{77}$$

$$= \pi a^2 \int u_\nu \, d\nu \, Q_{\rm pr}(\nu) \ , \quad Q_{\rm pr}(\nu) \equiv Q_{\rm abs} + Q_{\rm sca}(1 - \langle \cos\theta \rangle) \ . \tag{78}$$

where $Q_{\rm abs} \equiv C_{\rm abs}/\pi a^2$, $Q_{\rm sca} \equiv C_{\rm sca}/\pi a^2$, and $\langle \cos\theta \rangle$ is the mean of the cosine of the scattering angle.

We are generally interested in values of $Q_{\rm abs}$ and $Q_{\rm pr}$ averaged over the spectrum of the radiation.

We are often content to approximate the radiation from a star by a blackbody spectrum. In neutral regions of the interstellar medium we are frequently interested in averages over a blackbody spectrum with a cutoff at 13.6 eV, and it is useful to calculate spectrum-averaged values

$$\langle Q \rangle_T \equiv \frac{\int_0^{\nu_{\rm max}} d\nu \, B_\nu(T) Q(\nu)}{\int_0^{\nu_{\rm max}} d\nu \, B_\nu(T)} \quad , \tag{79}$$

where $B_\nu(T)$ is the Planck function (54) and $h\nu_{\rm max} = 13.6\,{\rm eV}$.

We will also be interested in Q values averaged over the interstellar radiation field, for which we use the estimate of Mathis, Mezger, & Panagia (1983).

$$\langle Q \rangle_{\rm ISRF} \equiv \frac{\int_0^{\nu_{\rm max}} d\nu \, u_\nu^{\rm ISRF} Q(\nu)}{\int_0^{\nu_{\rm max}} d\nu \, u_\nu^{\rm ISRF}} \quad . \tag{80}$$

Suppose that at some point in the interstellar medium the starlight has a net energy flux $c\Delta u_{\rm rad}$. By summing over the Skymap Star Catalog (Slater & Hashmall 1992), the Tycho Catalog (ESA 1997), and the all-sky ultraviolet observations of the S2/68 experiment on the TD-1A satellite, Weingartner &

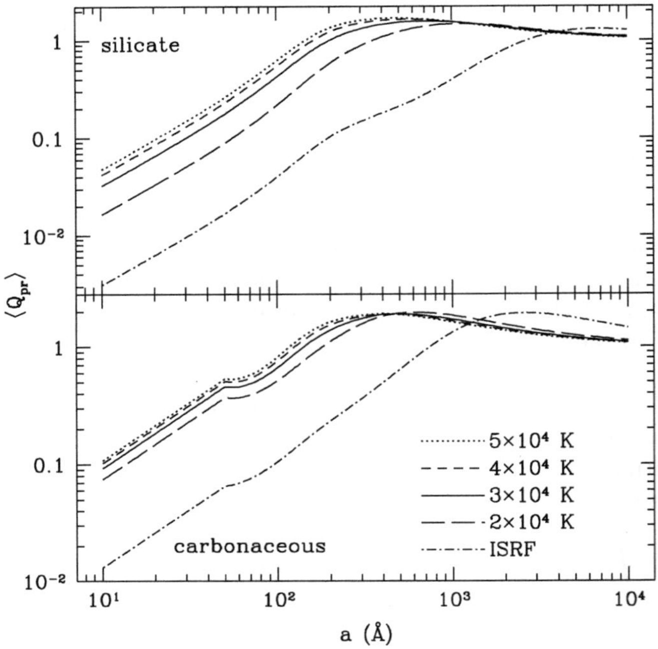

Fig. 31. Radiation pressure efficiency factor $\langle Q_{\rm pr} \rangle$ for neutral carbonaceous and silicate grains, averaged over the interstellar radiation field (ISRF) and blackbody spectra with indicated color temperatures, cut off at 13.6 eV. Taken from Weingartner & Draine (2001b).

Draine (2001b) estimated the anisotropy in the starlight background at V, B, and 4 ultraviolet bands. In Fig. 32 we show the net dipole moment of the radiation field at these 6 wavelengths.

5.4 Recoil from Photoelectric Emission and Photo-desorption

Radiation pressure is, of course, just the transfer of photon momentum to the grain. However, radiation can serve to energize other processes which can exert a thrust on the grain: photoelectric emission and photo-desorption. To see that these might be important, it is sufficient to compare the momentum of an $h\nu \approx 10\,\rm eV$ photon with the momentum of the photo-electron it might produce, if the photo-electron kinetic energy $= f \times h\nu$:

$$\frac{p(\text{electron})}{p(\text{photon})} = \frac{(2m_e f h\nu)^{1/2}}{h\nu/c} = 320\, f^{1/2} \left(\frac{10\,\rm eV}{h\nu}\right)^{1/2} \quad . \tag{81}$$

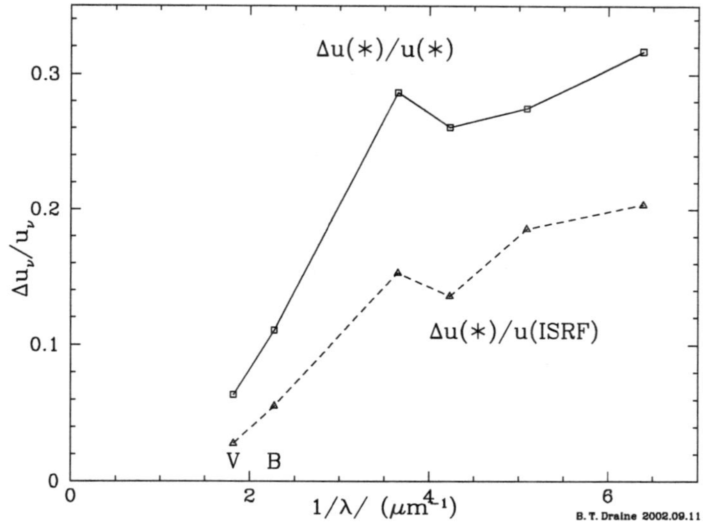

Fig. 32. Dipole fraction for the interstellar radiation field. Upper curve is the ratio of the measured dipole moment to the measured starlight (the measurements are incomplete). Lower curve is the ratio of the measured dipole moment to the total estimated ISRF. The true fraction is probably between these two curves. After Weingartner & Draine (2001b).

Similarly, suppose that a photon caused an H_2 molecule to be desorbed with kinetic energy $f \times h\nu$:

$$\frac{p(\mathrm{H}_2)}{p(\mathrm{photon})} = \frac{(4m_\mathrm{H} f h\nu)^{1/2}}{h\nu/c} = 1.9 \times 10^4 f^{1/2} \left(\frac{10\,\mathrm{eV}}{h\nu}\right)^{1/2} \quad . \tag{82}$$

Therefore if an appreciable fraction of absorbed photons produce photoelectrons or photodesorb atoms or molecules, there could be a significant recoil on the grain.

For a spherical grain in an isotropic radiation field, the photoelectrons would be emitted isotropically, with zero net thrust. Suppose, however, that we have a unidirectional radiation field (e.g., light from a single star). In this case we might expect a higher rate of photo-electron emission from the "bright side" than from the "dark side" of the grain. However, very small grains do not have a "dark side" – they are effectively transparent to the incident radiation, so one must try to estimate how this depends on grain size. How can we estimate the bright side/dark side asymmetry? A simple model is to assume that the rate of photoelectric emission from a point on the grain surface is proportional to $|\mathbf{E}|^2$ just inside the surface. Using Mie theory to calculate the electric field intensity $\mathbf{E}$ in a spherical grain, one can then evaluate the relative rates of photoelectric emission over the grain surface. Using this approach, Weingartner & Draine (2001b) have investigated the net thrust due to escaping photoelectrons – equal and opposite to the net momentum which the escaping electrons have when they reach infinity. The photoelectric yield, and the energy of the photoelectrons, is of course affected by the charge on the grain, and therefore depends on the environmental conditions. The analysis is complicated by the fact that if the grain is charged, photoelectrons with nonzero angular momentum will escape to infinity in directions different from the direction they are travelling when they emerge from the grain surface (until it reaches infinity, the electric field of the photoelectron exerts a force on the grain).

Figure 33 shows the results of Weingartner & Draine (2001b) for the ratio of the photoelectric thrust to the radiation pressure force, as a function of grain size. We see that this is an order-unity correction for grains with $a \approx 100$Å in the CNM.

For photoelectric emission, it is reasonable to assume that the grain is "spherically symmetric" prior to arrival of the photon – there are just as many electrons available for photo-ejection on the bright side as on the dark side of the grain. For photo-desorption, however, this is not a reasonable assumption.

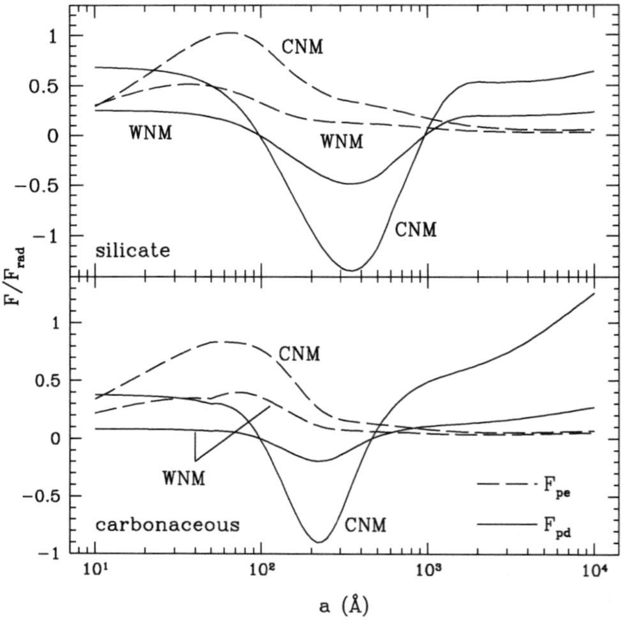

Fig. 33. Photoelectric thrust and photo-desorption thrust relative to radiation pressure for carbonaceous and silicate grains as a function of grain radius a. Curve labelled "CNM" is for cold diffuse clouds with $n_e = 0.03\,\mathrm{cm}^{-3}$, $T = 100\,\mathrm{K}$, and the standard ISRF. Note that the photo-desorption force can be negative for grains with radii $100\text{Å} \lesssim a \lesssim 500\text{Å}$. From Weingartner & Draine (2001b).

As we will see in §6 below, grains are expected to be spinning rapidly, with their principal axis of largest moment of inertia expected to be aligned with the angular momentum. Because of this gyroscopic stabilization, one may expect that if photo-desorption is a rapid process, then the "bright" side of the grain will have fewer adsorbed molecules ready to be photo-desorbed, an effect which will depend on the orientation of the grain's spin axis with the direction of radiation anisotropy.

If photo-desorption is rapid, the thrust will be limited by the need to resupply the grain surface with new molecules to be photo-desorbed. The story has a number of additional complications:

- The result will depend on the wavelength-dependence of the photo-desorption cross sections for the adsorbed species. This is essentially unknown. Weingartner & Draine simply assume that the photo-desorption rate for a molecule on the grain surface is proportional to $|\mathbf{E}|^2$ just outside the grain surface, integrated from $6 - 13.6\,\mathrm{eV}$.
- Adsorbed species may diffuse over the grain surface, so that the "bright side" could be replenished from the "dark side".
- If photo-desorption is rapid, the surface coverage of adsorbates will depend on the orientation of the grain spin axis relative to the direction of starlight anisotropy.
- The spinning grain can undergo "polar flips" – the "thermal flipping" process described by Lazarian & Draine (1999a).
- If the grain is moving, the grain motion will alter the rates at which molecules are resupplied to the different parts of the grain surface: the "bright side" may suffer a reduction in the rate of arrival of atoms or molecules from the gas.
- H atoms which are not photo-desorbed might react on the grain surface to form H_2, and could recoil from the grain with substantial kinetic energy. This process might happen more rapidly on the dark side where the H concentration is higher, thus acting to oppose the thrust due to photo-desorption.

With so many uncertainties, it is not possible to reach definite quantitative estimates for the photo-desorption force, but Weingartner & Draine (2001b) evaluate one plausible set of assumptions; results are shown in Fig. 33 for grains in the "warm neutral medium" and the "cold neutral medium". It is curious that there is a range of grain sizes (100–400Å for carbonaceous grains, 100 – 1000Å for silicate grains) where they find that the photo-desorption force is *negative* – this occurs because for this range of sizes, interference effects cause $|\mathbf{E}|^2$ to be larger on what would have been expected to be the dark side of the grain if the grain were opaque and geometric optics were applicable. Note that the larger grains do behave as "expected" for macroscopic targets.

The "bottom line" is that photo-desorption can make a significant contribution to the net force on the grain, which (coincidentally) is expected to be comparable in magnitude to the radiation pressure force and the photoelectric thrust, but which can in principle be of the opposite sign, depending on detailed physics of photo-desorption which are not known at this time.

5.5 Drift Velocities for Interstellar Grains

For the special case where the radiation anisotropy is parallel to the magnetic field direction, we can obtain the steady-state grain drift velocity by balancing the drag force against the radiation-related forces:

$$F_{\rm drag} = F_{\rm rad} + F_{\rm pe} + F_{\rm pd} \quad , \tag{83}$$

where $F_{\rm pe}$ and $F_{\rm pd}$ are the forces due to photoelectric emission and photodesorption. If we assume subsonic motion, and neglect Coulomb drag, this can be written

$$\frac{v_{\rm drift}}{\sqrt{kT/m_{\rm H}}} = \langle Q_{\rm pr}\rangle \left(\frac{\Delta u_{\rm rad}}{n_{\rm H}kT}\right) \frac{1 + F_{\rm pe}/F_{\rm rad} + F_{\rm pd}/F_{\rm rad}}{\sum_i (n_i/n_{\rm H})\sqrt{m_i/m_{\rm H}}} \quad , \tag{84}$$

so we see that the grain will drift with a "Mach number" which is approximately equal to $\langle Q_{\rm pr}\rangle$ times the ratio of the anisotropic radiation pressure to the gas pressure. Since $Q_{\rm pr}$ reaches values of order unity (see Fig. 31), we may expect peak Mach numbers or order 0.2 or so – the ratio of the anisotropic radiation energy density to the gas pressure in the diffuse ISM.

In Fig. 34 we show drift speeds calculated by Weingartner & Draine for carbonaceous and silicate grains in the WNM and CNM. The drift speeds are clearly a strong function of grain size, and appear to be negligible for small $a \lesssim 100$Å grains. For larger grains the drift velocities are modest but potentially significant since they could be sustained for long times: in the CNM, a radiation anisotropy of only $\Delta u_{\rm rad}/u_{\rm rad} = 0.1$ would give a $0.3\,\mu$m carbonaceous grain a drift speed of ~ 0.04km s^{-1}, so in 10 Myr it could drift 0.4 pc, and in the WNM the drift speeds are about an order of magnitude larger. This could in principle lead to removal of the large grains from some gas elements and concentration in others.

6 Rotational Dynamics of Interstellar Dust

The rotational dynamics of interstellar grains constitutes a fascinating story, not yet fully told. It is intimately tied to the long-standing problem of interstellar grain alignment, which we will discuss in §7.

6.1 Brownian Rotation?

Early discussions of the rotational dynamics of interstellar grains generally assumed the grains to be rigid spheres, which appeared to capture the essential physics, and discussed the scattering of impinging atoms from the grain

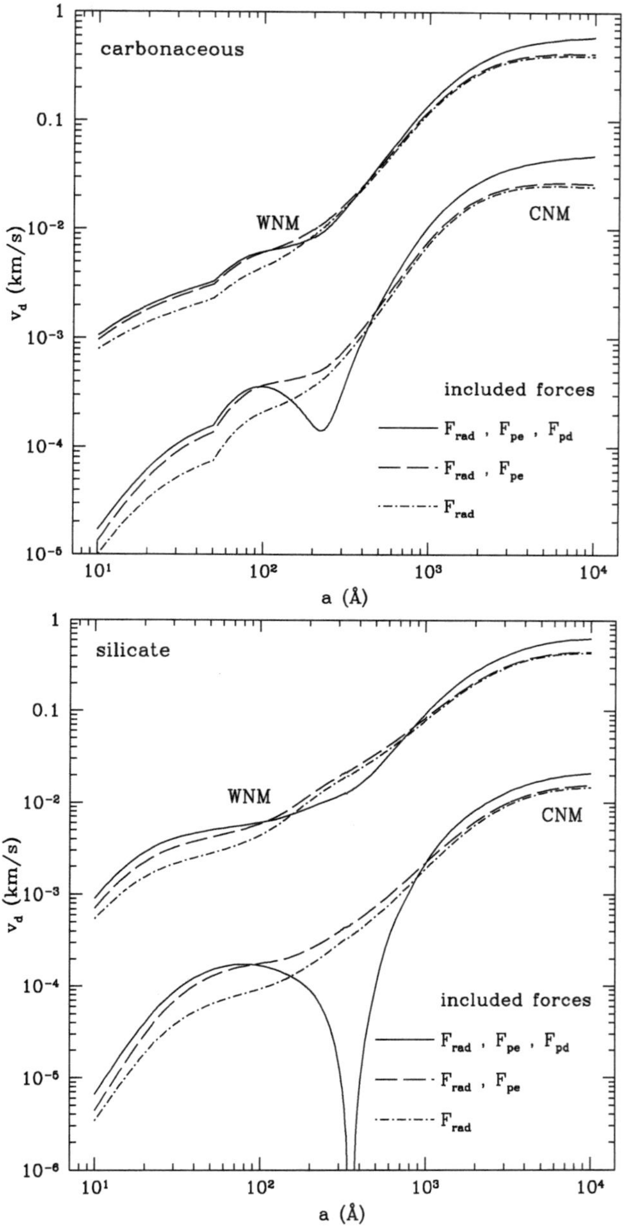

Fig. 34. Drift velocities for carbonaceous and silicate grains, as a function of grain size, in the CNM and WNM, for a radiation anisotropy $\Delta u_{rad}/u_{ISRF} = 0.1$. Taken from Weingartner & Draine (2001b). The minimum in drift velocity for $a \approx 250$Å carbonaceous grains or $a \approx 350$Å silicate grains in the CNM is due to the reversal of the photo-desorption force for grains in this size range (see Fig. 33).

surface. As far as grain rotation was concerned, it seemed obvious that the grain would simply undergo a form of Brownian motion, exchanging linear and angular momentum with the gas. Once statistical equilibrium was attained, a rigid grain (with purely elastic scattering of atoms from its surface) would have an expectation value of $kT_{\rm gas}/2$ for the kinetic energy in each of its degrees of freedom: three translational and three rotational. Thus the r.m.s. translational velocity of the grain would be

$$\langle v^2\rangle^{1/2} = \left(\frac{3kT}{M}\right)^{1/2} = \left(\frac{9kT}{4\pi\rho a^3}\right)^{1/2} \tag{85}$$

$$= 1.82\frac{\rm cm}{\rm s}\left(\frac{T}{100\,{\rm K}}\right)^{1/2}\left(\frac{3\,{\rm g\,cm^{-3}}}{\rho}\right)^{1/2}\left(\frac{10^{-5}\,{\rm cm}}{a}\right)^{3/2} , \tag{86}$$

which is negligible under all foreseeable circumstances. The r.m.s. rotation rate

$$\frac{\langle\omega^2\rangle^{1/2}}{2\pi} = \frac{1}{2\pi}\left(\frac{3kT}{2Ma^2/5}\right)^{1/2} = \frac{1}{2\pi}\left(\frac{45kT}{8\pi\rho a^5}\right)^{1/2} \tag{87}$$

$$= 4.6\times10^4{\rm Hz}\left(\frac{T}{100\,{\rm K}}\right)^{1/2}\left(\frac{3\,{\rm g\,cm^{-3}}}{\rho}\right)^{1/2}\left(\frac{10^{-5}\,{\rm cm}}{a}\right)^{5/2} \tag{88}$$

is more impressive, though it simply corresponds to an equatorial velocity equal to the r.m.s. translational velocity (86) times a factor $\sqrt{5/2} = 1.58$.

6.2 Suprathermal Rotation

Following the discovery of grain alignment in 1949, theoretical studies of the rotational dynamics were carried out to try to understand the alignment process, including the seminal paper by Davis & Greenstein (1951). It was recognized that, since interstellar grains were likely to have vibrational temperatures differing from the gas temperature, Equation (88) needed to be modified, and it was understood that for a spherical grain the temperature T appearing in Eq. (88) should be some weighted mean of the gas temperature and the grain temperature, since collisions of atoms with the grain surface will not be perfectly elastic.

However, in 1975 – 24 years after Davis & Greenstein's study – Purcell (1975) realized that interstellar grains could act as "pinwheels". After all, a simple pinwheel weighting many grams can readily be put into rotation with a rotational kinetic energy many orders of magnitude greater than kT by the air molecules which excite it. Purcell (1979) showed, in fact, that there were 3 distinct physical processes acting on interstellar grains which could each, acting alone, cause a grain to rotate with a rotational energy $(1/2)I\omega^2 \gg kT$:

1. The effective "accommodation coefficient" will probably vary over the grain surface, due to chemical inhomogeneities and/or geometrical irregularities. Such variations in accommodation coefficient on a non-spherical

grain will lead to a systematic torque (i.e., one which whose time-average is not expected to be zero) if the gas temperature and grain temperature differ.

2. Photoelectrons ejected from the grain may, on average, carry away angular momentum, resulting in a systematic torque.
3. H_2 formation on the grain surface, with the nascent H_2 molecules ejected with a significant kinetic energy, is expected to result in a large systematic torque, particularly if the H_2 formation occurs at a relatively small number of "active sites" on the grain surface.

Purcell showed that of the three processes, the H_2 formation process was likely to be the most important in regions where the H is atomic.

6.2.1 Suprathermal Rotation Driven by H_2 Formation

It is instructive to examine the argument for suprathermal rotation driven by H_2 formation at active sites on the grain surface.

To simplify the argument, let us consider a region where only atomic H is present (no He), and consider a cube of sides $2b \times 2b \times 2b$, with volume $8b^3$, surface area $24b^2$, and effective radius $a_{\rm eff} = (6/\pi)^{1/3}b = 1.24b$.

Choose a site at random on the grain surface, and assume that H_2 molecules are formed at that site at a rate $\dot{N}_1$. For simplicity, suppose that the newly-formed H_2 molecules leave normal to the surface with kinetic energy $E_{\rm H_2}$. The expectation value for the square of the moment arm is then just $(2/3)b^2$, and the expectation value for the mean square torque exerted by this site is

$$\Gamma_1^2 = \dot{N}_1^2[m_{\rm H_2}v_{\rm H_2}]^2 \times \frac{2}{3}b^2 \tag{89}$$

$$= \frac{8}{3}\dot{N}_1^2 m_{\rm H} E_{\rm H_2} b^2 \quad . \tag{90}$$

Now suppose that there are a total of $N_{\rm site}$ such sites on the grain surface, and suppose that a fraction γ of the H atoms which impinge on the surface are converted to H_2 (so there are $\gamma/2$ H_2 formation events per arriving H atom). In a thermal gas, the rate per area at which atoms collide with a convex surface is just $n(kT/2\pi m)^{1/2}$, so

$$\dot{N}_1 = \frac{\gamma}{2N_{\rm site}} n({\rm H}) \left(\frac{kT}{2\pi m_{\rm H}}\right)^{1/2} 24b^2 \quad . \tag{91}$$

Since the sites are assumed to be randomly located, the torques add like a random walk, so that the expectation value for the square of the total torque is just

$$\langle \Gamma_{\rm ex}^2 \rangle = N_{\rm site}\Gamma_1^2 \tag{92}$$

$$= \frac{192}{\pi}\frac{1}{N_{\rm site}}\gamma^2 n({\rm H})^2 E_{\rm H_2} kT\, b^6 \quad . \tag{93}$$

The rotating grain will experience gas drag. For slow rotation rates ω, it is not difficult to show that the drag torque is

$$\Gamma_{\rm drag} = -\frac{80}{3} n({\rm H}) \left(\frac{m_{\rm H} kT}{2\pi} \right)^{1/2} b^4 \omega \quad . \tag{94}$$

To find $\langle \omega^2 \rangle$, we now set $\langle \Gamma_{\rm ex}^2 \rangle = \langle \Gamma_{\rm drag}^2 \rangle$ to obtain

$$\langle \omega^2 \rangle = \frac{27}{50} \frac{\gamma^2}{N_{\rm site}} \frac{E_{\rm H_2}}{m_{\rm H}} \frac{1}{b^2} \quad , \tag{95}$$

or a rotational kinetic energy, relative to the thermal value,

$$\frac{I\langle \omega^2 \rangle}{3kT} = \frac{3}{25} \frac{M}{m_{\rm H}} \frac{\gamma^2}{N_{\rm site}} \frac{E_{\rm H_2}}{kT} \tag{96}$$

$$= 2.0 \times 10^{11} \frac{\gamma^2}{N_{\rm site}} \left(\frac{b}{10^{-5}\,{\rm cm}} \right)^3 \left(\frac{\rho}{3\,{\rm g\,cm^{-3}}} \right) \left(\frac{E_{\rm H_2}}{\rm eV} \right) \left(\frac{100\,{\rm K}}{T} \right) . \tag{97}$$

where $M = 8b^3\rho$ is the grain mass. Now this is no small number! The number of active sites is uncertain. If we suppose that there is a surface area s^2 per active site, then $N_{\rm site} = 24b^2/s^2$, and

$$\frac{I\langle \omega^2 \rangle}{3kT} = 8.4 \times 10^5 \gamma^2 \left(\frac{s}{10{\rm \AA}} \right)^2 \left(\frac{b}{10^{-5}\,{\rm cm}} \right) \left(\frac{\rho}{3\,{\rm g\,cm^{-3}}} \right) \left(\frac{E_{\rm H_2}}{\rm eV} \right) \left(\frac{100\,{\rm K}}{T} \right) . \tag{98}$$

The r.m.s. rotation rate is

$$\frac{\langle \omega^2 \rangle^{1/2}}{2\pi} = 3.6 \times 10^5 {\rm Hz}\, \gamma \left(\frac{s}{10{\rm \AA}} \right) \left(\frac{10^{-5}\,{\rm cm}}{b} \right) \left(\frac{E_{\rm H_2}}{1\,{\rm eV}} \right) \quad , \tag{99}$$

so that we might have MHz rotational frequencies for $b < 3 \times 10^{-6}$ cm! It is apparent that systematic torques can play a major role in grain dynamics.

Each H_2 formation site can be thought of as being like a small rocket thruster attached to the grain surface: the systematic torque due to H_2 formation is fixed in body coordinates, so long as the H_2 formation sites do not change. The kinetics of H_2 formation on grain surfaces is poorly understood at this time, and it is not certain how long-lived the active sites are likely to be.

6.2.2 Radiative Torques Due to Starlight

The three torques identified by Purcell do indeed appear to be individually capable of driving grains to suprathermal rotation. In fact, there appears to be an additional physical process which can compete with the H_2 formation torque for grains larger than $\sim 10^{-5}$ cm: torques exerted on interstellar grains by starlight.

We discussed above the "radiation pressure" force which anisotropic starlight could exert on a grain. It turns out that starlight can also exert a torque on grains, and these torques can be dynamically important!

How can we address this problem? Analytic progress is difficult, because the only cases for which we have analytic solutions are cases where radiative torques vanish: (1) grains of complex shape but small compared to the wavelength; (2) spheres; (3) spheroids. In the first case the torques vanish because only the dipole response of the grain is important, and the dipole radiation does not carry angular angular momentum. In the case of spheres and spheroids, there is no torque because of the symmetry of the target.

In order to obtain a nonzero torque we must have an asymmetric target. Draine & Weingartner (1996) discussed use of the discrete dipole approximation to calculate radiative torques. The DDSCAT code (Draine & Flatau 2000) now provides the capability to compute torques. For an asymmetric target illuminated by a beam of unpolarized radiation, the direction and magnitude of the torque depends on the orientation of the target relative to the beam. It is useful to define a "torque efficiency vector" $\mathbf{Q}_\Gamma$ defined so that the torque on the grain is given by

$$\mathbf{\Gamma}_{\rm rad} = \pi a_{\rm eff}^2 u_{\rm rad} \frac{\lambda}{2\pi} \mathbf{Q}_\Gamma \quad , \tag{100}$$

where the effective radius $a_{\rm eff}$ is the radius of an equal volume sphere. We will generally want to average $\mathbf{Q}_\Gamma$ over some spectrum; Draine & Weingartner (1996) calculate

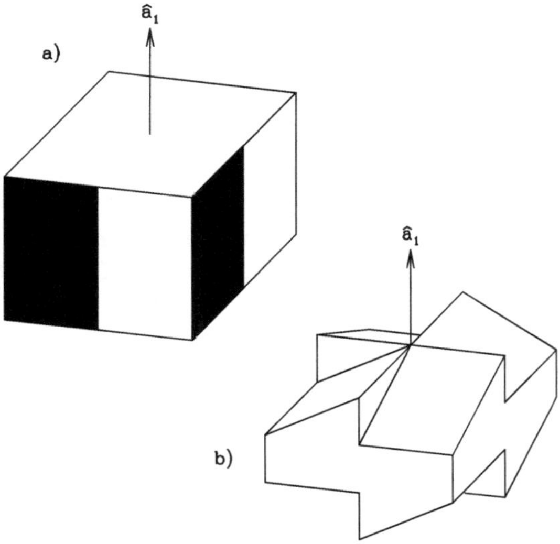

Fig. 35. Macroscopic examples of targets which would be subject to radiative torques. From Draine & Weingartner (1996).

$$\langle \mathbf{Q_\Gamma} \rangle \equiv \frac{\int \mathbf{Q_\Gamma} \lambda u_\lambda \, d\lambda}{\int \lambda u_\lambda \, d\lambda} \quad , \tag{101}$$

$$\bar{\lambda} \equiv \frac{\int \lambda u_\lambda \, d\lambda}{\int u_\lambda \, d\lambda} \quad , \tag{102}$$

for u_λ taken to be the MMP ISRF (for which $\bar{\lambda} = 1.202\,\mu\text{m}$). With these definitions, the torque

$$\mathbf{\Gamma}_{\text{rad}} = \pi a_{\text{eff}}^2 u_{\text{rad}} \frac{\bar{\lambda}}{2\pi} \langle \mathbf{Q_\Gamma} \rangle \quad . \tag{103}$$

We recall from classical mechanics that an arbitrary object has three principal axes, $\hat{\mathbf{a}}_1$, $\hat{\mathbf{a}}_2$, $\hat{\mathbf{a}}_3$, and the moment of inertia tensor is diagonal in a coordinate system aligned with these three axes. Let the eigenvalues of the moment of inertia tensor be $I_1 \geq I_2 \geq I_3$.

For purposes of discussion, let us assume that $I_1 > I_2 = I_3$. If the grain has angular momentum $\mathbf{J}$, then the rotational kinetic energy is

$$E_{\text{rot}} = \frac{(J\cos\beta)^2}{2I_1} + \frac{(J\sin\beta)^2}{2I_2} \tag{104}$$

$$= \frac{J^2}{2I_1} + J^2 \frac{(I_1 - I_2)}{I_1 I_2} \sin^2\beta \quad , \tag{105}$$

with the rotational kinetic energy minimized for $\beta = 0$. As we will see below, when J is large we expect the grain to remain close to this minimum energy state, with $\sin^2\beta \ll 1$.

The grain spins rapidly, so we are concerned with the torque after averaging over rotations of the grain around the axis $\hat{\mathbf{a}}_1$. This rotation-averaged torque therefore depends only on a single angle – the angle Θ between $\hat{\mathbf{a}}_1$ and the direction of the incident radiation.

We will see below that we expect a spinning dust grain to precess around the direction of the local magnetic field $\mathbf{B}$, so what we really need is the torque averaged over grain rotations *and* over precession of the axis $\hat{\mathbf{a}}_1$ around the magnetic field direction. For a given grain and spectrum of radiation, this will now depend on two angles: the angle ψ between $\mathbf{B}$ and the direction of the radiation, and the angle ξ between $\hat{\mathbf{a}}_1$ and $\mathbf{B}$.

For purposes of spinup (or spindown), we are interested in the projection of the torque efficiency vector along the spin axis, i.e., the axis of largest moment of inertia: $\mathbf{Q_\Gamma}(\hat{\mathbf{\Theta}}) \cdot \hat{\mathbf{a}}_1$. Draine & Weingartner (1996, 1997) define $H(\xi, \phi) = \langle \mathbf{Q_\Gamma} \cdot \hat{\mathbf{a}}_1 \rangle$ averaged over rotation around $\hat{\mathbf{a}}_1$ and precession of $\hat{\mathbf{a}}_1$ around $\mathbf{B}$. They calculate $H(\xi, \phi)$ for three grain shapes with effective radii $a_{\text{eff}} = 0.2\,\mu\text{m}$. The numerical values of H of course depend on ξ and ϕ, but a typical value of $|H|$ would be $|H| \approx 0.005$ (see Fig. 5 of Draine & Weingartner 1997), so that the spinup torque would be

$$\langle \mathbf{\Gamma}_{\text{rad}} \cdot \hat{\mathbf{a}}_1 \rangle = H \bar{\lambda} \Delta u_{\text{rad}} \pi a_{\text{eff}}^2 \quad . \tag{106}$$

It is of interest to compare this to the spinup torque due to H_2 formation. We can rewrite Eq. (93) replacing b by $(\pi/6)^{1/3}a_{\rm eff}$, and project onto a random direction $\hat{\mathbf{a}}_1$ to obtain

$$\langle(\mathbf{\Gamma}_{\rm ex}\cdot\hat{\mathbf{a}}_1)^2\rangle^{1/2}=\frac{1}{3}\left(\frac{2}{3\pi}\right)^{1/2}\left(\frac{6}{\pi}\right)^{1/3}\gamma\left(\frac{E_{\rm H_2}}{kT}\right)^{1/2}s\,n({\rm H})kT\,\pi a_{\rm eff}^2\quad. \qquad (107)$$

Thus the ratio of the radiative spinup torque (106) to the r.m.s. value of the H_2 formation torque is

$$\frac{\langle\mathbf{\Gamma}_{\rm rad}\cdot\hat{\mathbf{a}}_1\rangle}{\langle(\mathbf{\Gamma}_{\rm ex}\cdot\hat{\mathbf{a}}_1)^2\rangle^{1/2}}=\frac{H\bar{\lambda}}{0.191\gamma(E_{\rm H_2}/kT)^{1/2}s}\,\frac{\Delta u_{\rm rad}}{n({\rm H})kT} \qquad (108)$$

$$=\frac{2.92}{\gamma}\left(\frac{H}{0.005}\right)\left(\frac{T}{100\,{\rm K}}\right)^{1/2}\left(\frac{\rm eV}{E_{\rm H_2}}\right)^{1/2}\left(\frac{10{\rm \AA}}{s}\right)\frac{\Delta u_{\rm rad}}{n({\rm H})kT}\quad. \qquad (109)$$

Adopting the 10% anisotropy estimated by Weingartner & Draine (2001b), we have $\Delta u_{\rm rad}\approx 8.64\times10^{-14}\,{\rm erg\,cm^{-3}}$. With $n({\rm H})=30\,{\rm cm^{-3}}$, and $T=100\,{\rm K}$, we have $\Delta u_{\rm rad}/n({\rm H})kT=0.21$. Therefore with an H_2 formation efficiency $\gamma<0.5$ the radiative torque would exceed the H_2 formation torque if the other parameters have nominal values (in particular, $s=10{\rm \AA}$). Many of the numbers are uncertain; the important point is that for $a_{\rm eff}=0.2\,\mu{\rm m}$ grains the radiative torque has a magnitude which is comparable to our estimate for the H_2 formation torque.

7 Alignment of Interstellar Dust

Polarization of starlight was discovered in 1949. It was immediately deduced that the only plausible mechanism was "linear dichroism" of the interstellar medium due to aligned dust grains, but understanding the mechanism responsible for this alignment has frustrated astrophysicists for half a century. In fact, the story of the efforts of astrophysicists to understand grain alignment is humbling. It is not just because it has taken us so long to solve the problem – the humbling lesson is that several very important and relatively simple physical effects were overlooked for decades. It was not that these physical effects were difficult to understand, or that they involved physics which has only recently been understood. On the contrary – the physics involved would have been familiar to well-trained physicist in 1950. It was simply that important processes were overlooked:

- The Rowland effect: a charged, spinning dust grain will develop a magnetic moment due to its circulating charge (Martin 1971).
- The Barnett effect: a spinning dust grain with unpaired electron spins will spontaneously magnetize (Dolginov & Mytrophanov 1976).
- Suprathermal rotation due to dust-gas temperature differences (Purcell 1975, 1979).

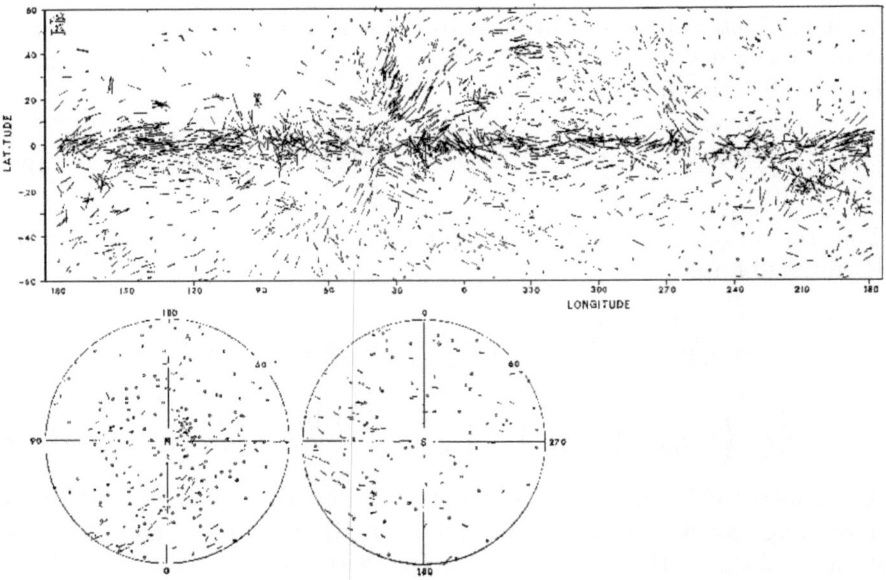

Fig. 36. Linear polarization of light from 1800 stars. From Mathewson & Ford 1970.

- Suprathermal rotation due to photoelectric emission (Purcell 1975, 1979).
- Suprathermal rotation due to H_2 formation (Purcell 1975, 1979).
- Viscoelastic dissipation of rotational kinetic energy due to time-varying stresses in a grain which is not rotating around a principal axis (Purcell 1979)
- "Barnett dissipation" of rotational kinetic energy due to the electron spin system (Purcell 1979).
- Dissipation of rotational kinetic energy due to the nuclear spin system (Purcell 1979).
- Suprathermal rotation due to starlight torques (Draine & Weingartner 1996).
- Fluctuation phenomena associated with Barnett dissipation (Lazarian & Roberge 1997; Lazarian & Draine 1999a) and coupling to the nuclear spins (Lazarian & Draine 1999b).

Have we thought of all the important processes? Or are there other important phenomena to which our eyes are still closed?

7.1 The Motion of Spinning Grains

As seen above, dust grains spin very rapidly. Consider for the moment one spinning grain. If the angular momentum **J** does not happen to be along one of the grain's principal axes, then the grain's angular velocity ω will not be parallel to **J**, and will nutate around **J**, with a nutation rate that will be smaller than (but comparable to) the rotation rate.

Now imagine observing a single grain as it spins at kHz rotation frequencies. Because the rotation and nutation are both rapid, we will quickly obtain a rotation-nutation average of the grain orientation. During this observation, the angular momentum **J** is conserved, and defines a preferred direction. Now consider light propagating in some particular direction. Because of the rotation-nutation averaging, the grain will tend to polarize the light with electric vector either exactly perpendicular to **J**, or exactly parallel to **J**, depending on whether the grain on average has its "long" axis perpendicular to **J**, or parallel to **J**.

This is for one grain. If a population of grains shows a tendency to polarize light, then the individual grain angular momenta cannot be randomly-oriented – there must be a tendency for the angular momenta **J** to be aligned with some preferred direction in space.

The problem of grain alignment is to account for the alignment of grain angular momenta **J**.

Many alignment mechanisms have been proposed over the years. Some directly involve the galactic magnetic field (Davis & Greenstein 1951). Other proposed mechanisms rely on so-called "mechanical alignment" due to gas (Gold 1952) or cosmic rays (Salpeter & Wickramasinghe 1969) streaming relative to the grains. Harwit (1970) suggested that grain angular momenta might be aligned as a result of the angular momentum absorbed from starlight photons, if the starlight is anisotropic. It was only in 1971 that Martin (1971) pointed out that even in the case of nonmagnetic alignment processes, the observed alignment would be either exactly parallel or exactly perpendicular to the magnetic field, simply as a result of the magnetic moment on a spinning grain.

7.2 Grain Magnetic Moments and Precession Around B

Grain in diffuse regions tend to be electrically charged (see Fig. 30). If charged, a spinning dust grain will develop a magnetic moment due to the "Rowland effect": the rotating charge on the grain constitutes a current circulating around a loop, and there will be a resulting magnetic field and magnetic moment interior to the loop. Recall that a current I flowing in a loop with area A generates a magnetic moment $m = IA/c$. The mean of $x^2 + y^2$ over the surface of a sphere is just $(2/3)a^2$, so the magnetic moment due to charge Ze distributed uniformly over the surface of a spinning sphere is

$$\mu = \frac{\omega Ze}{2\pi c}\pi\frac{2a^2}{3} = \frac{\omega Zea^2}{3c} \quad . \tag{110}$$

If ω is at an angle $\theta > 0$ relative to $\mathbf{B}$, the spinning sphere will precess with a precession frequency

$$\omega_p = \frac{\mu B \sin\theta}{I\omega \sin\theta} = \frac{5}{8\pi}\frac{ZeB}{\rho a^3 c} \tag{111}$$

$$= 1.01 \times 10^{-4}\,\mathrm{yr}^{-1} Z \left(\frac{B}{3\mu\mathrm{G}}\right)\left(\frac{3\,\mathrm{g\,cm}^{-3}}{\rho}\right)\left(\frac{10^{-5}\,\mathrm{cm}}{a}\right)^3 \quad . \tag{112}$$

We have seen above (Fig. 30) that in a CNM environment, we might expect an $a = 10^{-5}$ cm grain to have $\langle Z \rangle \approx 10$, for a precession period ~ 6000 yr. This time is short compared to the other dynamical times: the gas drag time (72), or (as we shall see below) the time for grain alignment to occur.

We can therefore conclude that regardless of what process is responsible for alignment of interstellar grains, each grain will have a time-averaged angular momentum distribution which must be symmetric around $\mathbf{B}$. Therefore the interstellar dust mixture will tend to polarize starlight either exactly parallel to $\mathbf{B}$ or exactly perpendicular to $\mathbf{B}$, for any conceivable alignment mechanism!

7.3 Barnett Effect

Dolginov & Mytrophanov (1976) pointed out that the "Barnett effect" would result in grain magnetic moments much larger than due to the Rowland effect. The Barnett effect can be stated very simply: an uncharged object rotating with angular velocity ω tends to spontaneously magnetize, with a magnetization $M = \chi\omega/\gamma$, where χ is the susceptibility, and γ is the "gyromagnetic ratio" for the material – the ratio of (magnetic moment)/(angular momentum) for the electron orbitals or electron spins responsible for the magnetizability of the material. For electron spins, $\gamma = \mu_B/(\hbar/2) = e/2m_e c$, where $\mu_B \equiv e\hbar/2m_e c$ is the Bohr magneton.

The spontaneous magnetization is equal to that which would be produced in a stationary sample in a (fictitious) applied magnetic field $H_{Be} = \omega/\gamma = 2m_e c\omega/e$ – the so-called "Barnett equivalent field".

At first sight the Barnett effect seems surprising, but it is not difficult to understand. Consider an unmagnetized sample (equal numbers of "spin-up" and "spin-down" electrons), with angular momentum J, spinning at angular velocity ω. At constant total angular momentum, the angular velocity of the sample can be reduced, with consequent reduction of the rotational kinetic energy $I\omega^2/2$, if some of the angular momentum is put into the electron spin system. The electrons have magnetic moments, so with unequal numbers of electrons in the two spin states, the sample is now magnetized. The reduction in rotational kinetic energy is accompanied by heating of the grain, with an increase in entropy. The effect was observed in the laboratory by Barnett (1915).

For normal paramagnetic materials, the magnetic susceptibility is just $\chi = n_s \mu_B^2 / kT_g$, where n_s is the number density of unpaired electrons, and T_g is the grain temperature. The magnetic moment due to the Barnett effect is thus

$$\mu = \frac{4\pi a^3}{3} \frac{n_s \mu_B^2}{kT} \frac{2m_e c}{e} \omega \quad . \tag{113}$$

The ratio of the "Barnett effect moment" to the "Rowland effect moment" is

$$\frac{\mu(\text{Barnett})}{\mu(\text{Rowland})} = \frac{2\pi a n_s \hbar^2}{Z k T_g m_e} \tag{114}$$

$$= \frac{2.8 \times 10^5}{Z} \left(\frac{a}{10^{-5}\,\text{cm}} \right) \left(\frac{20\,\text{K}}{T_g} \right) \left(\frac{n_s}{10^{22}\,\text{cm}^{-3}} \right) \quad . \tag{115}$$

An unpaired electron density $n_s \approx 10^{22}\,\text{cm}^{-3}$ is typical for normal paramagnetic materials, so it is clear that the Barnett effect is very important for interstellar grains, and will result in precession periods of order days or months rather than the thousands of years for the Rowland effect.

7.4 Alignment by Paramagnetic Dissipation: Davis-Greenstein Mechanism

Davis and Greenstein (1951) described a mechanism which could produce grain alignment. Consider a grain spinning with angular velocity ω in a static magnetic field $\mathbf{B}_0$. Let θ be the angle between $\mathbf{B}_0$ and ω.

The grain material has a magnetic susceptibility χ, and it will try to magnetize in response to an applied magnetic field. In grain body coordinates, the applied field has a static component parallel to the grain rotation axis, with magnitude $B_\parallel = B_0 \cos\theta$, and a component of magnitude $B_\perp = B_0 \sin\theta$ which appears to be rotating, with angular velocity ω relative to the grain. The rotating field $\mathbf{B}_\perp$ is perpendicular to ω.

It seems clear that the grain material will develop a magnetization $\mathbf{M}$ which will have a static component $\mathbf{M}_\parallel = \chi_0 B_\parallel$ parallel to ω, and a component $\mathbf{M}_\perp$ which rotates in body coordinates (but is stationary in inertial coordinates), with magnitude $M_\perp = |\chi(\omega)| B_\perp$, where $\chi(\omega)$ is the (complex) magnetic susceptibility for a magnetic field rotating at frequency ω. In body coordinates, the rotating magnetization $\mathbf{M}_\perp$ will "lag" behind the rotating field $\mathbf{B}_\perp$ – there will be an "in-phase" component $\chi_1(\omega) B_\perp$ and an "out-of-phase" component $\chi_2(\omega) B_\perp$.

The galactic magnetic field $\mathbf{B}_0$ will exert a torque on the grain magnetic moment $V\mathbf{M}$. The torque is obviously $\perp$ to $\mathbf{B}_0$. A bit of calculation shows that the out-of-phase component of the magnetization results in a torque of magnitude $V\chi_2 B_\perp B_0 = V\chi_2 B_0^2 \sin\theta$ which acts to leave the component of the angular momentum $\parallel$ to B_0 unchanged (the torque, after all, must be $\perp$ $\mathbf{B}_0$) but acts to reduce the component $J_\perp = J\sin\theta$ perpendicular to $\mathbf{B}_0$:

$$\frac{d}{dt} J_\perp = -V\chi_2 B_0^2 \sin\theta \quad . \tag{116}$$

At low frequencies the imaginary component of the magnetic susceptibility varies linearly with frequency: $\chi_2 = K\omega$. With $\omega = J/I = J_{\|}/I\cos\theta$, noting that $J_{\|} = \text{const}$, we obtain

$$\frac{d}{dt}J_{\perp} = J_{\|}\frac{d}{dt}\tan\theta = -J_{\|}\frac{VKB_0^2}{I}\tan\theta \quad , \tag{117}$$

with solution

$$\tan\theta = \tan(\theta_0)e^{-t/\tau_{\rm DG}} \quad , \tag{118}$$

$$\tau_{\rm DG} \equiv \frac{I}{VKB_0^2} \quad . \tag{119}$$

The magnetic susceptibilities of paramagnetic materials are quite well understood at the frequencies of interest (see, e.g., Morrish 1965), with

$$K \approx 1.25 \times 10^{-13}\,{\rm s}\left(\frac{20\,{\rm K}}{T}\right) \quad . \tag{120}$$

With this value, and $I \approx (2/5)\rho V a^2$, the Davis-Greenstein alignment time

$$\tau_{\rm DG} = \frac{2\rho a^2}{5KB_0^2} \tag{121}$$

$$= 1.2 \times 10^6\,{\rm yr}\left(\frac{a}{10^{-5}\,{\rm cm}}\right)^2\left(\frac{\rho}{3\,{\rm g\,cm^{-3}}}\right)\left(\frac{T}{20\,{\rm K}}\right)\left(\frac{5\mu{\rm G}}{B_0}\right)^2 . \tag{122}$$

Thermally-rotating grains are disaligned by random collisions with gas atoms on a time scale $\tau_{\rm drag}$. The ratio of the alignment time to the disalignment time is

$$\frac{\tau_{\rm DG}}{\tau_{\rm drag}} = 2.8\left(\frac{n_{\rm H}}{30\,{\rm cm^{-3}}}\right)\left(\frac{T_{\rm g}}{20\,{\rm K}}\right)\left(\frac{5\mu{\rm G}}{B_0}\right)^2\left(\frac{T}{100\,{\rm K}}\right)^{1/2}\left(\frac{a}{10^{-5}\,{\rm cm}}\right), \tag{123}$$

where we have assumed the hydrogen to be atomic. Thus we see that even for $a = 10^{-5}$ cm grains and a relatively high value of the magnetic field strength, alignment is slow compared to disalignment, and the ratio is linear in the grain radius a. The Davis-Greenstein alignment mechanism favors alignment of *smaller* grains, but this is a problem, since small grains are observed to NOT be aligned, while large grains are (Kim & Martin 1995). As we will see below, some important physics has been overlooked.

7.5 Effect of Suprathermal Rotation

Consider now a grain rotating suprathermally. The Davis-Greenstein alignment time is independent of the rotation rate, since the torque is proportional to the rate of rotation. On the other hand, the time scale for random collisions to change the direction of the angular momentum is approximately

$(J/J_{\rm th})^2\tau_{\rm drag}$, where $J_{\rm th}$ is the angular momentum expected for Brownian rotation. So if the grain rotation is suprathermal, with $J/J_{\rm th} \gg 1$, we can virtually ignore the disaligning effects of random collisions, and Davis-Greenstein paramagnetic dissipation will bring the grain into alignment with $\mathbf{B}_0$ in a few million years. As we have seen in §6.2, suprathermal rotation appears to be expected. Could it for some reason be suppressed in small grains?

7.6 Thermal Flipping vs. Suprathermal Rotation

Why aren't small grains aligned? From Eq. (98) it certainly appears that H_2 formation should be able to drive grains as small as 10^{-6} cm into suprathermal rotation, and yet there is strong evidence that grains smaller than 10^{-5} cm are not aligned (Kim & Martin 1995).

Lazarian & Draine (1999a) have proposed an explanation for this – the process of "thermal flipping" may prevent small grains from achieving suprathermal rotation.

The key issue is that the grain does NOT behave like a "rigid body" subject only to external torques – the grain has internal vibrational degrees of freedom. The grain rotational kinetic energy depends on the angle β between principal axis $\hat{\mathbf{a}}_1$ and $\mathbf{J}$ through Eq. (105):

$$E_{\rm rot} = \frac{J^2}{2I_1} + J^2\frac{(I_1 - I_2)}{I_1 I_2}\sin^2\beta \quad . \tag{124}$$

In the absence of external torques (and ignoring the small amount of angular momentum in the spin system), J is fixed, but the rotational kinetic energy can change if energy is exchanged with the vibrational degrees of freedom of the grain. Thus we expect the angle β to fluctuate. Suppose, for example, that we start at $\beta = 0$. Thermal fluctuations will cause β to "explore" the region $0 \leq \beta < \pi/2$. If

$$J^2\frac{(I_1 - I_2)}{I_1 I_2} \lesssim kT \quad , \tag{125}$$

then thermal fluctuations could allow β to reach $\pi/2$: the maximum possible rotational kinetic energy. At this point β can go either way, with equal probabilities, so the grain may "flip over" to the state $\pi/2 < \beta \leq \pi$.

Now consider the effects of H_2 formation torques. These torques are *fixed* in body coordinates. If the grain "flips" from $\beta = 0$ to $\beta = \pi$, the torques change sign in inertial coordinates! As a result, if the grain flips frequently, the H_2 torques can be averaged out to zero, and have no significant effect on the grain dynamics, so that the grain rotation remains essentially thermal! Lazarian & Draine (1999a) refer to this as "thermal trapping".

The question now comes down to estimating the rate of flipping, and how the rate depends on grain size. Lazarian & Draine (1999a) discussed the role which the Barnett effect (i.e., the electron spin system) can play in coupling the grain rotation to the vibrational modes, and concluded that flipping would be rapid enough to result in thermal trapping for grain sizes $a \lesssim 0.1\,\mu$m. This seemed to successfully account for the absence of grain alignment for $a \lesssim 0.1\,\mu$m.

However, further analysis of grain dissipational processes showed that the *nuclear* spin system could be important at the lower rotational rates appropriate to larger grains (Lazarian & Draine 1999b). It was found that the coupling of grain rotation to the nuclear spin system would lead to rapid thermal flipping even for grains as large as $\sim 1\,\mu$m!

Thermal flipping therefore now appears able to prevent H_2 formation torques from being able to achieve superthermal rotation even for the $a \approx 0.2\,\mu$m grains which are observed to be aligned! How is grain alignment to be achieved?

7.7 Effect of Radiative Torques

So long as the grain surface is not modified, the H_2 formation torques are fixed in body coordinates, and thermal flipping will cause these torques to average to zero if the grain spends equal amounts of time in the two "flip states" (i.e., $\beta < \pi/2$ and $\beta > \pi/2$). However, the grain is also subject to other systematic torques, the most important of which is the radiative torque due to anisotropic starlight. This torque is *not* fixed in body coordinates, so that even if the grain spends equal amounts of time in the two flip states, there is a secular change in angular momentum due to the radiative torques. Radiative torques are unimportant for small $a \lesssim 0.05\,\mu$m grains, but very important for $a \gtrsim 0.1\,\mu$m grains. It therefore appears that "thermal trapping" will prevent small ($a \lesssim 0.05\,\mu$m) grains from achieving suprathermal rotation, but that larger ($a \gtrsim 0.1\,\mu$m) grains *can* achieve suprathermal rotation, following which they can be gradually aligned by the Davis-Greenstein mechanism.

But life turns out to be more complicated. Draine & Weingartner (1997) found that radiative torques act not only to spin up grains, but also to change their alignment. The characteristic time scale for radiative torques to change the grain alignment turns out to be just the drag time $\tau_{\rm drag}$, which is shorter than the Davis-Greenstein alignment time, so that the radiative torques appear able to dominate over the systematic effects of the Davis-Greenstein alignment torque.

Study of three different grain shapes and different angles between the magnetic field direction and the starlight anisotropy direction showed that under some conditions the starlight torques brought about alignment (the aligned state was an "attractor"), while under other conditions the grain did not settle down in any stationary state. There appeared to be an overall tendency for grain alignment to take place, but the sample of 3 grain shapes was insufficient to reach general conclusions.

Further study of this problem is underway, and it is hoped that in a year or two we will have definite conclusions on the role of radiative torques in the grain alignment problem. But for the moment the situation appears very promising – we seem to finally understand the absence of alignment of small grains, and it appears that the physics we now understand will explain the alignment of the larger ones, with radiative torques due to starlight playing a critical role.

7.8 Alignment and Disalignment in Different Regions

There is evidence that the degree of grain alignment varies from one region to another. Observations of polarization of starlight indicate that the $a \gtrsim 0.1\,\mu$m grains in diffuse clouds are substantially aligned, but JHK observations of quiescent dark clouds (Goodman et al. 1995) indicate that the dust grains in the dark interiors are not aligned. How can we understand this?

There are several possible plausible answers:

- The inner dark regions have insufficient starlight for the radiative torques to achieve grain alignment.
- The inner dark regions are molecular, so that H_2 formation torques cannot drive suprathermal rotation.
- Photoelectric emission is suppressed in the inner dark regions.

It is also possible, of course, that either the magnetic field structure is completely tangled in the interior, or the grains in the central regions are for some reason more spherical, but these do not seem plausible.

It is also interesting to note that in some dark clouds the grains are reasonably well aligned: for example, the M17 molecular cloud (Fig. 37) shows linear polarization (averaged over the 22″ beam) as large as 5%. Note, though, that although this cloud is "dark" (i.e., we cannot see into it at optical wavelengths) it is full of stars – the infrared spectrum indicates a grain temperature $T \approx 45\,$K, and the grains must be heated by starlight: the mean starlight intensity must be $\sim (45/20)^6 \approx 10^2$ times greater than the ISRF in order to heat the grains to this temperature! It therefore appears likely that starlight torques could drive the grains in this cloud to suprathermal rotation rates.

Finally, a recent study by Padoan et al. (2001) concluded that submm polarization maps of quiescent cores were consistent with a model where the grains in the outer parts of the cloud with $A_V \lesssim 3$ mag were aligned, while grains in regions with $A_V > 3$ mag were randomly oriented.

All of the above appear consistent with the idea that starlight torques are an essential part of the grain alignment process. Grains deep in quiescent dark clouds are unable to align because starlight torques are weak or absent, whereas starlight torques are present are do produce alignment within star-forming clouds.

7.9 Summary of the Grain Alignment Story

The grain alignment story has been long and complicated, with a number of false turns. The discussion above has skipped some of the important physics

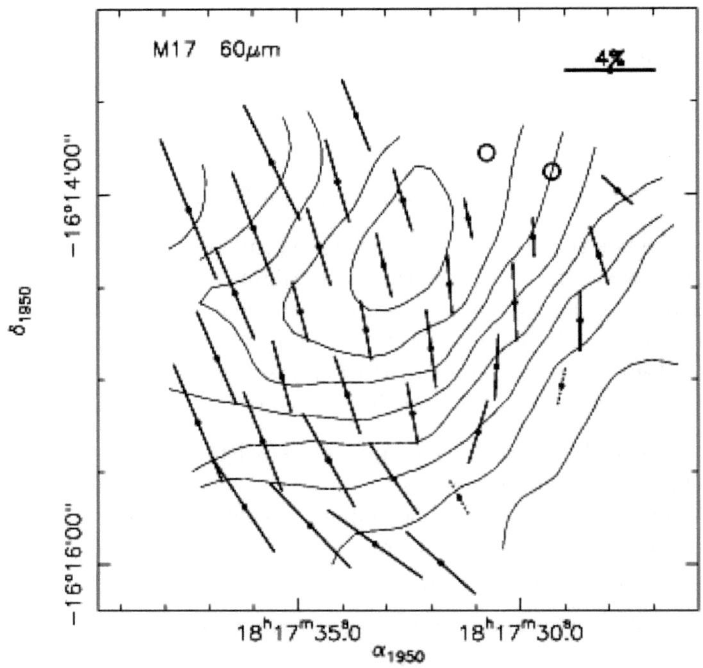

Fig. 37. 60 μm linear polarization toward M17. Flux contours are at 20%, 30%,. . . ,90% of the peak flux. From Dotson et al (2000).

involving coupling of the grain rotation to the electron and nuclear spin systems (see Lazarian & Draine 1999a,b) as well as the important question of the grain dynamics during "crossovers", when reversals of the systematic torques on the grain cause the grain angular momentum to become small – during this time the grain is very vulnerable to disalignment (see, e.g., Lazarian & Draine 1997).

To summarize: the important elements of the grain alignment story now appear to be as follows:

- Grains are subject to 2 important systematic torques:
  - H_2 formation torques,
  - radiative torques due to starlight.
- For grains with rotational kinetic energies less than a few times kT_d, thermal fluctuations allow the grain to sample different orientations of the principal axes relative to the instantaneous angular momentum **J**, leading to "thermal flipping" of the grain between the two "flip states".
- If the thermal flipping is sufficiently rapid, the grain will spend 50% of the time in each flip state, in which case the H_2 formation torque, which is fixed in body coordinates, will integrate to zero in inertial coordinates: the H_2 formation torque will be unable to drive the grain to suprathermal rotation.

- Starlight torques are not fixed in body coordinates, and hence do not integrate to zero even if the grain spends equal amounts of time in the two flip states.
- For grains with $a \gtrsim 0.1\,\mu$m in regions with sufficiently strong starlight, the systematic torque due to starlight will be able to drive the grain to suprathermal rotation.
- Once a grain achieves suprathermal rotation, thermal flipping will cease, the grain will settle into a state where the principal axis $\hat{\mathbf{a}}_1$ is either parallel or antiparallel to $\mathbf{J}$, and all systematic torques will operate simultaneously.
- Grains in suprathermal rotation are essentially immune from disalignment by random collisions. Alignment will result from the combined effects of Davis-Greenstein alignment and starlight torques.
- Under some circumstances, starlight torques do not result in strong alignment. However, numerical experiments suggest that, averaged over irregular grain shapes, starlight torques lead to overall alignment.
- The fact that starlight torques are important only for grains with $a \gtrsim 0.1\,\mu$m appears to account for the observation that $a \lesssim 0.05\,\mu$m grains in diffuse clouds are not aligned, while $a \gtrsim 0.1\,\mu$m grains appear to be well-aligned.

8 Evolution of the Grain Population

The interstellar grain population in a galaxy is the result of a rich and complicated mix of processes.

We would like to know the composition of interstellar dust, how and where it forms, and the extent to which individual grains are either homogeneous or conglomerate. We would like to know what the size distribution of the different types of grains is, how and where the size distribution is fashioned, and the manner in which the size distribution varies from one region to another. If we understood the balance of processes acting in our part of the Milky Way, perhaps we could deduce how grains might differ in the inner or outer parts of the Milky Way, and in the LMC, SMC, and other galaxies.

Unfortunately, the overall grain evolution problem seems to me to be beyond our grasp at this time – I don't think we understand the parts well enough to justify trying to put them together as a whole. At this time, we have to try to clarify the individual elements of the story.

8.1 Dust Formation in Stellar Outflows

There is abundant evidence that dust grains form in some stellar outflows: we see infrared emission from dust around red giants, around carbon stars, and in planetary nebulae. In all these cases the dust condensed out of material which at an earlier time was completely gaseous. Observations show that different types of stellar outflows form different types of dust.

The dust around red giants with O/C > 1 shows a 10 μm emission feature which is due to the Si-O stretching mode in amorphous silicate material; around some OH/IR stars the 10 μm feature appears in absorption (e.g., Demyk et al. 2000), along with emission features revealing the presence of crystalline silicates.

The dust around carbon stars, on the other hand, does not show the 10 μm feature, which is consistent with the fact that we do not expect silicates to form in an atmosphere with O/C < 1. Some carbon stars (e.g. IRC+10216) instead show a 11.3 μm emission feature which is attributed to SiC (e.g., Blanco et al. 1998).

It should also be mentioned that we find presolar grains in meteorites whose isotopic composition clearly indicates formation in outflows from evolved stars (Hoppe & Zinner 2000); these include nanodiamonds, SiC, graphite, and Al_2O_3. Some of these grains may have formed in supernova ejecta.

So there is no question that stellar outflows contribute dust to the interstellar medium. However, this does *not* necessarily imply that the bulk of the dust in the interstellar medium was formed in stellar outflows.

The physics and chemistry of dust formation in stellar outflows is complex: the material is generally far from thermodynamic equilibrium, and the outflows themselves are often (perhaps always?) hydrodynamically complex – neither steady nor spherically symmetric. We can not yet reliably describe the details of the grain formation process.

8.2 Grain Growth in the ISM

8.2.1 Accretion in the ISM

Dust grains in the interstellar medium can grow, since atoms from the gas can stick to them. We can calculate the lifetime against accretion for an atom or ion in the gas phase:

$$\tau^{-1} = A_{\rm i}^{-1/2} s_{\rm i} \left(\frac{8kT}{\pi m_{\rm H}}\right)^{1/2} \int da\, \pi a^2 \frac{dn_{\rm d}}{da} D(a) \quad , \tag{126}$$

where $A_{\rm i}$ is the mass number of the ion, $s_{\rm i}$ is the sticking coefficient, and $n_{\rm d}(a)$ is the number density of grains with radii less than a. The enhancement factor $D(a)$ gives the collision rate relative to what it would be in the absence of electrostatic effects:

$$D(a) = \sum_{Z_{\rm d}} f(Z_{\rm d}, a) B(Z_{\rm d}, a) \quad , \tag{127}$$

where

$$B(Z_{\rm d}, a) = \begin{cases} \exp\left(-Z_{\rm d} Z_{\rm i} e^2/akT\right) & \text{for } Z_{\rm g} Z_{\rm i} > 0 \\ \left(1 - \frac{Z_{\rm d} Z_{\rm i} e^2}{akT}\right) & \text{for } Z_{\rm g} Z_{\rm i} < 0 \\ 1 + \left(\frac{\pi Z_{\rm i}^2 e^2}{2akT}\right)^{1/2} & \text{for } Z_{\rm g} = 0 \end{cases} \quad . \tag{128}$$

Figure 38 shows the collisional enhancement factor $D(a)$ calculated by Weingartner & Draine (1999) for carbonaceous and silicate grains in various environments. Enhancement factors $D \gtrsim 10^2$ are found for $a \lesssim 10^{-7}$ cm grains in the CNM. As a result, an ion (e.g., SiII, TiII, FeII) would have a lifetime against accretion of $\sim 1 \times 10^5$ yr in the CNM if the grain size distribution includes the ultra-small grain population of Fig. 10. Therefore we expect that those metals which can "stick" to the small grains will be strongly depleted from the gas phase; presumably those elements which are not strongly depleted (e.g., Na, K, S) for some reason do not remain on the ultra-small grains after they collide.

8.2.2 Mantle Formation

As discussed in §1.8, grains in dark molecular clouds are apparently coated with ice "mantles". The overall mantle composition is dominated by H_2O ice. It is not known whether the H_2O molecules arrived from the gas phase and then froze onto the grain, or whether the H_2O formed by surface chemistry. In most clouds the gas phase H_2O abundance is very low, and the observed mantles would not be able to form by accretion, suggesting that O and H atoms arriving at the grain react there to form H_2O (Jones & Williams 1984).

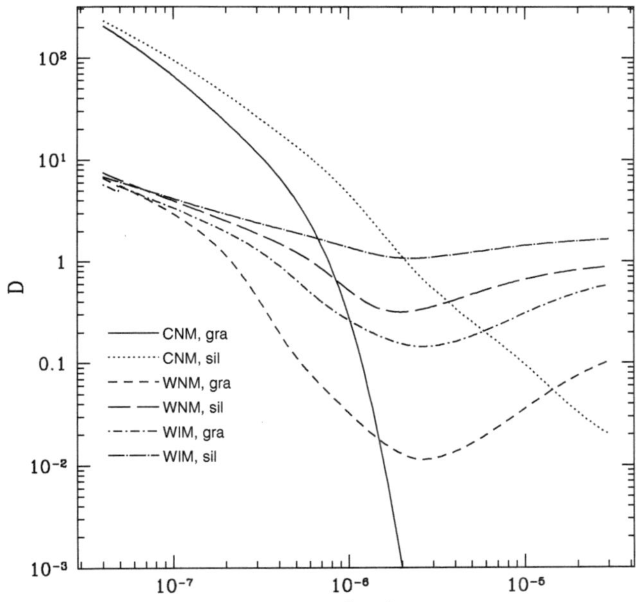

Fig. 38. Collision enhancement factor for positively charged ion (see text). From Weingartner & Draine (1999).

However, it has recently been proposed that the H_2O could be produced in the gas phase during occasional passage of a shock wave, and then quickly accreted onto the grain surfaces (Bergin et al. 1999). Other molecules (e.g., CO) may be mixed with the H_2O, or in some cases condensed separately, as the absorption profile in some cases requires that the CO not be mixed with polar molecules. This could result if the H_2O was deposited while the grain was too warm to retain CO, with the CO later freezing out when the conditions changed and the grain became colder.

8.2.3 Coagulation

The grain population in normal diffuse clouds has a very large population of very small grains, required to explain the rapid rise in extinction at short wavelengths. In dense regions, however, the extinction at short wavelengths appears to be reduced. This can only happen if some of the small grains are either destroyed altogether, or simply coagulated onto big grains. The latter seems more likely.

What is the time scale for such coagulation? This depends on the velocity differences between different grains. Grain-grain velocities can be the result of differential drift velocities (see §5.5), hydrodynamic turbulence (see, e.g., Draine 1985), or MHD turbulence (Lazarian & Yan 2002).

Grains larger than $\sim 10^{-5}\,\mathrm{cm}$ contribute a geometric cross section per H $\Sigma \approx 0.5 \times 10^{-21}\,\mathrm{cm}^2$. If the characteristic grain-grain velocity difference is Δv_d, then the grain-grain coagulation time would be

$$t_\mathrm{coag} = [n_\mathrm{H} \Sigma \Delta v_\mathrm{d}]^{-1} = 2 \times 10^8\,\mathrm{yr} \left(\frac{30\,\mathrm{cm}^{-3}}{n_\mathrm{H}}\right) \left(\frac{0.1\mathrm{km\,s}^{-1}}{\Delta v_\mathrm{d}}\right) \quad , \qquad (129)$$

so that grain coagulation will not be very important in diffuse clouds, but could proceed in dense regions.

8.3 Photolysis

When grains with ice mantles are returned to diffuse regions and exposed to ultraviolet radiation, it appears that most of the material is driven off, but the ultraviolet radiation photolyzes the ice to produce an organic refractory residue (e.g., Greenberg et al. 2000). This residue may be responsible for the interstellar $3.4\,\mu\mathrm{m}$ "aliphatic C-H" absorption band.

8.4 Photo-desorption

The process of photo-desorption is potentially very important, but is unfortunately not well understood because of limited laboratory studies. Photo-desorption is a quantum process whereby absorption of a single photon can lead to ejection of an atom or molecule from a solid. For example, an absorbed photon might cause an atom or molecule on a substrate to be excited to a repulsive electronic state which would then be ejected from the substrate.

Photo-desorption was discussed by Draine & Salpeter (1979b), who argued that a molecule in a surface monolayer might have a photo-desorption cross section as large as $\sim 10^{-18}\,\mathrm{cm}^2$.

If an atom or molecule adsorbed on a grain surface has a cross section $\sigma_{\mathrm{pd}} = 10^{-18}\sigma_{-18}\,\mathrm{cm}^2$ for $8 < h\nu < 13.6\,\mathrm{eV}$ photons, the photo-desorption rate for an adsorbed atom or molecule would be $\sim 6 \times 10^{-11}\sigma_{18}\,\mathrm{s}^{-1}$ in the diffuse ISM, since the density of $8 - 13.6\,\mathrm{eV}$ photons is $\sim 2.0 \times 10^{-3}\,\mathrm{cm}^{-3}$ in the MMP ISRF. Laboratory studies show that H_2O ice has a photo-desorption cross section $\sigma_{\mathrm{pd}} \approx 8 \times 10^{-18}\,\mathrm{cm}^2$ at $\lambda = 1215\text{Å}$ (Westley et al. 1995a, 1995b), so that it seems likely that at least some species will have $\sigma_{-18} \gtrsim 1$, with photo-desorption rate $\gtrsim 6 \times 10^{-11}\,\mathrm{s}^{-1}$. Note that this estimated photo-desorption rate is comparable to photo-dissociation rates for diatomic molecules in the interstellar radiation field.[2]

O is the most abundant species after H and He. The probability per unit time that an O atom will arrive at a single surface site (of area $\sim 10^{-15}\,\mathrm{cm}^2$) is $\sim 2 \times 10^{-13}\,\mathrm{s}^{-1}$ in the CNM. It is therefore apparent that any species with $\sigma_{18} \gtrsim 0.01$ will be photo-desorbed before accretion of anything chemically interesting (other than H) can take place on top of it.

We conclude that photo-desorption most likely plays a dominant role in determining what kind of accretional growth takes place on a grain. It is no surprise that the noble gases (He, Ne, Ar) do not deplete, but the fact that some chemically-reactive elements (e.g., Na, K, S) appear to undergo minimal depletion, at least in the diffuse ISM, may be due to photo-desorption.[3] Laboratory studies of photo-desorption from carbonaceous or silicate substrates would be of great value to clarify why some elements deplete and others do not in the diffuse ISM.

8.5 Grain Destruction in Shock Waves

From time to time a shock wave will pass through an element of interstellar gas, compressing, heating, and accelerating the gas (Draine & McKee 1993), and creating conditions under which grain destruction can occur (Draine 1995). Shock waves can be driven into cold cloud material ahead of an ionization/dissociation front (Bertoldi & Draine 1996), as the result of energetic outflows from newly-formed stars, or as the result of the explosion of a nearby supernova.

It is easiest to study the shock in the "shock frame" where the shock front is stationary: in the upstream direction the grains and gas move together

[2] Roberge et al. (1991) estimate photo-dissociation rates $8.6 \times 10^{-10}\,\mathrm{s}^{-1}$ and $3.5 \times 10^{-10}\,\mathrm{s}^{-1}$ for CH and OH, respectively.

[3] The only other depletion-suppressing process would be chemisputtering by H atoms: an impinging H atom might react with the adsorbed atom, and the reaction energy might eject the resulting hydride.

toward the shock with a speed equal to the shock speed v_s. The fluid is then decelerated at the shock front, with the post-shock gas velocity $v_g \approx v_s/4$ for a strong shock. The post-shock gas is initially hot. If the shock is strong, and we can neglect the effects of a possible magnetic precursor, the gas temperature will rise to

$$T_s \approx \frac{3}{16k}\mu v_s^2 = 2300 \times \text{K} \left(\frac{\mu}{m_\text{H}}\right) \left(\frac{v_s}{10\text{km s}^{-1}}\right)^2 \quad , \tag{130}$$

where μ is the mass per free particle ($\mu/m_\text{H} \approx 1.4/1.1 = 1.27$ for cold atomic clouds, and $\mu/m_\text{H} = 1.4/0.6 = 2.33$ for molecular gas, where we assume He/H=0.1).

The grains move across the shock front into the post-shock fluid, where they initially have velocity $v_s - v_g \approx (3/4)v_s$ relative to the gas. The grains now begin to decelerate by gas drag, but are also acted on by the magnetic field $\mathbf{B}$ and electric field $\mathbf{E} = -\mathbf{v}_g \times \mathbf{B}/c$ (evaluated in the shock frame). As a result, the grain population develops grain-grain velocity differences which can be of order the shock speed v_s, or even larger if conditions are favorable for "betatron" acceleration of the dust as the density and magnetic field increase in cooling post-shock gas (Spitzer 1976; Draine & Salpeter 1979b).

8.5.1 Grain-Grain Collisions

In the absence of betatron acceleration, a grain will slow down by a factor $1/e$ when it has collided with its own mass of gas. If all grains were identical, then the chance that a grain would run into another grain before appreciable slowing-down would be of order the dust-to-gas mass ratio of ~ 0.01, and we can therefore expect that a typical grain (one representative of the bulk of the grain mass) will have a probability of ~ 0.01 of collision with another comparable grain with a grain-grain velocity difference $\sim v_s$.

The physics of grain-grain collisions has been discussed by Tielens et al. (1994). For refractory grains one expects virtually complete destruction if the energy in the center-of-mass system is sufficient to vaporize both grains. For identical grains, mean atomic mass 20 amu, and binding energy per atom of 5.7 eV (these numbers are appropriate to $MgFeSiO_4$ silicate) complete vaporization could occur in a head-on collision between identical grains with a velocity difference of 15 km s^{-1}, so we may expect vaporization of perhaps of $\sim 1\%$ of the grains in a hydrodynamic shock with shock speed $v_s \gtrsim 20\text{km s}^{-1}$. At lower grain-grain collision speeds, shattering may occur. Thus we should expect that several percent of the large dust grains will be shattered in shocks with $v_s \gtrsim 10\text{km s}^{-1}$. Grain-grain collisions may be responsible for maintaining the population of small grains in the interstellar medium.

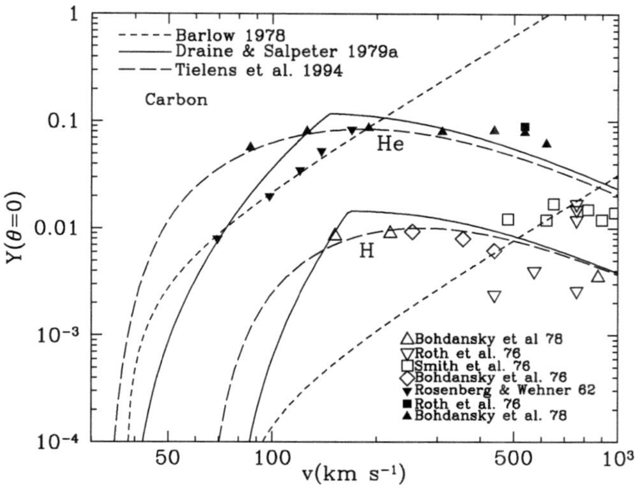

Fig. 39. Sputtering yields for H and He on carbon estimated by Barlow (1978), Draine & Salpeter (1979a), and Tielens et al. (1994), and compared with laboratory data.

8.5.2 Sputtering

In a fast shock, the grain material can be eroded by the process of "sputtering", where individual atoms or ions from the gas collide with the grain, occasionally ejecting one of the grain atoms. In interstellar gas, the sputtering rate depends mainly on the sputtering "yields" $Y(E)$ for H and He projectiles incident on silicate or carbonaceous target materials. In ionized gas the sputtering rates will be affected by the grain charge. An expression for the rate of sputtering of a moving grain, and estimates for sputtering yields Y, are given by Draine & Salpeter (1979a).

In the initial post-shock region where the gas is hot, the sputtering is due to a combination of thermal energy and the motion of the grain through the gas with velocity $\sim 3v_s/4$. If the gas cools before the grain slows down, then subsequent sputtering is entirely due to the rapid motion of the grain through the gas, with He atoms having kinetic energies 4 times larger than H atoms. The sputtering continues until the grain slows to the point where the kinetic energy of the impinging He atoms is too low for appreciable sputtering.

Detailed models of grain destruction in shock waves are in general agreement that shock speeds $v_s \gtrsim 100\mathrm{km\,s^{-1}}$ result in substantial grain destruction, with essentially complete grain destruction for $v_s > 200\mathrm{km\,s^{-1}}$ in the case of a radiative shock – see Fig. 41 (Draine & Salpeter 1979b; Jones et al. 1994).

These estimates for grain destruction appear to be in agreement with the observation that high-velocity gas (which has been shocked) tends to have enhanced gas phase abundances of elements, such as Si, which are normally

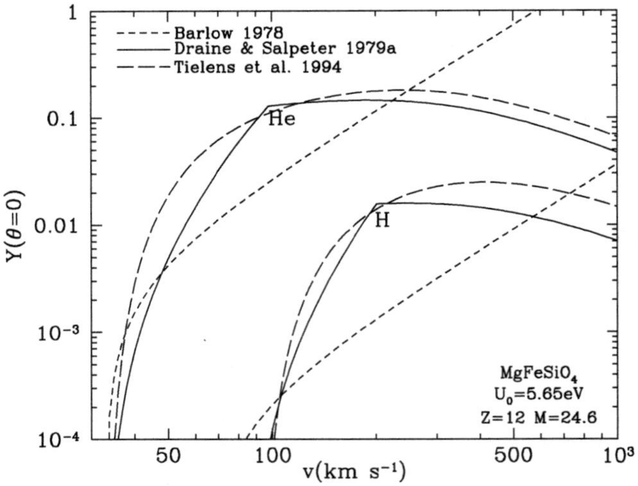

Fig. 40. Sputtering yields for H and He on silicate material estimated by Barlow (1978), Draine & Salpeter (1979a), and Tielens et al. (1994).

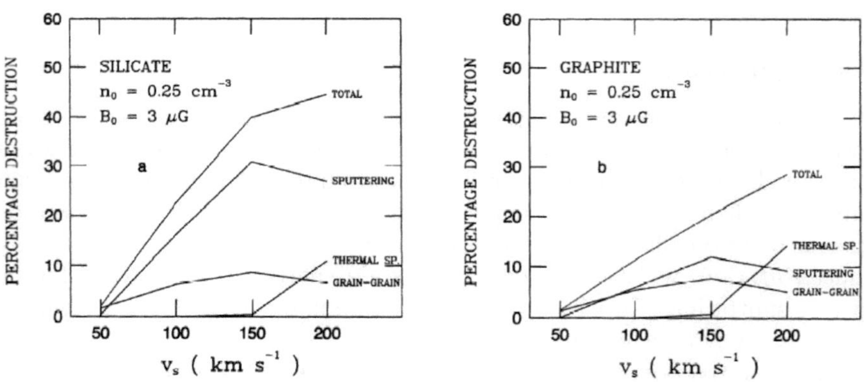

Fig. 41. Fractional grain destruction calculated for graphite and silicate grains in interstellar shock waves in low density gas. Figure from Jones et al. (1994).

heavily depleted in interstellar gas, an effect first noted for Ca by Routly & Spitzer (1952). Figure 42 shows this effect for Si.

8.6 Thermal Sublimation

Ice mantles on grains can be removed if the grains become too warm. What is the critical grain temperature for ice mantle removal?

The probability per unit time for a surface molecule to desorb is approximately

$$\frac{dP}{dt} = \nu_0 e^{-U_0/kT} \quad , \tag{131}$$

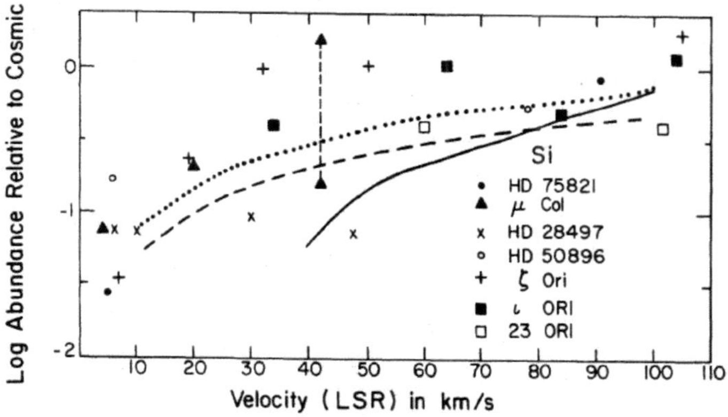

Fig. 42. Observed gas-phase Si abundances: evidence for refractory grain destruction in high-velocity shocks. *Solid*, *dashed*, and *broken lines* are theoretical models discussed by Cowie (1978). Figure from Cowie (1978).

where U_0 is the binding energy of the molecule, and $\nu_0 \approx 10^{13}\,\mathrm{s}^{-1}$ is a characteristic vibration frequency. If there are N_{mono} monolayers of ice on the grain and we want it to survive a time Δt, the grain temperature should not exceed

$$T_{\mathrm{sub}} = \frac{U_0/k}{\ln(\nu_0 \Delta t / N_{\mathrm{mono}})} \tag{132}$$

Some heating phenomena are brief (e.g., a stellar flare or a supernova) but we are usually interested in longer time scales. For example, a massive star in or near the cloud might have a lifetime $\Delta t \approx 5 \times 10^6\,\mathrm{yr}$. We can write

$$T_{\mathrm{sub}} = \frac{U_0/54k}{1 + 0.018 \ln[(\Delta t/10^6\,\mathrm{yr})(10^3/N_{\mathrm{mono}})]} \tag{133}$$

Thus, a H_2O ice mantle ($U_0 = 0.5\,\mathrm{eV}$) 10^3 monolayers thick will survive $10^6\,\mathrm{yr}$ provided $T < 108\,\mathrm{K}$. Therefore we should not be surprised to find H_2O ice in clouds where the grain temperature is 80 K (as in the molecular cloud OMC-1), but it would be a great surprise to find H_2O ice if the grain temperature were, say, 120K.

8.7 Gamma Ray Bursts

Gamma ray bursts may destroy dust grains out to a substantial distance. While this is probably uninteresting in terms of evolution of the overall grain population in galaxies, it may nevertheless be of observational interests because it affects the observable properties of gamma-ray bursts. This is of considerable interest because there is evidence that gamma-ray bursts are associated with star-forming regions in galaxies, and yet the optical afterglows from a number of bursts show little evidence for reddening.

Grain destruction near GRBs has been discussed by Waxman & Draine (2000), Fruchter, Krolik & Rhoads (2001), and Draine & Hao (2002). There are two separate mechanisms which could be important. Thermal sublimation (as discussed above, but now with larger binding energies $U_0/k \approx 7 \times 10^4\,\mathrm{K}$ and shorter times $\Delta t \approx 10\,\mathrm{s}$ for the optical transient, so the grain temperature can rise to $\sim 3000\,\mathrm{K}$. Draine & Hao find that for a reasonable (but very uncertain) estimate of the GRB peak optical luminosity, dust grains can be destroyed out to distances of $\sim 5\,\mathrm{pc}$.

A more exotic grain destruction mechanism may operate; if effective, it could destroy grains out to greater differences. This is the mechanism of "Coulomb explosion" (Waxman & Draine 2000). The grains will be quite highly charged due to the intense X-ray irradiation from the GRB. If the grains are sufficiently weak, the electrostatic stresses on the grain might conceivably tear it apart. Fruchter et al give arguments in favor of this outcome, in which case GRBs could destroy grains out to quite large distances.

8.8 Time Scales

When all grain destruction processes are considered, it is found that sputtering in shock waves dominates the overall destruction of refractory grains. The frequency with which grains are overtaken by such shocks depends on the frequency of the supernovae which drive these blastwaves, and on the structure of the interstellar medium, since material in dense regions is "sheltered" from the effects of a passing blastwave. The problem is complex, and therefore there are many places where different assumptions could alter the quantitative conclusion, but Draine & Salpeter (1979b) estimated a lifetime $\sim 1.3 \times 10^8\,\mathrm{yr}$ for silicate material in a diffuse cloud, while Jones et al. (1994) find a lifetime $2.2 \times 10^8\,\mathrm{yr}$ for silicate grains.

Draine & Salpeter (1979b) stress that this relatively short lifetime was impossible to reconcile with the idea that "interstellar grains are stardust": nearly all of the Si in the interstellar medium is in solid form (e.g., see Fig. 9, showing that only 2.5% of the Si is in the gas phase in the cloud toward ζOph). The "turnover" time of the interstellar medium is equal to the ratio of (mass in gas)/(rate of mass going into stars), or $\sim 5\times10^9 M_\odot/5 M_\odot\,\mathrm{yr}^{-1} \approx 1\times10^9\,\mathrm{yr}$. Suppose that all the Si leaving stars entered the ISM in solid form, but there was no grain growth in the ISM. A grain destruction time $\sim 2 \times 10^8\,\mathrm{yr}$ – only ~20% of the turnover time – would imply that ~ 80% of the Si would be found in the gas phase. However, observations of diffuse clouds typically show $< 10\%$ of the Si in the gas gas phase (see, e.g., Fig. 9).

The ineluctable conclusion is that *most* of the Si atoms in grains must have joined those grains in the ISM through some process of grain growth. The kinetics of accretion implies that only in denser regions can accretion rates be large enough to remove Si from the gas fast enough to maintain the low observed gas-phase abundances. This in turn requires that there must be fairly rapid exchange of mass between the dense and less dense phases of the ISM. A more quantitative discussion of the required mass exchange rates may be found in Draine (1990) and Weingartner & Draine (1999).

9 Effects of Dust on Interstellar Gas

9.1 Photoelectric Heating

Photoelectrons emitted from dust grains depart with nonzero kinetic energy. This is often the dominant mechanism for heating interstellar gas, whether in diffuse clouds or in photo-dissociation regions.

Photoelectric heating rates have been re-estimated recently by Bakes & Tielens (1994) and Weingartner & Draine (2001c). For starlight with the spectrum of the ISRF, Weingartner & Draine find a heating rate per H nucleon $\Gamma/n_{\mathrm{H}} \approx 6\times10^{-26}\,\mathrm{erg\,s^{-1}}$ for conditions characteristic of cool H I clouds. This heating rate is much larger than other heating mechanisms, such as ionization by cosmic rays or X-rays.

9.2 H_2 Formation and Other Chemistry

The H_2 molecule is central to interstellar chemistry. The rovibrational lines can be an important coolant for gas at temperatures $100 \lesssim T \lesssim 2000\,\mathrm{K}$.

The principal gas-phase formation processes are via H^-

$$\mathrm{H} + e^- \rightarrow \mathrm{H}^- + h\nu \quad ; \quad \mathrm{H}^- + \mathrm{H} \rightarrow \mathrm{H}_2 + e^- \tag{134}$$

or 3-body collisions:

$$\mathrm{H} + \mathrm{H} + X \rightarrow \mathrm{H}_2 + X + \Delta E \tag{135}$$

where the third body X could be H, He, or another H_2. The H^- formation pathway (134) is slow because of the generally low abundance of H^- in diffuse gas, and the 3 body process (135) requires very high densities to become important. In the interstellar medium of the Milky Way, the dust abundance is such that H_2 formation is overwhelmingly due to grain surface catalysis. The grain catalysis process was first proposed by Gould & Salpeter (1963), and the basic picture has not changed:

1. An H atom arrives at a grain surface and "sticks".
2. The H atom diffuses over the grain surface until it becomes trapped by either chemisorption or physisorption.
3. A second H atom arrives at the surface, sticks, diffuses over the surface, and encounters the trapped H atom.
4. The two H atoms react to form H_2, releasing $\sim 4.5\,\mathrm{eV}$ of energy, and ejecting the H_2 from the grain surface into the gas.

Part of the 4.5 eV goes into overcoming the binding of the two H atoms to the surface, part goes into vibrational excitation of the grain lattice ("heat"), part appears as kinetic energy of the H_2, and part appears in vibrational and rotational energy of the newly-formed H_2 molecule.

In the above picture, the H_2 formation rate is

$$\frac{dn(\mathrm{H}_2)}{dt} = R n_{\mathrm{H}} n(\mathrm{H}) \tag{136}$$

$$R = \frac{1}{2}\left(\frac{8kT}{\pi m_{\mathrm{H}}}\right)^{1/2} \int da \, \frac{1}{n_{\mathrm{H}}}\frac{dn_{\mathrm{d}}}{da}\pi a^2 \gamma(a) \tag{137}$$

$$= 7.3 \times 10^{-17}\,\mathrm{cm}^3\,\mathrm{s}^{-1}\left(\frac{T}{100\,\mathrm{K}}\right)^{1/2} \frac{A\langle\gamma\rangle}{10^{-21}\,\mathrm{cm}^2\mathrm{H}^{-1}} \tag{138}$$

$$A \equiv \int da \, (1/n_{\mathrm{H}})(dn_{\mathrm{d}}/da)\pi a^2 \tag{139}$$

$$\langle\gamma\rangle \equiv \frac{1}{A}\int da \, (1/n_{\mathrm{H}})(dn_{\mathrm{d}}/da)\pi a^2 \gamma(a) \tag{140}$$

where $\gamma(a)$ is the fraction of H atoms colliding with a grain of radius a which are converted to H_2; $\gamma < 1$ since not all impinging H atoms will "stick", and some of those which stick may later be removed from the surface by some process other than H_2 formation. Gould & Salpeter argued for γ of order unity under interstellar conditions.

Dust models which reproduce the observed extinction of starlight have

$$A \equiv \int da \, \frac{1}{n_{\mathrm{H}}}\frac{dn_{\mathrm{d}}}{da}\pi a^2 \gtrsim 10^{-21}\,\mathrm{cm}^2 \tag{141}$$

For example, the dust model of Weingartner & Draine (2001a) has $A = 6.7 \times 10^{-21}\,\mathrm{cm}^2/\mathrm{H}$. The H_2 formation "rate coefficient" R can be determined from ultraviolet observations of H_2 in diffuse regions; Jura (1975) found $R \approx 3 \times 10^{-17}\,\mathrm{cm}^3\,\mathrm{s}^{-1}$, which implies $\langle\gamma\rangle \approx 0.06$. It appears, then, that the very small grains which dominate the grain surface area (in particular, the PAHs) must not be efficient catalysts for H_2 formation.

How might H_2 formation on the smallest grains be suppressed? The time scale between photon absorption events is

$$\tau_{\mathrm{abs}} \approx 10^6\,\mathrm{s}(10\text{Å}/a)^3 \quad , \tag{142}$$

while the time scale between arrival of H atoms is

$$\tau_{\mathrm{H}} \approx 8 \times 10^6\,\mathrm{s}\left(\frac{10\text{Å}}{a}\right)^2\left(\frac{30\,\mathrm{cm}^{-3}}{n_{\mathrm{H}}}\right)\left(\frac{100\,\mathrm{K}}{T}\right)^{1/2} \tag{143}$$

so it is possible that an H atom physisorbed on a very small grain may be removed when the grain is heated by a photon absorption event, so that only rarely would an arriving H atom find another H atom with which to recombine. The efficacy of this process depends on the binding energy of the H atom.

9.3 Grains as Sources of Complex Molecules

Observational studies of star-forming regions sometimes find molecular abundances in the gas phase which are difficult to understand in the context of pure gas-phase chemistry. An extreme example of this is the observation of $D_2CO/H_2CO = 0.003$ toward Orion (Turner 1990), or $D_2CO/H_2CO = 0.03 - 0.16$ toward the low-mass protostar IRAS 16293-2422 (Ceccarelli et al 2001). Such extreme deuteration seems impossible to envision in the gas phase, but could occur by chemistry on grain mantles during a precollapse phase, with the molecules put into the gas phase when the dust is warmed by energy from the protostar. There seems little doubt that at least some regions have molecular abundances which are heavily influenced by grain surface chemistry.

9.4 Ion Recombination on Dust Grains

In diffuse clouds, the ion/neutral fraction elements with ionization potentials $< 13.6\,\mathrm{eV}$ is indicative of the relative rates for photoionization of starlight vs. recombination of ions with electrons. It is often assumed that the only channel for neutralization of metal ions X^+ is radiative recombination, $X^+ + e^- \rightarrow X + h\nu$. However, collisions with neutral or negatively charged grains can be more important than radiative recombination for neutralization of metal ions for conditions typical of the diffuse interstellar medium (Weingartner & Draine 2001d). Weingartner & Draine provide estimates for the effective "rate coefficient" for ion neutralization via collisions with dust grains. Recombination on dust grains is particularly effective for protons in H I regions, and therefore is involved in regulating the free electron density.

9.5 Coupling Neutral Gas to Magnetic Fields

In gas of low fractional ionization, a significant fraction of the "free" charge present may be located on dust grains. The fraction of the free charge residing on the grains depends very much on the numbers of very small dust grains, but appears likely to become significant in regions of low fractional ionization $n_e/n_\mathrm{H} \lesssim 10^{-6}$ (Draine & Sutin 1987).

Coupling of the magnetic field to the gas is due to the force exerted on moving charges; in regions of low fractional ionization, the neutral gas will be unaffected by the magnetic field unless the charged species are collisionally coupled to the neutral atoms and molecules. Because dust grains have relatively large physical cross sections, the charge trapped on dust grains can be important at coupling the magnetic field to the neutral gas in MHD shocks in gas with fractional ionization $n_e/n_\mathrm{H} \lesssim 10^{-7}$ (Draine 1980; Draine, Roberge & Dalgarno 1983).

9.6 Dust Grains as Magnetometers?

Magnetic fields are dynamically important in the ISM, but difficult to observe remotely. Nature has been kind enough to strew microscopic magnetometers – dust grains – throughout the interstellar medium. Unfortunately we have not yet figured out how these magnetometers work, so that even if we could measure their degree of alignment, we would not be able to determine the magnetic field strength. But we *can* use dust grains as "compasses" to indicate the *direction* of the magnetic field.

As discussed above, dust grains have their angular momenta $\mathbf{J}$ systematically aligned with the magnetic field $\mathbf{B}_0$, at least in regions where sufficient starlight is present. Here we review the main features of this alignment:

- It is virtually certain that the sense of alignment is to have $\mathbf{J} \parallel \mathbf{B_0}$ (to within a sign – parallel and antiparallel are equally favored).
- Suppose the grain has 3 principal axes $\hat{\mathbf{a}}_1$, $\hat{\mathbf{a}}_2$, and $\hat{\mathbf{a}}_3$, with moments of inertia $I_1 \geq I_2 \geq I_3$. If the grain has rotational kinetic energy $E_{\rm rot} > (3/2)kT_{\rm gr}$, then the grain will tend to have its shortest axis $\hat{\mathbf{a}}_1 \parallel \mathbf{J}$ so as to minimize its rotational kinetic energy at constant $\mathbf{J}$. In regions where the grains are cooler than the gas (the most common situation), this is expected to be true for both large grains (which rotate suprathermally) and small grains (which rotate thermally).
- Thus we expect the grains to have their "long axes" $(\hat{\mathbf{a}}_2, \hat{\mathbf{a}}_3) \perp \mathbf{B_0}$.
- Starlight extinction will be greatest for starlight with $\mathbf{E} \parallel$ the long axes, and therefore we expect the transmitted starlight to have $\mathbf{E} \parallel \mathbf{B}_0$.
- Far-infrared and submm emission will be polarized $\parallel$ to the grains' long axes, and hence we expect the FIR/submm emission to be polarized with $\mathbf{E} \perp \mathbf{B}_0$.

Therefore we can obtain a map of the magnetic field by either measuring the polarization of starlight for many stars, or by mapping the polarized far-infrared or submm emission from the cloud. CCDs and infrared detectors make possible extensive measurements of starlight polarization, even in dense regions, and FIR/submm arrays (e.g., SCUBA on the JCMT) make possible measurements of polarized submm emission.

Grains are likely to have a fairly high polarizing efficiency in the FIR/submm. Define

$$C_{\rm pol} \equiv \frac{1}{2}\left[C(\mathbf{E} \parallel \hat{\mathbf{a}}_2) + C(\mathbf{E} \parallel \hat{\mathbf{a}}_3)\right] - C(\mathbf{E} \parallel \hat{\mathbf{a}}_1) \tag{144}$$

$$C_{\rm ran} \equiv \frac{1}{3}\left[C(\mathbf{E} \parallel \hat{\mathbf{a}}_1) + C(\mathbf{E} \parallel \hat{\mathbf{a}}_2) + C(\mathbf{E} \parallel \hat{\mathbf{a}}_3)\right] \tag{145}$$

If the grains were spherical, there would of course be no polarization in extinction or emission. Figure 43 shows how the polarizing efficacy depends on the axial ratio of the dust grains. The grains are assumed to be either oblate

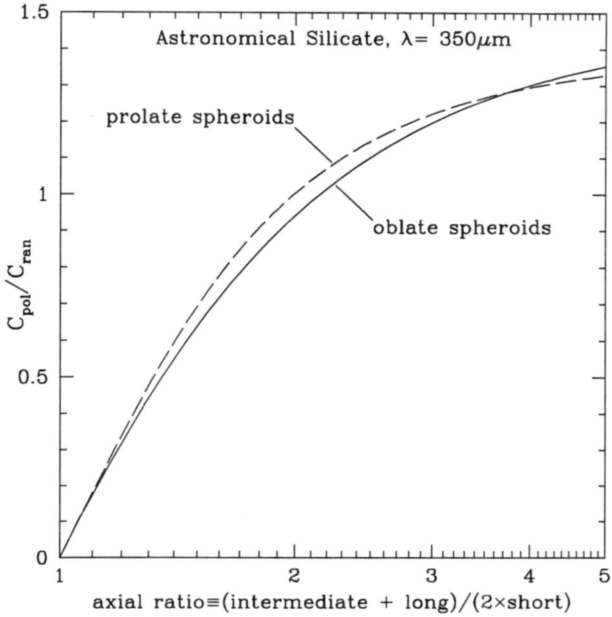

Fig. 43. Polarization efficiency $C_{\rm pol}/C_{\rm ran}$ for spheroidal dust grains (oblate and prolate) as a function of the axial ratio. From Padoan et al. (2001).

or prolate spheroids, and the dielectric function of the grains is taken to be that of "astronomical silicate" at 350 μm.

For perfect spinning alignment, the maximum polarization is obtained for a sightline $\perp \mathbf{B}_0$:

$$P_{\rm max} = \frac{C_2 + C_3 - 2C_1}{C_2 + C_3 + 2C_1} = \frac{C_{\rm pol}/C_{\rm ran}}{1 - \frac{1}{6}\frac{C_{\rm pol}}{C_{\rm ran}}} \tag{146}$$

Suppose the grain shape has (intermediate+long/(2×(short) = 1.4 (e.g., axial ratios 1 : 1.2 : 1.6). Then $C_{\rm pol}/C_{\rm ran} \approx 0.5$ for astronomical silicate. Perfect alignment would give polarization in emission $P = 0.5/(1 - 0.08) = 0.55$. However, the largest observed polarizations are $\sim 10\%$, implying either

- imperfect alignment
- less elongation
- disorder in the magnetic field on short length scales.

Unfortunately, it will be difficult to determine the relative importance of these three effects. For perfectly aligned grains in a uniform magnetic field perpendicular to the line-of-sight, $P = 0.1$ would be produced by grains with axial ratio of only ~ 1.05 – very minimal elongation.

10 Concluding Remarks

There are many unanswered questions in the astrophysics of interstellar dust grains in cold clouds. The questions are interesting in their own right, but one is also driven to answer them by the need to understand dust grains well enough to use them as diagnostics of interstellar conditions, and to understand the effects which dust grains have on the dynamics and evolution of interstellar gas. With the advent of powerful infrared and submillimeter observing facilities – including *SIRTF, SOFIA, ALMA, Herschel,* and *Planck* – we can anticipate that there will be progress on the astrophysics of dust, driven by the challenge of understanding the new data. An exciting decade lies before us!

I thank Robert Lupton for making available the SM software package. This work was supported in part by NSF grant AST 9988126.

References

1. Agladze, N.I., Sievers, A.J., Jones, S.A., Burlitch, J.M., & Beckwith, S.V.W. 1996: *Ap. J.*, **462**, 1026-1040
2. Allamandola, L.J., Hudgins, D.M., & Sandford, S.A. 1999: *Ap. J. Lett.*, **511**, L115-L119
3. Arendt, R.G., et al. 1998: *Ap. J.*, **508**, 74-105
4. Asano, S., & Sato, M. 1980: *Appl. Opt.*, **19**, 962
5. Asano, S., & Yamamoto, G. 1975: *Appl. Opt.*, **14**, 29
6. Ashcroft, N.W., & Mermin, N.D. 1976: Solid State Physics (New York: Holt, Rinehart, and Winston)
7. Bakes, E.L.O., & Tielens, A.G.G.M. 1994: *Ap. J.*, **427**, 822-838
8. Barlow, M.J. 1978: *M.N.R.A.S.*, **183**, 367-395
9. Barnett, S.J. 1915: *Phys. Rev.* **6**, 239-270
10. Bergin, E.A., Neufeld, D.A., & Melnick, G.J. 1999: *Ap. J. Lett.*, **510**, L145-148
11. Bertoldi, F., & Draine, B.T. 1996: *Ap. J.*, **458**, 222-232
12. Blanco, A., Borghesi, A., Fonti, S., & Orofino, V. 1998: *Astr. Ap.*, **330**, 505-514
13. Bohdansky, J., Roth, J., Sinha, M.K. 1976: in *Proc. 9th Symp. on Fusion Technology* (London: Pergamon), p. 541
14. Bohdansky, J., Bay, H.L, & Ottenberger, W. 1978: *J. Nucl. Mat.*, **76**, 163
15. Bohren, C.F., & Huffman, D.R. 1983: Absorption and Scattering of Light by Small Particles (New York: Wiley)
16. Bouchet, P., Lequeux, J., Maurice, E., Prévot, L., & Prévot-Burnichon, M.L. 1985: *Astr. Ap.*, **149**, 330-336
17. Boulanger, F., Boissel, P., Cesarsky, D., Ryter, C. 1998: *Astr. Ap.*, **339**, 194-200
18. Boulanger, F., & Pérault, M. 1988: *Ap. J.*, **330**, 964-985
19. Cardelli, J.A., Clayton, G.C., & Mathis, J.S. 1989: *Ap. J.*, **345**, 245-256
20. Cardelli, J.A., Mathis, J.S., Ebbets, D.C., & Savage, B.D. 1993: *Ap. J. Lett.*, **402**, L17-20

21. Ceccarelli, C., Loinard, L, Castets, A., Tielens, A.G.G.M., Caux, E., Lefloch, B., & Vastel, C. 2001: *Astr. Ap.*, **372**, 998-1004
22. Cesarsky, D., Lequeuex, J., Abergel, A., Perault, M., Palazzi, E., Madden, S., & Tran, D. 1996: *Astr. Ap.*, **315**, L305-308
23. Clayton, G.C., Anderson, C.M., Magalhaes, A.M., Code, A.D., Nordsieck, K.H. et al. 1992: *Ap. J. Lett.*, **385**, L53-57
24. Cowie, L.L. 1978: *Ap. J.*, **225**, 887-892
25. Crinklaw, G., Federman, S.R., & Joseph, C.L. 1994: *Ap. J.*, **424**, 748-753
26. Dale, D.A., Helou, G., Contursi, A., Silberman, N.A., & Kolhatkar, S. 2001: *Ap. J.*, **549**, 215-227
27. Davis, L., & Greenstein, J.L. 1951: *Ap. J.*, **114**, 206-240
28. Debye, P. 1909: *Ann. Phys. (Leipzig)*, **30**, 57
29. Demyk, K., Dartois, E., Wiesemeyer, H., Jones, A.P., & d'Hendecourt, L. 2000: *Astr. Ap.*, **364**, 170-178
30. de Oliveira-Costa, A., Kogut, A., Devlin, M.J., Netterfield, C.B., Page, L.A., & Wollach, E.J. 1997: *Ap. J.*, **482**, L17-20
31. de Oliveira-Costa, A., Tegmark, M., Devlin, M.J., Haffner, L.M., Herbig, T., Miller, A.D., Page, L.A., Reynolds, R.J., & Tufte, S.L. 2000: *Ap. J.*, **542**, L5-8
32. de Oliveira-Costa, A., Tegmark, M., Gutiérrez, C.M., Jones, A.W., Davies, R.D., Lasenby, A.N., Rebolo, R., & Watson, R.A. 1999: *Ap. J.*, **527**, L9-12
33. de Oliveira-Costa, A., Tegmark, M., Page, L.A., & Boughn, S.P. 1998: *Ap. J.*, **509**, L9-12
34. Dolginov, A.Z., & Mytrophanov, I.G. 1976: *Ap. Sp. Sci.*, **43**, 291-317
35. Dotson, J.L., Davidson, J., Dowell, C.D., Schleuning, D.A., Hildebrand, R.H. 2000: *Ap. J. Suppl.*, **128**, 335-370
36. Draine, B.T. 1978: *Ap. J. Suppl.*, **36**, 595-619
37. Draine, B.T. 1980: *Ap. J.*, **241**, 1021-1038
38. Draine, B.T. 1981: *Ap. J.*, **245**, 880-890
39. Draine, B.T. 1985: in Protostars and Planets II, ed. D.C. Black & M.S. Matthews (Tucson: Univ. of Arizona Press), 621-640
40. Draine, B.T. 1988: *Ap. J.*, **333**, 848-872
41. Draine, B.T. 1989a: in Interstellar Dust, Proceedings of IAU Symp. 135, ed. L.J. Allamandola & A.G.G.M. Tielens, (Dordrecht: Kluwer), 313-327
42. Draine, B.T. 1989b: in ESA, Infrared Spectroscopy in Astronomy, 93-98
43. Draine, B.T. 1990: in *The Evolution of the Intestellar Medium.* ed. L. Blitz, Astr. Soc. Pacific. Conf. Series, **12**, 193-205
44. Draine, B.T. 1995: in *Shocks in Astrophysics*, ed. T.J. Millar, *Ap. Sp. Sci.*, **233**, 111-123
45. Draine, B.T. 2000: in *Light Scattering by Nonspherical Particles: Theory, Measurements, and Applications*, ed. M.I. Mishchenko, J.W. Hovenier, & L.D. Travis (San Diego: Academic Press), 131-145
46. Draine, B.T., & Flatau, P.J. 1994: *J. Opt. Soc. Am.* **A**, **11**, 1491-1499
47. Draine, B.T., & Flatau, P.J. 2000: http://xxx.lanl.gov/abs/astro-ph/0008151
48. Draine, B.T., & Goodman, J. 1993: *Ap. J.*, **405**, 685-697
49. Draine, B.T., & Hao, L. 2002: *Ap. J.*, **569**, 780-791
50. Draine, B.T., & Lazarian, A. 1998a: *Ap. J. Lett.*, **494**, L19-22
51. Draine, B.T., & Lazarian, A. 1998b: *Ap. J.*, **508**, 157-179
52. Draine, B.T., & Lazarian, A. 1999a: *Ap. J.*, **512**, 740-754 (DL99a)

53. Draine, B.T., & Lazarian, A. 1999b: in Microwave Foregrounds, eds. A. de Oliveira-Costa & M. Tegmark, A.S.P. Conf. Series **181** (San Francisco: Astr. Soc. Pac.), 133-149
54. Draine, B.T., & Lee, H.-M. 1984: *Ap. J.*, **285**, 89–108
55. Draine, B.T., & Li, A. 2001: *Ap. J.*, **551**, 807-824
56. Draine, B.T., & Malhotra, S. 1993: *Ap. J.*, **414**, 632-645
57. Draine, B.T., & McKee, C.F. 1993: *Ann. Rev. Astr. Ap.*, **31**, 373-432
58. Draine, B.T., Roberge, W.G., & Dalgarno, A. 1983: *Ap. J.*, **264**, 485-507
59. Draine, B.T., & Salpeter, E.E. 1979a: *Ap. J.*, **231**, 77-94
60. Draine, B.T., & Salpeter, E.E. 1979b: *Ap. J.*, **231**, 438-455
61. Draine, B.T., & Sutin, B. 1987: *Ap. J.*, **320**, 803-817
62. Draine, B.T., & Tan, J.C. 2002: astro-ph/0208302
63. Draine, B.T., & Weingartner, J.C. 1996: *Ap. J.*, **470**, 551-565
64. Draine, B.T., & Weingartner, J.C. 1997: *Ap. J.*, **480**, 633-646
65. Dufour, R. J. 1984: in IAU Symp. 108, Structure and Evolution of the Magellanic Clouds, ed. S. van den Bergh & K.S. de Boer (Dordrecht: Reidel), 353-360
66. Duley, W.W., Scott, A.D., Seahra, S., & Dadswell, G. 1998: *Ap. J. Lett.*, **503**, L183-185
67. Erickson, W.C. 1957: *Ap. J.*, **126**, 480-492
68. ESA, 1997: The Hipparcos and Tycho Catalogues (ESA SP-1200)
69. Farafonov, V.G., Voshchinnikov, N.V., & Somsikov, V.V. 1996: *Appl. Opt.*, **35**, 5412-5426
70. Ferrara, A., & Dettmar, R.-J. 1994: *Ap. J.*, **427**, 155-159
71. Finkbeiner, D.P., Davis, M., & Schlegel, D.J. 1999: *Ap. J.*, **524**, 867-886
72. Finkbeiner, D.P., Schlegel, D.J., Frank, C., & Heiles, C. 2002: *Ap. J.*, **566**, 898-904
73. Fitzpatrick, E.L. 1999: *P. A. S. P.*, **111**, 63-75
74. Fitzpatrick, E.L., Massa, D. 1986: *Ap. J.*, **307**, 286-294
75. Fruchter, A., Krolik, J.H., & Rhoads, J.E. 2001: *Ap. J.*, **563**, 597-610
76. Gillett, F.C., Forrest, W.J., Merrill, K.M., Soifer, B.T., & Capps, R.W. 1975: *Ap. J.*, **200**, 609-620
77. Gillett, F.C., Jones, T.W., Merrill, K.M., & Stein, W.A. 1975: *Astr. Ap.*, **45**, 77-81
78. Gilra, D.P. 1972: in The Scientific Results of OAO-2, ed. A.D. Code (NASA SP-310), p. 295
79. Gold, T. 1952: *M.N.R.A.S.*, **112**, 215-218
80. Goodman, A.A., Jones, T.J., Lada, E.A., & Myers, P.C. 1995: *Ap. J.*, **448**, 748-765
81. Goodman, J.J., Draine, B.T., & Flatau, P.J. 1991: *Optics Lett.*, **16**, 1198-1200
82. Gordon, K.D., & Clayton, G.C. 1998: *Ap. J.*, **500**, 816-824
83. Gould, R.J., & Salpeter, E.E. 1963: *Ap. J.*, **138**, 393-407
84. Greenberg, J.M., Gillette, J.S., Muñoz, C., Guillermo, M., Mahajan, T.B., Zare, R.N., Li, A., Schutte, W.A., de Groot, M., & Mendoza-Gómez, C. 2000: *Ap. J. Lett.*, **531**, L71-73
85. Hall, J.S. 1949: *Science*, **109**, 166
86. Hall, J.S., & Mikesell, A.H. 1949: *A. J.*, **54**, 187-188
87. Harwit, M. 1970: *Bull. Astron. Inst. Czech.*, **21**, 204-210
88. Hayakawa, S. 1970: *Prog. Theor. Phys.*, **43**, 1224

89. Heger, M.L. 1922: *Lick Obs. Bull.*, **10**, 146-147
90. Hiltner, W.A. 1949a: *Science*, **109**, 165
91. Hiltner, W.A. 1949b: *Ap. J.*, **109**, 471-478
92. Hoppe, P., & Zinner, E. 2000: *J. Geophys. Res.*, **105**, 10371-10386
93. Jackson, J.D. 1975: Classical Electrodynamics (New York: Wiley), p. 154
94. Jenniskens, P., & Desert, F-X. 1994: *Astr. Ap. Suppl.*, **106**, 39-78
95. Joblin, C., d'Hendecourt, L., Leger, A., & Maillard, J.P. 1990: *Nature*, **346**, 729-731
96. Jones, A.P., Tielens, A.G.G.M., Hollenbach, D.J., & McKee, C.F. 1994: *Ap. J.*, **433**, 797-810
97. Jones, A.P., & Williams, D.A. 1984: *M.N.R.A.S.*, **209**, 955-960
98. Jura, M. 1975: *Ap. J.*, **197**, 575-580
99. Kerr, T.H., Hibbins, R.E., Fossey, S.J., Miles, J.R., Sarre, P.J. 1998: *Ap. J.*, **495**, 941-945
100. Kim, S.-H., & Martin, P.G. 1995: *Ap. J.*, **444**, 293-305
101. Kogut, A., Banday, A.J., Bennett, C.L., Górski, K.M., Hinshaw, G., & Reach, W.T. 1996a: *Ap. J.*, **460**, 1-9
102. Kogut, A., Banday, A.J., Bennett, C.L., Górski, K.M., Hinshaw, G., Smoot, G.F., & Wright, E.L. 1996b: *Ap. J.*, **464**, L5-L9
103. Landau, L. D., & Lifshitz, E.M. 1960: Electrodynamics of Continuous Media (New York: Pergamon)
104. Lazarian, A., & Draine, B.T. 1997: *Ap. J.*, **487**, 248-258
105. Lazarian, A., & Draine, B.T 1999a: *Ap. J. Lett.*, **516**, L37-40
106. Lazarian, A., & Draine, B.T. 1999b: *Ap. J. Lett.*, **520**, L67-70
107. Lazarian, A., & Draine, B.T. 2000: *Ap. J. Lett.*, **536**, L15-18
108. Lazarian, A., & Roberge, W.G. 1997: *Ap. J.*, **484**, 230-237
109. Lazarian, A., & Yan, H. 2002: *Ap. J. Lett.*, **566**, L105-108
110. Li, A., & Draine, B.T. 2001: *Ap. J.*, **554**, 778-802
111. Li, A., & Draine, B.T. 2002: *Ap. J.*, **576**, 762-772
112. Martin, C., Arakawa, E.T., Callcott, T.A., & Warmack, R.J. 1987: *J. Electr. Spectrosc. Rel. Phenom.*, **42**, 171-175
113. Martin, P.G. 1971: *M.N.R.A.S.*, **153**, 279-285
114. Mathewson, D.S., & Ford, V.L. 1970: *Mem.R.A.S.*, **74**, 139
115. Mathis, J.S., Cohen, D., Finley, J.P., & Krautter, J. 1995: *Ap. J.*, **449**, 320-329
116. Mathis, J.S., Mezger, P.G., & Panagia, N. 1983: *Astr. Ap.*, **128**, 212-229 (MMP)
117. McFeely, F.R., Cartier, E., Yarmoff, J.A., & Joyce, S.A. 1990: *Phys. Rev.* **B**, **42**, 5191
118. Mennella, V., Brucato, J.R. Colangeli, L., & Palumbo, P. 1999: *Ap. J. Lett.*, **524**, L71-74
119. Merrill, P.W. 1934: *P. A. S. P.*, **46**, 206-207
120. Mie, G. 1908: *Ann. Phys. (Leipzig)*, **25**, 377
121. Mishchenko, M.I., Hovenier, J.W., & Travis, L.D. 2000: in Light Scattering by Nonspherical Particles: Theory, Measurements, and Applications (San Diego: Academic Press), 3-27
122. Mishchenko, M.I., Travis, L.D., & Macke, A. 2000: in Light Scattering by Nonspherical Particles: Theory, Measurements, and Applications (San Diego: Academic Press), 147-172

123. Morrish, A.H. 1965: The Physical Principles of Magnetism (New York: IEEE Press)
124. Morton, D.C. 1975: *Ap. J.*, **197**, 85-115
125. Onaka, T., Yamamura, I., Tanabé, T., Roellig, T.L., & Yuen, L. 1996: *P. A. S. J.*, **48**, L59-63
126. Overbeck, J.W. 1965: *Ap. J.*, **141**, 864-886
127. Padoan, P., Goodman, A., Draine, B.T., Juvela, M., Nordlund, A, & Rögnvaldsson, Ö.E. 2001: *Ap. J.*, **559**, 1005-1018
128. Pendleton, Y.J., & Allamandola, L.J. 2002: *Ap. J. Suppl.*, **138**, 75-98
129. Purcell, E.M. 1969: *Ap. J.*, **158**, 433-440
130. Purcell, E.M. 1975: in The Dusty Universe (New York: Neale Watson), 155-167
131. Purcell, E.M. 1979: *Ap. J.*, **231**, 404-416
132. Roberge, W.G., Jones, D., Lepp, S., & Dalgarno, A. 1991: *Ap. J. Suppl.*, **77**, 287-297
133. Roche, P.F., & Aitken, D.K. 1984: *M.N.R.A.S.*, **208**, 481-492
134. Rogers, C., & Martin, P.G. 1979: *Ap. J.*, **228**, 450-464
135. Rosenberg, D., & Wehner, G.K. 1962: *J. Appl. Phys.*, **33**, 1842
136. Roth, J., Bohdansky, J., Pschenrieder, W., & Sinha, M.K. 1976: *J. Nucl. Mat.*, **63**, 222
137. Routly, P.M., & Spitzer, L. 1952: *Ap. J.*, **115**, 227-243
138. Salpeter, E.E., & Wickramasinghe, N.C. 1969: *Nature*, **222**, 442
139. Sarre, P.J., Miles, J.R., Kerr, T.H., Hibbins, R.E., Fossey, S.J., & Somerville, W.B. 1995: *M.N.R.A.S.*, **277**, L41-43
140. Savage, B.D., Cardelli, J.A., & Sophia, U.J. 1992: *Ap. J.*, **401**, 706-723
141. Scott, A., Duley, W.W., 1996: *Ap. J. Lett.*, **472**, L123-125
142. Serkowski, K. 1973: in IAU Symp. No. 52, Interstellar Dust and Related Topics, eds. J. M. Greenberg and H. C. van de Hulst (Dordrecht: Reidel), 145-152
143. Slater, M., & Hashmall, J. 1992: NASA/Goddard Space Flight Center Document 554-FDD-89/001R3UD1
144. Slysh, V.I. 1969: *Nature*, **224**, 159
145. Smith, J.N., Meyer, C.H., Layton, J.K. 1976: *Nucl. Tech.*, **29**, 318
146. Smith, R.K., & Dwek, E. 1998: *Ap. J.*, **503**, 831-842
147. Sofia, U.J., & Meyer, D.M. 2001: *Ap. J.*, **554**, L221-224 (and erratum: *Ap. J.*, **558**, L147)
148. Spitzer, L. 1976: *Comments Astrophys.*, **6**, 177-187
149. Stanimirovic, S., Stavely-Smith, L., van der Hulst, J.M., Bontekoe, T.J.R., Kester, D.J.M., & Jones, P.A. 2000: *M.N.R.A.S.*, **315**, 791-807
150. Stecher, T.P., & Donn, B. 1965: *Ap. J.*, **142**, 1681-83
151. Stein, S.E., & Brown, R.L. 1991: *J. Am. Chem. Soc.*, , **113**, 787-
152. Tanaka, M., et al. 1996: *P. A. S. J.*, **48**, L53-57
153. Tielens, A.G.G.M., McKee, C.F., Seab, C.G., & Hollenbach, D.J. 1994: *Ap. J.*, **431**, 321-340
154. Trumpler, R.J. 1930: *P. A. S. P.*, **42**, 214-227
155. Turner, B.E. 1990: *Ap. J. Lett.*, **362**, L29-33
156. van de Hulst, H.C. 1957: Light Scattering by Small Particles (New York: Dover)

157. Verstraete, L., Léger, A., d'Hendecourt, L., Défourneau, D., & Dutuit, O. 1990: å**237**, 436-444
158. Voshchinnikov, N.V., & Farafonov, V.G. 1993: *Ap. Sp. Sci.*, **204**, 19-86
159. Watson, W.D. 1973: *J. Opt. Soc. Am.*, **63**, 164-
160. Waxman, E., & Draine, B.T. 2000: *Ap. J.*, **537**, 796-802
161. Weingartner, J.C., & Draine, B.T. 1999: *Ap. J.*, **517**, 292-298
162. Weingartner, J.C., & Draine, B.T. 2001a: *Ap. J.*, **548**, 296-309 (WD01)
163. Weingartner, J.C., & Draine, B.T. 2001b: *Ap. J.*, **553**, 581-594
164. Weingartner, J.C., & Draine, B.T. 2001c: *Ap. J. Suppl.*, **134**, 263-281
165. Weingartner, J.C., & Draine, B.T. 2001d: *Ap. J.*, **563**, 842-852
166. Westley, M.S., Baragiola, R.A., Johnson, R.E., & Baratta, G.A. 1995a: *Nature*, **373**, 405
167. Westley, M.S., Baragiola, R.A., Johnson, R.E., & Baratta, G.A. 1995b: *Plan. Sp. Sci.*, **43**, 1311-1315
168. Witt, A.N., Smith, R.K., & Dwek, E. 2001: *Ap. J. Lett.*, **550**, L201-205
169. Wolff, M.J., Clayton, G.C., Kim, S.-H., Martin, P.G., & Anderson, C.M. 1997: *Ap. J.*, **478**, 395-402
170. Wright, E.L., Mather, J.C., Bennett, C.L., Cheng, E.S., Shafer, R.A., et al. 1991: *Ap. J.*, **381**, 200-209

From left to right, Bruce Draine, Andrew Blain and Françoise Combes, almost done with the week of lectures, are still listening carefully to the audience last questions.

Index

MIX
Papier aus verantwortungsvollen Quellen
Paper from responsible sources
FSC® C105338

If you have any concerns about our products,
you can contact us on
ProductSafety@springernature.com

In case Publisher is established outside the EU,
the EU authorized representative is:
Springer Nature Customer Service Center GmbH
Europaplatz 3, 69115 Heidelberg, Germany

Printed by Libri Plureos GmbH
in Hamburg, Germany